Calcium Transport and Intracellular Calcium Homeostasis

NATO ASI Series

Advanced Science Institutes Series

A series presenting the results of activities sponsored by the NATO Science Committee, which aims at the dissemination of advanced scientific and technological knowledge, with a view to strengthening links between scientific communities.

The Series is published by an international board of publishers in conjunction with the NATO Scientific Affairs Division

A **Life Sciences** B **Physics**	Plenum Publishing Corporation London and New York
C **Mathematical and Physical Sciences** D **Behavioural and Social Sciences** E **Applied Sciences**	Kluwer Academic Publishers Dordrecht, Boston and London
F **Computer and Systems Sciences** G **Ecological Sciences** H **Cell Biology**	Springer-Verlag Berlin Heidelberg New York London Paris Tokyo Hong Kong Barcelona

Series H: Cell Biology Vol. 48

Calcium Transport and Intracellular Calcium Homeostasis

Edited by

Danielle Pansu

Ecole Pratique des Hautes Etudes
and INSERM U 45
Hôpital E. Herriot
69437 Lyon Cedex 03, France

and

Felix Bronner

Dept. of Biostructure
and Function
University of Connecticut
Health Center
Farmington, CT 06032, USA

Springer-Verlag Berlin Heidelberg New York
London Paris Tokyo Hong Kong Barcelona
Published in cooperation with NATO Scientific Affairs Division

Proceedings of the NATO Advanced Research Workshop on Calcium Transport and Intracellular Calcium Homeostasis held in Lyon, France, March 4–7, 1990

ISBN-13: 978-3-642-83979-5 e-ISBN-13: 978-3-642-83977-1
DOI: 10.1007/978-3-642-83977-1

Softcover reprint of the hardcover 1st edition 1990

2131/3140-543210 – Printed on acid-free-paper

CONTENTS

Contributors

D.C. Ahrens, Department of Biochemistry, Roche Institute of Molecular Biology, Roche Research Center, Nutley, NJ 07110 USA.

O. Alvarez, Department of Physiology, UCLA School of Medicine, 10833 Le Conte Ave, Los Angeles, CA 90024-1751 USA.

R.J.J.M. Bakens, Department of Physiology, University of Nijmegen, 6500 HB Nijmegen, the Netherlands.

A. Bar, Institute of Animal Science, Agricultural Research Organization, the Volcani Center, Bet Dagan, Israel.

C. Bellaton, Laboratoire de Physiologie des Echanges Minéraux, Ecole Pratique des Hautes Etudes, Pavillon H Bis, Hôpital E. Herriot, 69437 Lyon cédex 03, France.

M. Bichara, Faculté des Cordeliers, Laboratoire de Physiologie et d'Endocrinologie Rénales, 15-21 rue de l'Ecole de Médecine Rénale, 75270 Paris cédex 06, France.

R.J.M. Bindels, Department of Physiology, University of Nijmegen, 6500 HB Nijmegen, The Netherlands.

M.P. Blaustein, Department of Physiology, University of Maryland School of Medicine, 660 West Redwood St., Baltimore, MD 21201, USA.

J.P. Bonjour, Division de Pathophysiologie clinique, Département de Médecine, Hôpital Cantnal Universitaire, 24 rue Micheli-du-Crest, 1211 Genève 4, Suisse.

A.B. Borle, Department of Physiology, University of Pittsburgh School of Medicine, Pittsburgh, PA 15261, USA.

R. Bouillon, Lab. Expl. Medicine & Endocrinology, University Hospital St Rafael, Catholic University of Leuven, B-3000 Leuven, Belgium.

S. Bova, Department of Physiology, University of Maryland School of Medicine, 660 West Redwood St., Baltimore, MD 21201, USA.

A. Brehier, Unité de Recherche INSERM U 120, 44 Chemin de Ronde, 78110 Le Vesinet, France.

F. Bronner, Department of Biostructure and Function, University of Connecticut Health Center, Farmington, CT 06032, USA.

D.E. Bruns, Department of Pathology, University of Virginia Medical Center, Charlottesville, VA 22908, USA.

M.E. Bruns, Department of Pathology, University of Virginia Medical Center, Charlottesville, VA 22908, USA.

M. Canonaco-Friedrich, Department of Biochemical Pharmacology, University of Innsbruck, 1 Peter-Mayer Str., 16020 Innsbruck, Austria.

C.J. Cao, Department of Biological Chemistry, University of Maryland School of Medicine, 660 West Redwood St., Baltimore, MD 21201, USA.

E. Carafoli, Department of Biochemistry, Swiss Federal Institute of Technology, 16 Universität Str., CH 8092 Zürich, Switzerland.

M. Case, Department of Physiological Sciences, Stopford Building Oxford Road, Manchester M13 9PT UK.

S. Chandra, Department : Section of Physiology, N.Y. State College of Veterinary Medicine, Cornell University, 717 Veterinary Research Tower, Ithaca, NY 14853, USA.

J. Cheon, Department of Biochemistry, Roche Institute of Molecular Biology, Roche Research Center, Nutley, NJ 07110, USA.

S. Christakos, Departement of Biochemistry, University of Medicine and Dentistry, Newark, NJ 07103, USA.

K. Christensen, Veterans Administration Medical Center and Department of Internal Medicine, University of Iowa College of Medicine, Iowa City, IA 52240, USA.

H.S. Cross, Department of General and Experimental Pathology, University of Vienna, 13 Wahringer Str., A-1090 Vienna, Austria.

C. Douglass, Department of Biological Chemistry, University of Maryland School of Medicine, 660 West Redwood St., Baltimore, MD 21201, USA.

G.P. Downey, Department of Cell Biology, Hospital of Sick Children, 555 University Ave, Toronto, Ontario M5G 1X8, Canada.

T. Drüeke, Unité de Recherche INSERM U 90, Hôpital Necker, 149 Rue de Sèvres, 75743 Paris cédex 15, France.

J.M. Dupret, Unité de Recherche INSERM U 120, 44 Chemin de Ronde, 78110 Le Vésinet, France.

J.T. Durkin, Department of Biochemistry, Roche Institute of Molecular Biology, Roche Research Center, Nutley, NJ 07110, USA.

G. Eisenman, Department of Physiology, UCLA School of Medicine, 10833 Le Conte Ave, Los Angeles, CA 90024-1751, USA.

A.E. Evangelopoulos, National Hellenic Research Foundation, Biological Research Center, 48 Vassileos Constantinou Ave, Athens 11635, Greece.

J.P. Felix, Laboratory of Membrane Biochemistry, Merck Institute for Therapeutic Research, P.O.Box 2000, Rahway, NJ 07095, USA.

N. Fratzl-Zelman, Department of General and Experimental Pathologie, University of Vienna, 13 Währinger Str., A-1090 Vienna, Austria.

F. Friedrich, Department of Physiology University of Innsbruck, 3 Fritz-Pregl Str., A-6010 Innsbruck, Austria.

C.S. Fullmer, Department : Section of Physiology, N.Y. State College of Veterinary Medicine, Cornell University, 717 Veterinary Research Tower, Ithaca, NY 14853, USA.

W. Furuya, Department of Cell Biology, Hospital of Sick Children, 555 University Ave, Toronto, Ontario M5G 1X8, Canada.

M.L. Garcia, Laboratory of Membrane Biochemistry, Merck Institute for Therapeutic Research, P.O.Box 2000, Rahway, NJ 07095, USA.

R. Gill, Deparment of Biochemistry, University of Medicine and Dentistry, Newark, NJ 07103, USA.

H. Glossmann, Department of Biochemical Pharmacology, University of Innsbruck, 1 Peter-Mayer Str., A6020 Innsbruck, Austria.

W.F. Goldman, Department of Physiology, University of Maryland School of Medicine, 660 West Redwood St., Baltimore, MD 21201, USA.

R. Goldshleger, Department of Biological Chemistry, Institute of Life Sciences, The Hebrew University of Jerusalem, Jerusalem 91904, Israel.

S. Grinstein, Department of Cell Biology, Hospital of Sick Children, 555 University Ave, Toronto, Ontario M5G 1X8, Canada.

M. Harding, A.F.R.C. Institute of Animal Physiology and Genetics Research, Babraham Hall, Cambridge, CB2 4 AT, UK.

A. Hartog, Department of Physiology, University of Nijmegen, 6500 HB Nijmegen, The Netherlands.

C.W. Heizmann, Department of Pediatrics, Division of Clinical Chemistry, University of Zürich, 75 Steinwies Str., Ch 8032 Zürich, Switzerland.

U. Hennessen, Unité de Recherche INSERM U 90, Hôpital Necker, 149 Rue de Sèvres, 75743 Paris cédex 15, France.

A. Herrmann-Frank, Department of Biochemistry 231A, University of North Carolina, Chapel Hill, NC 27514, USA.

T.R. Hinds, Department of Pharmacology, University of Washington SJ - 30, Seattle, WA 97201, USA.

S. Hurwitz, Institute of Animal Science, Agricultural Research Organization, The Volcani Center, Bet Dagan, Israel.

A.M. Iacopino, Department of Biochemistry, University of Medicine and Dentistry, Newark, NJ 07103, USA.

L.C. Isaacson, Department of Physiology, University of Cape Town Medical School, Observatory, 7925, RSA.

G.J. Kaczorowski, Laboratory of Membrane Biochemistry, Merck Institute for Therapeutic Research, P.O.-Box 2000, Rahway, NJ 07095, USA.

U. Karbach, Medizinische Klinik Innenstadt, Universität München, Ziemssenst, 1.8000, München 2, RFA.

S.J.D. Karlish, Department of Biological Chemistry, Institute of Life Sciences, The Hebrew University of Jerusalem, Jerusalem 91904, Israel.

K. Klaushofer, Department of General and Experimental Pathology, University of Vienna, 13 Währinger Str., A-1090 Vienna, Austria.

V.F. King, Laboratory of Membrane Biochemistry, Merck Institute for Therapeutic Research, P.O.Box 2000, Rahway, NJ 07095, USA.

H.G. Kraus, Department of Biochemical Pharmacology, University of Innsbruck, 1 Peter-Mayr Str., A6020 Innsbruck, Austria.

R.H. Kretsinger, Department of Biology, University of Virginia, Charlottesville, VA 22901, USA.

F. L'Horset, Unité de Recherche INSERM U 120, 44 Chemin de Ronde, 78110 Le Vésinet, France.

B. Lacour, Unité de Recherche INSERM U 90, Hôpital Necker, 149 Rue de Sèvres, 75743 Paris cédex 15, France.

F.A. Lai, Department of Biochemistry 231 A, University of North Carolina, Chapel Hill, NC 27514, USA.

F. Lang, Department of Physiology University of Innsbruck, 3 Fritz-Pregl Str., A-6010 Innsbruck, Austria.

K.R. Lau, Department of Physiological Sciences, Stopford Building Oxford Road, Manchester M13 9PT, UK.

D.E.M. Lawson, A.F.R.C. Institute of Animal Physiology and Genetics Research, Babraham Hall, Cambridge, CB2 4AT, UK.

S. Lee, Department of Biochemistry, University of Medicine and Dentistry, Newark, NJ 07103, USA.

E. Lehner, Department of General and Experimental Pathology, University of Vienna, 13 Währinger Str., A-1090 Vienna, Austria.

H. Li, Dept. of Biochemistry, University of Medicine and Dentistry, Newark, NJ 97103, USA.

A. Lindner, Department of Pharmacology, University of Washington SJ - 30, Seattle, Wa 97201, USA.

T. Lockwich, Department of Biological Chemistry, University of Maryland School of Medicine, 660 West Redwood St., Baltimore, MD 21201, USA.

N. Lomri, Unité de Recherche INSERM U 120, 44 chemin de Ronde, 78110 Le Vésinet, France.

G. Meissner, Department of Biochemistry 231 A, University of North Carolina, Chapel Hill, NC 27514, USA.

R.J. Miller, Department of Pharmacology and Physiological Sciences, University of Chicago, 947 E 88 St. Chicago, IL 60637, USA.

G. Morrison, Departement : Section of Physiology, N.Y. State College of Veterinary Medicine, Cornell University, 717 Veterinary Research Tower, Ithaca, NY 14853, USA.

E. Muir, A.F.R.C. Institute of Animal Physiology and Genetics Research, Babraham Hall, Cambridge, CB2 4AT, UK.

H. Mykannen, Department : Section of Physiology, N.Y. State College of Veterinary Medicine, Cornell University, 717 Research Tower, Ithaca, NY 14853, USA.

I. Nemere, Department of Biochemistry, University of California, Riverside, CA 92521, USA.

Y. Nys, Station de Recherche de Nutrition, INRA, Nouzilly, 37380 Monnaie, France

M. Paillard, Faculté des Cordeliers, Laboratoire de Physiologie et d'Endocrinologie Rénales, 15-21 rue de l'Ecole de Médecine 75270 Paris Cédex 06.

D. Pansu, Laboratoire de Physiologie des Echanges Minéraux, Ecole Pratique des Hautes Etudes, Pavillon Hbis, Hôpital E. Herriot, 69437 Lyon cédex 03, France.

M. Paulmichl, Department of Physiology University of Innsbruck, 3 Fritz -Pregl Str., A-6010 Innsbruck, Austria.

C. Perret, Unité de Recherche INSERM U 120, 44 Chemin de Ronde, 78110 Le Vésinet, France.

P. Peterlik, Department of General and Experimental Pathology, University of Vienna, 13 Währinger Str., A-1090 Vienna, Austria.

O.H. Petersen, Physiological Laboratory, University of Liverpool, Brownlow Hill, P.O.Box 147, Liverpool L69 3BX, UK.

J. Pfeilschifter, Department of Physiology University of Innsbruck, 3 Fritz-Pregl Str., A-6010 Innsbruck, Austria.

K.G. Pote, Department of Biology, University of Virginia, Charlottesville, VA 22901, USA.

J.P. Reeves, Department of Biochemistry, Roche Institute of Molecular Biology, Roche Research Center, Nutley, NJ 07110, USA.

R. Rizzoli, Division de Pathophysiologie Clinique, Département de Médecine, Hôpital Cantonal Universitaire, 24 rue Micheli-du-Crest, 1211 Genève 4, Suisse.

W. Ronnenberg, Veterans Administration Medical Center and Department of Internal Medicine, University of Iowa College of Medicine, Iowa City, IA 52240, USA.

J.C.J. Saunders, Department of Physiology, University of Cape Town Medical School, Observatory, 7925 RSA.

H.P. Schedl, Veterans Administration Medical Center and Department of Internal Medicine, University of Iowa College of Medicine, Iowa City, IA 5240, USA.

F. Scheffauer, Department of Biochemical Pharmacology, University of Innsbruck, 1 Peter-Mayer Str., A6020 Innsbruck, Austria.

A.E. Shamoo, Department of Biological Chemistry, University of Maryland School of Medicine, 660 West Redwood St., Baltimore, MD 21201, USA.

J.L. Shevell, Laboratory of Membrane Biochemistry, Merck Institute for Therapeutic Research, P.O.Box 2000, Rahway, NJ 07095, USA.

T.J.B. Simons, Biomedical Sciences Division, King's College, Strand, London WC2R 2LS; UK.

R.S. Slaughter, Laboratory of Membrane Biochemistry, Merck Institute for Therapeutic Research, P.O. Box 2000, Rahway, NJ 07095, USA.

T.G. Sotiroudis, National Hellenic Research Foundation, Biological Research Center, 48 Vassileos Constantinou Ave, Athens 11635, Greece.

W.D. Stein, Department of Biological Chemistry, Institute of Life Sciences, The Hebrew University of Jerusalem, Jerusalem 91904, Israel.

J. Striessnig, Department of Biochemical Pharmacology, University of Innsbruck, 1 Peter-Mayer Str., A6020 Innsbruck, Austria.

A.N. Taylor, Department of Anatomy, Baylor College of Dentistry, 3302 Gaston Ave, Dallas, TX 75246, USA.

M. Thomasset, Unité de Recherche INSERM U 120, 44 Chemin de Ronde, 78110 Le Vésinet, France.

J.A.H. Timermanns, Department of Physiology, University of Nijmegen, 6500 HB Nijmegen, The Netherlands.

N. Tolosa de Talamoni, Department : Section of Physiology, N.Y. State College of Veterinary Medicine, Cornell University, 717 Veterinary Research Tower, Ithaca, NY 14853, USA.

S. Trudel, Department of Cell Biology, Hospital of Sick Children, 555 University Ave, Toronto, Ontario M5G 1X8, Canada.

E. van Lceuwen, Department of Physiology, University of Nijmegen, 6500 HB Nijmegen, The Netherlands.

C.H. van Os, Department of Physiology, University of Nijmegen, 6500 HB Nijmegen, The Netherlands.

J. Verhaeghe, Lab. Expl. Medicine & Endocrinology, University Hospital St Rafael, Catholic University of Leuven, B-3000 Leuven, Belgium.

F.F. Vincenzi, Department of Pharmacology, University of Washington SJ - 30 Seattle, WA 97201, USA.

J.R.F. Walters, Gastroenterology Unit, Postgraduate Medical School, Hammersmith Hospital Ducane Road, London W12 OHS, UK.

R.H. Wasserman, Department : Section of Physiology, N.Y. State College of Veterinary Medicine, Cornell University, 717 Veterinary Research Tower, Ithaca, NY 14853, USA.

M.M. Weiser, Division of Gastroenterology and Nutrition, State University of New York at Buffalo and Buffalo General Hospital, 100 High St., Buffalo, NY 14203, USA.

H. Weiss, Department of Physiology University of Innsbruck, 3 Fritz-Pregl Str., A-6010 Innsbruck, Austria.

P. Wilson, A.F.R.C. Institute of Animal Physiology and Genetics Research, Babraham Hall, Cambridge, CB2 4AT, UK.

E.E. Windhager, Department of Physiology, Cornell University Medical College, 1300 York Ave, New York, NY 10021, USA.

E. Woell, Department of Physiology University of Innsbruck, 3 Fritz-Pregl Str., A-6010 Innsbruck, Austria.

L. Xu, Department of Biochemistry 231 A, University of North Carolina, Chapel Hill, NC 27514, USA.

X.J. Yuan, Department of Physiology, University of Maryland School of Medicine, 660 West Redwood St., Baltimore, MD 21201, USA.

C. Zech, Department of Biochemical Pharmacology, University of Innsbruck, 1 Peter-Mayer Str., A6020 Innsbruck, Austria.

J. Zelinski, Division of Gastroenterology and Nutrition, State University of New York at Buffalo and Buffalo General Hospital, 100 High St., Buffalo, NY 14203, USA.

G. Zernig, Department of Biochemical Pharmacology, University of Innsbruck, 1 Peter-Mayer Str., A6020 Innsbruck, Austria.

V.G. Zevgolis, National Hellenic Research Foundation, Biological Research Center, 48 Vassileos Constantinou Ave, Athens 11635, Greece.

Participants

C. Bellaton
M. Bichara
R.J.M. Bindels
M.P. Blaustein
J.P. Bonjour
R. Bouillon
A.B. Borle
F. Bronner
M.E.H. Bruns
E. Carafoli
R.M. Case
S. Christakos
M.O. Christen
T.B. Drüeke
Y. Dupuis
G. Eisenman
H. Glossmann
S. Grinstein
C.W. Heizmann
S. Hurwitz
L.C. Isaacson
G.J. Kaczorowski
U. Karbach
R.H. Kretsinger
F. Lang
D.E.M. Lawson
G. Meissner
R.J. Miller
K. Nemere
Y. Nys
D. Pansu
M. Peterlik
O.H. Petersen
J.P. Reeves
C. Roche
S. Roche
T.J.B. Simons
H.P. Schedl
A.E. Shamoo
T.G. Sotiroudis
W.D. Stein
A.N. Taylor
M. Thomasset
S. Uhlrich
F.F. Vincenzi
J.R.F. Walters
R.H. Wasserman
M.M. Weiser
E.E. Windhager

PREFACE

The crucial role played by calcium as a cellular messenger has become increasingly evident, as has the recognition that cells spend much energy in maintaining the cytosolic concentration of this cation both constant and low. It is thought they do this to avoid precipitating phosphate, needed as a source of bond energy and to modulate protein structure. Moreover, since calcium that does enter the cell must be disposed with, processes that utilize calcium have evolved, e.g. secretion, contraction, signaling, to name just some.

New knowledge concerning the processes of cellular calcium entry, extrusion and the fate of intracellular calcium has accumulated in recent years. Much has also been learned about calcium transport by and across epithelial cells. It seems logical to think that the processes of calcium entry, extrusion and intracellular handling are similar in all cells. We have therefore assembled in one volume overviews and research reports of transport and cellular calcium regulation so as to explore similarities and differences between cells that utilize calcium for metabolic purposes and those whose primary function is transport.

Channels seem to play a major role in the regulation of cellular calcium entry. Electrical impulses or hormonal messengers initiate a cascade that leads to increased calcium entry. In epithelial transporting cells, on the other hand, neither the existence nor the regulation of calcium channels is currently known. Section I of this book describes the various calcium channels, their function and provides a first report of the existence of such channels in the transporting epithelium of the renal tubule.

Once calcium enters the cell, it must be either bound or extruded. Section II describes the plasma calcium ATPase, a major calcium extrusion pump, and a novel possibility of how it may be upregulated. The same Section also contains a description of the muscle Ca ATPase, a pump that functions in intracellular calcium shuttling. Included is also a study that identifies an important component of the Na/K ATPase, the portion of the pump molecule that involves coupling between ATP hydrolysis and the cation sites.

The role of the Na/Ca exchanger in cellular Ca extrusion is less obvious. There is uncertainty about the relative importance of the exchanger and of the Ca ATPase; in some cells, like those of the duodenum, the exchanger seems to play a minor role, whereas in others, as in heart cells, its role is quite significant. Section III examines the nature, structure and function of the Na/Ca exchanger in cardiac and vascular cells.

Sodium/calcium exchange seems to play an important role in intracellular sodium and proton regulation. These processes are examined in detail in Section IV, both in terms of how various cells regulate their intracellular calcium and how this regulation modulates other processes. As might be expected, many but not all intracellular events are calcium-dependent. The last chapter in Section IV addresses the intriguing possibility that intracellular calcium may be inappropriately high in hypertension, a major disease category.

If cells have developed mechanisms for keeping calcium out and intracellular calcium low, how do specialized cells cope with the necessity of transporting calcium? This question becomes even more acute if one remembers that the extracellular calcium concentration is some 10,000 times higher than its intracellular concentration. Thus, calcium that enters the cell drops along a steep electrochemical gradient and, after crossing the cell interior, must be again raised against a steep electrochemical gradient. In Section V it is pointed out that the probable rate at which calcium moves from the luminal pole of the cells of the duodenum or distal convoluted tubule to the basolateral pole is close to 100 times slower than the experimentally determined active transport rate. To overcome this it is proposed that with a sufficient number of intracellular calcium binding sites, either on cytosolic, small, moveable calcium-binding proteins or perhaps also on lysosomes, an adequate rate of intracellular calcium movement can be assured. Thus it seems that the evolution of mammalian calcium transporting cells involved the synthesis and utilization of a cytosolic calcium-binding protein, the product of the hormonal action of vitamin D. Other reports in this section deal with the regulation of transepithelial calcium transport and the importance of the paracellular route, probably not subject to direct regulation by vitamin D. One paper attempts to integrate by way of modeling cellular events into what happens at the tissue and organisms levels.

Section VI deals with the exciting field of calcium-binding proteins, their characteristics and molecular structure. It provides a first analysis of calcium binding by a molecule somewhat simpler in structure than those that have been characterized by a binding loop (EF hand) flanked by two helices. It also contains descriptions of an enzyme and an enzyme-like molecule that have interesting calcium-binding properties.

The next part of the book, Section VII, deals with the calbindin gene and its regulation. Calbindin is the name given to the vitamin D-induced cytosolic calcium-binding proteins found in abundance in calcium transporting cells and to which a calcium-ferrying role has been attributed. The section also localizes and explores roles of these proteins in uterus, brain and tooth enamel.

Section VIII of the book discusses possible or potential defects of calcium transport as encountered in some pathological states such as diabetes, hypertension, lead poisoning, or hypercalcemia due to a parathyroid hormone-like peptide. One chapter discusses hormonal effects on the development of calcium transport.

The book is the outcome of a NATO and INSERM-sponsored advanced research workshop that was held in Lyon in March, 1990. The Workshop represented a sequel to three previous Workshops, held in Vienna in 1981, 1984, and 1987, which analyzed calcium and phosphate transport across biomembranes. By 1990 the amount of information on calcium alone had become so large that the phosphate portion was sequestered and constitutes a separate Harden Conference, to be held in Wales in September, 1990. The Lyon Workshop brought together scientists from many countries and diverse disciplines and stimulated interdisciplinary questions and collaborative researches. We hope this book will serve a similar purpose for the wider biological community and want to thank the authors and publishers for helping in this endeavor.

May, 1990

Lyon, France — Danielle Pansu
Farmington, CT, USA — Felix Bronner

ACKNOWLEDGMENTS

Financial support of the Workshop was generously provided by the following organizations:

INTERNATIONAL

North Atlantic Treaty Organization (NATO)

FRANCE

Institut National de la Santé et de la Recherche Médicale (INSERM)
Ecole Pratique des Hautes Etudes (EPHE)
Unité d'Enseignement et de Recherche (UER) de Biologie Humaine. Université Claude Bernard Lyon 1
Conseil Général du Rhône
Ville de Lyon

Laboratoire IMEDEX	Lyon
Laboratoire LATEMA	Suresnes - Paris
Laboratoire SANDOZ	Rueil-Malmaison
Laboratoires THYLMER	Dijon

USA

CIBA-GEIGY Corporation	New Jersey
ICI Pharmaceutical Group	Delaware
MARION Laboratories	Missouri
SHERING-PLOUGH Research Division	New Jersey

SECTION I

CELLULAR CALCIUM ENTRY - CALCIUM CHANNELS

MODULATION AND FUNCTIONS OF NEURONAL Ca^{2+} PERMEABLE CHANNELS

Richard J. Miller
Department of Pharmacological and Physiological Sciences
University of Chicago
947 E. 58th Street
Chicago, IL 60637, U.S.A.

Ca^{2+} acts as an important second messenger molecule in virtually every cell type. $[Ca^{2+}]_i$ is normally kept at very low levels ~ 10^{-8} - 10^{-7} M. On the other hand the Ca^{2+} concentration of the extracellular milieu is approximately 10^{-3} M. Consequently there is a very large electrochemical gradient for Ca^{2+} ions across the plasma membrane of the cell. This membrane is normally very impermeable to Ca^{2+}. The Ca^{2+} permeability of the plasma membrane can be rapidly increased by opening a number of ion channels. These channels may be activated by changes in membrane potential, by agonists or, in some instances, by both. In neurons Ca^{2+} has a very large number of roles to play. Among these are the control of neuronal excitability and also the triggering of neurotransmitter release. Voltage sensitive Ca^{2+} channels and receptor operated Ca^{2+} channels have been studied extensively in nerve cells both from the peripheral and central nervous systems. In this article I shall review some of the properties of these channels, how they can be regulated physiologically and some of the consequences of this regulation.

TYPES OF NEURONAL Ca^{2+} CHANNELS

Voltage sensitive Ca^{2+} channels are present in virtually all neurons where they appear to regulate a number of functions. The best known of these is the control of neurotransmitter release. It is well established that when an action potential invades a nerve terminal the resulting depolarization leads to the opening of voltage sensitive Ca^{2+} channels, Ca^{2+} influx into the nerve terminal and the initiation of secretion. A key development in the study of neuronal Ca^{2+} channels was the observation that nerve cells contained more than one channel (Miller, 1987; Nowycky et al., 1985) type. These Ca^{2+} channels differ in both their biophysical and pharmacological properties (see Table I). L-type and T-type Ca^{2+} channels are also found in cells outside the nervous system. However N-type Ca^{2+} channels have only.been found in neurons. It is particularly interesting to ask which Ca^{2+} channels provide the Ca^{2+} influx for triggering the release of neurotransmitters. The answer to this question

NATO ASI Series, Vol. H 48
Calcium Transport and Intracellular Calcium Homeostasis
Edited by D. Pansu and F. Bronner

TABLE I. Electrical and pharmacological properties of the three types of vertebrate calcium channels in chick DRG neurons.

Channel type	**T**	**N**	**L**
Single-channel conductance (110 Ba)	~ 8ps	~ 13 pS	~ 25 pS
Single-channel kinetics	Late opening brief burst, inactivation	Long burst, inactivation	Almost no inactivation
Relative conductance	$Ba^{2+} = Ca^{2+}$	$Ba^{2+} > Ca^{2+}$	$Ba^{2+} > Ca^{2+}$
Inorganic ion block	$Ni^{2+} > Cd^{2+}$	$Cd^{2+} > Ni^{2+}$	$Cd^{2+} > Ni^{2+}$
ω-CgTx via block	Weak, reversible	Persistent	(Persistent)
Dihydropyridine sensitive?	Resistant	Resistant	Sensitive
Activation range (for 10 Ca)	Positive to -70 mV	Positive to -20 mV	Positive to -10 mV
Inactivation range (for 10 Ca)	-100 to -60 mV	-120 to -30 mV	-60 to -10 mV
Inactivation rate (0 mV, 10 Ca or (10 Ba)	Rapid (tau ~ 20-50 ms)	Moderate (tau ~ 50-80 ms)	Very slow (tau > 500 ms)

appears to be somewhat variable depending on the particular neuron in question. However, in many instances the release of neurotransmitters has been shown to be resistant to the effects of dihydropyridine drugs, but potently blocked by the toxin ω-conotoxin which is a powerful blocker of N-type Ca^{2+} channels (Hirning et al., 1988; Lipscombe et al., 1989). Thus, there are a number of circumstances where it appears to be N-type Ca^{2+} channels that regulate neurotransmitter secretion.

In addition to the voltage sensitive Ca^{2+} channels, the existence of one particular type of receptor operated channels that is highly Ca^{2+} permeable has also been demonstrated in nerve cells. This Ca^{2+} permeable channel is activated by the action of the excitatory neurotransmitter glutamate (Mayer and Miller, 1990). Indeed, glutamate can activate a number of different types of receptors on nerve cells. One particular glutamate receptor is activated by the archetypal glutamate agonist N-methyl-D-aspartatic acid (NMDA) and it is this channel which appears to be highly permeable to Ca^{2+}. The NMDA activated channel has a number of interesting properties in addition to being activated by glutamate. The first of these is that it is blocked by physiological concentrations of Mg^{2+} in a voltage dependent manner. Thus, at normal resting membrane potentials the NMDA activated channel is substantially blocked by Mg^{2+}. However, as the nerve cell depolarizes the Mg^{2+} block is relieved. Thus, under many circumstances the NMDA activated ion channel gives the impression of being a voltage gated channel. Another interesting property of the NMDA receptor is that very low concentrations of glycine act as a positive allosteric regulator of the system. Thus, in the concentration range 10 - 1000 nM, glycine greatly potentiates the agonist effects of glutamate or NMDA. The action of glutamate at synaptic NMDA receptors is thought to be most important for promoting Ca^{2+} influx into central neurons under physiological conditions. However, under some pathological conditions, such as those prevailing in stroke or epilepsy, abnormally large amounts of glutamate are released into the extracellular milieu. Excessive stimulation of glutamate receptors occurs and this leads to an abnormal increase in Ca^{2+} uptake by neurons which can lead to cell death. Thus, these NMDA receptors are extremely important mediators of both normal and pathological events in the central nervous system. It is thought likely that the structure of the NMDA receptor resembles one of the other multisubunit receptor linked ionophores such as the nicotinic, glycine or GABA-A receptor. Indeed, recent cloning of another member of the glutamate receptor family supports this contention (Hollman et al., 1989).

The regulation of voltage sensitive Ca^{2+} channels in neurotransmitter release

Voltage sensitive Ca^{2+} channels provide the influx of Ca^{2+} which triggers the release of neurotransmitters. In many cases this appears to be due to influx of Ca^{2+} through N-type Ca^{2+} channels. An important aspect of the communication between nerve cells is that the amount of neurotransmitter released by an action potential is subject to modulation through the action of neurotransmitters at presynaptic receptors. Thus, it is commonly observed that a neurotransmitter can inhibit either its own release or the release of another neurotransmitter through a presynaptic action (Starke, 1987; Kandel and Schwarz, 1982). This process is known as presynaptic inhibition. One way such presynaptic inhibition could potentially be exerted is by restricting the influx of Ca^{2+} into the nerve terminal during an action potential. This might be achieved if the action of a neurotransmitter at a presynaptic receptor site could lead directly to the inhibition of voltage sensitive Ca^{2+} channels in the nerve terminal. Recently this has become clear that this is actually a very commonly used molecular mechanism for producing presynaptic inhibition. It has now been observed in many nerve preparations that neurotransmitters can inhibit the Ca^{2+} current recorded under voltage clamp (Bean, 1989a; Tsien et al., 1988). The same neurotransmitters that block the Ca^{2+} current in neurons are also frequently observed to exert presynaptic inhibition on these same cells. A good example of such a system occurs in peripheral sensory (dorsal root ganglion, DRG) neurons of the rat. These cells utilize several transmitters including the peptide substance P. It has been found that certain substances such as opioids, neuropeptide Y or norepinephrine can inhibit the evoked release of substance P from these sensory neurons both in vivo and in vitro (Holz et al., 1989; Mudge et al., 1979; Walker et al., 1988). How do these various neurotransmitters inhibit the evoked release of substance P? Fig. 1 shows that in these

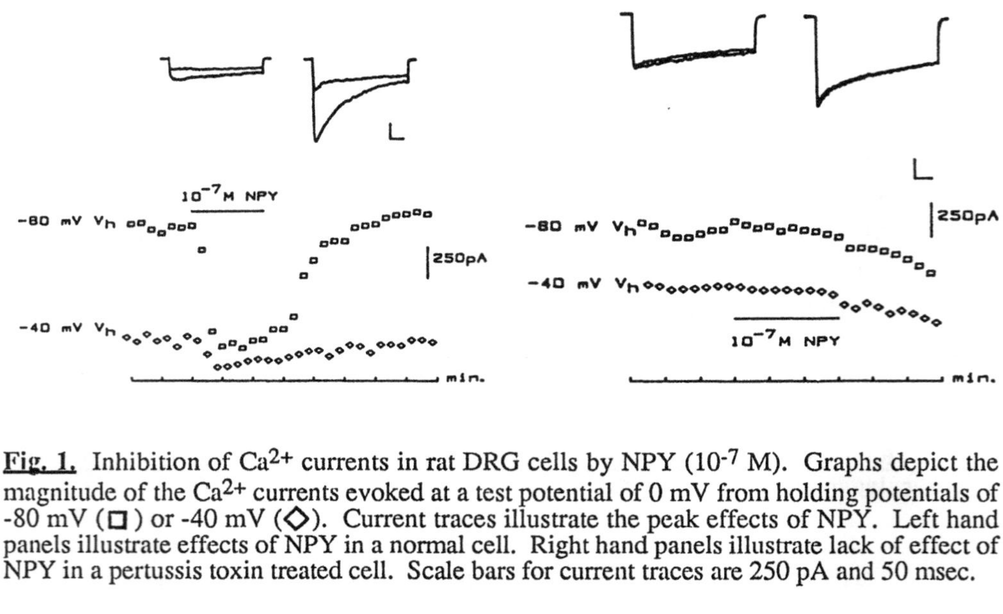

Fig. 1. Inhibition of Ca^{2+} currents in rat DRG cells by NPY (10^{-7} M). Graphs depict the magnitude of the Ca^{2+} currents evoked at a test potential of 0 mV from holding potentials of -80 mV (□) or -40 mV (◇). Current traces illustrate the peak effects of NPY. Left hand panels illustrate effects of NPY in a normal cell. Right hand panels illustrate lack of effect of NPY in a pertussis toxin treated cell. Scale bars for current traces are 250 pA and 50 msec.

cells inhibitory neurotransmitters can block the Ca^{2+} current recorded electrophysiologically. It is quite clear that this includes the portion of the Ca^{2+} current that is due to the activation of N-channels in these cells. Presumably the inhibition of voltage sensitive Ca^{2+} influx by such neurotransmitters is at least partially responsible for their ability to block the evoked release of neurotransmitter from these cells. An important question is how is this inhibition of Ca^{2+} influx produced? Fig. 1 also illustrates the fact that if dorsal root ganglion neurons are treated with pertussis toxin then inhibitory neurotransmitters no longer block the Ca^{2+} current in these cells (Dunlap et al., 1987). This implies that a pertussis toxin sensitive G-protein participates in the transduction mechanism linking the inhibitory receptor to the voltage sensitive Ca^{2+} channel. Further studies have illustrated that in pertussis toxin treated neurons, the coupling between the inhibitory receptor and the voltage sensitive Ca^{2+} channel can be reconstituted by perfusion of purified or recombinant G-protein α-subunits into the cells (Ewald et al., 1988). However not all α-subunits are equally effective. It has usually been observed that the α-subunit of the G-protein heterotrimer G_o is the most effective at coupling neurotransmitter receptors to voltage sensitive Ca^{2+} channels. This is certainly of interest as G_o is clearly the most abundant pertussis toxin substrate in the nervous system (Sternweis and Robishaw, 1984). Thus, this G-protein may carry out this transduction mechanism under physiological circumstances.

A further question of interest is whether the G-protein α-subunit normally acts upon the voltage sensitive Ca^{2+} channel directly or whether its effects are mediated through a diffusible second messenger molecule. In the case of modulation of presynaptic N-type Ca^{2+} channels, it seems likely that no diffusible second messenger molecule is involved. An experiment which illustrates this is shown in Fig. 2 (See also Forscher et al., 1988; Green and Cottrell, 1988). In this case, N-type Ca^{2+} channels were measured on on-cell patches from another kind of rat peripheral neuron, those from the rat myenteric plexus. It is quite clear that in these cells as well the neurotransmitter neuropeptide Y profoundly inhibits the Ca^{2+} current. However in the paradigm illustrated in Fig. 2 addition of neuropeptide Y outside the patch pipette failed to influence the activity of voltage sensitive Ca^{2+} channels in the on-cell patch indicating that no diffusible second messenger is involved. Similar kinds of results have also been obtained in experiments on sympathetic and sensory neurons as well (Green and Cottrell, 1988; Forscher et al., 1988; Lipscombe et al., 1989). Thus, in general, it appears that this kind of modulation does not require a diffusible second messenger molecule. It should also be mentioned that there are examples of the neurotransmitter induced modulation of other kinds of voltage sensitive Ca^{2+} channels in nerve cells including T and L-channels (Bean, 1989a). In these cases diffusible second messengers may sometimes be important. Diacylglycerol and activation of protein kinase C have often been suggested as being important in these instances (Rane et al., 1989). It is possible that in some cases neurotransmitter release can mediated by L-channels rather than N-channels (Perney et al., 1986). L-channels clearly do exist in nerve terminals in addition to N-channels (Lemos and Nowycky, 1989) and in some cases they may be the only channel type present (Lindgren and Moore, 1989). Furthermore, it has also been shown that in some cases neurons utilize more than one type of neurotransmitter. In such cases different conditions may be required for the release of the different substances and different Ca^{2+} channels may be involved.

These experiments therefore show that N-type Ca^{2+} channels in nerve cells can be modulated by inhibitory transmitters through a G-protein mediated mechanism. Further studies have revealed the mechanism by which these G-proteins alter the function of N-type Ca^{2+} channels. Bean, (1989b) has shown that inhibitory transmitters appear to produce a large shift in the voltage dependence of channel activation. Neurotransmitters appear to shift a proportion of the Ca^{2+} channels into a state where they only open at very strong depolarizations. Thus, for small to moderate depolarizations, the Ca^{2+} current appears to be reduced by the transmitter although for very strong depolarizations the Ca^{2+} channels are still

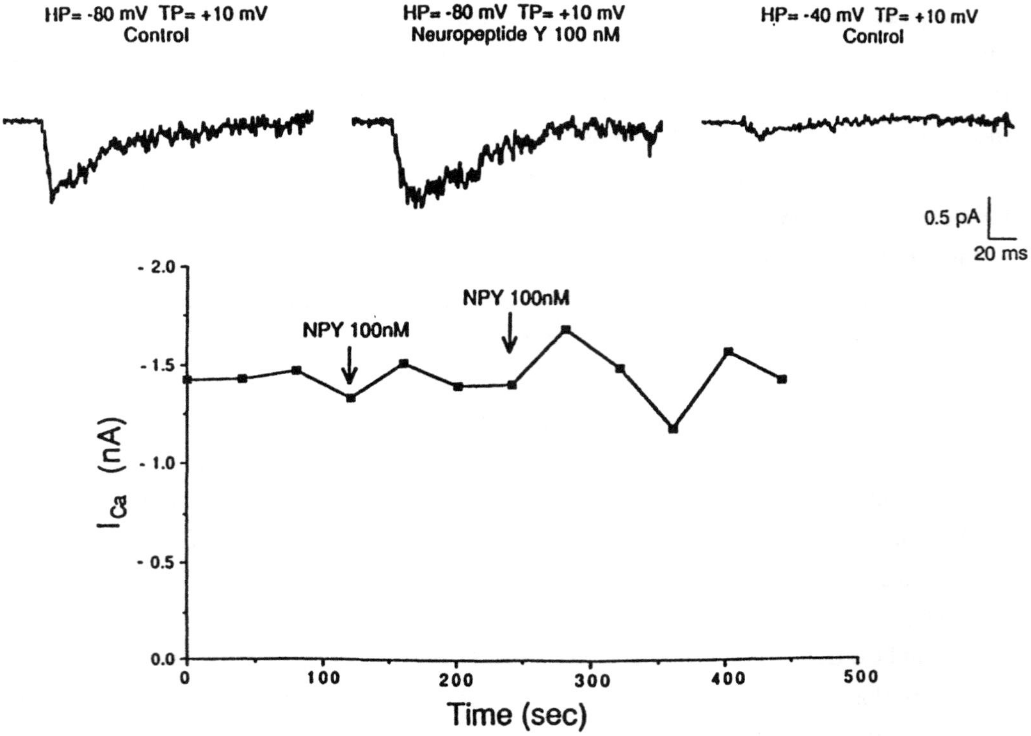

Fig. 2. No easily diffusible second messenger mediates the NPY inhibition of N-type Ca^{2+} channels in a cell attached multi-channel patch. Graph illustrates amplitude of peak currents as a function of time after the addition of NPY (100 nM) to the bath, outside the patch pipette. NPY added at each arrow. A mean current was obtained by averaging ten individual current records. Each time point on the graph represents the peak inward Ca^{2+} current measured from these mean currents. Representative mean currents obtained before and after the addition of NPY are shown in the top row (left, middle traces). The right mean current shows that the patch had mainly N-type Ca^{2+} channels; after changing to HP = -40 mV, almost all the current inactivated.

available for activation and inhibitory effects of the transmitter are lost. It has been suggested that N-type Ca^{2+} channels exist in two states which are in equilibrium with one another. Inhibitory neurotransmitters shift the channel equilibrium in favor of the state which is more difficult to activate. At the single channel level these inhibitory neurotransmitters also produce several effects on channel gating properties including a reduction in the mean open time of the channel, an increase in the latency to first opening and an increase in a number of null sweeps observed during an experiment (Lipscombe et al., 1989).

Although regulation of voltage sensitive Ca^{2+} channels of the type described is clearly important in the production of presynaptic inhibition, it is the actual Ca^{2+} signal within the nerve cell which is responsible for triggering the release of the neurotransmitter. It is therefore important to understand those factors which shape the $[Ca^{2+}]_i$ transient in the neuron which results from the influx of Ca^{2+} through voltage sensitive Ca^{2+} channels during an action potential. Clearly, in addition to Ca^{2+} entry other factors such as Ca^{2+} buffering systems will also be important in this regard. Fig. 3 illustrates recordings in which voltage sensitive Ca^{2+} currents and their accompanying Ca^{2+} transients have been simultaneously recorded in rat dorsal root ganglion cells under voltage clamp filled using the Ca^{2+} sensing dye fura-2. It can be seen that following a relatively short Ca^{2+} influx the amplitude of the $[Ca^{2+}]_i$ transient remains elevated for a considerable amount of time and is buffered relatively slowly. As the length of the voltage step becomes longer and Ca^{2+} influx increases in duration, the amplitude of the accompanying $[Ca^{2+}]_i$ transient increases, but eventually reaches an asymptote. Thus, it appears that as $[Ca^{2+}]_i$ becomes greater, the efficiency of Ca^{2+} buffering systems within the nerve cell also becomes greater. A similar phenomenon is observed if Ca^{2+} influx is elicited by firing trains of action potential rather than by voltage steps of increased duration (Fig. 3). Here again the size of the $[Ca^{2+}]_i$ transient obtained in response a train of action potentials increases with the length of the spike train, but eventually reaches an asymptote. Such behavior has some interesting consequences for the regulation of transmitter release by inhibitory neurotransmitters. Thus, it can be predicted that for relatively short spike trains or relatively short voltage steps, a decrease in Ca^{2+} influx produced by inhibition of presynaptic Ca^{2+} channels will lead to a corresponding decrease in the amplitude of the $[Ca^{2+}]_i$ transient. This will presumably decrease the secretion of the neurotransmitter. However, if the cell is stimulated strongly with a spike train of long duration then the inhibition of the Ca^{2+} channels can be "overcome" and inhibition of Ca^{2+} influx will not lead to a corresponding decrease in the amplitude of the Ca^{2+} transient. This observation is interesting in the light of a considerable amount of data in the literature showing that the effectiveness of presynaptic inhibition depends on the activation state of the neuron cell. Inhibitory transmitters that exert presynaptic inhibition tend to be much more effective when the nerve cell is firing at a fairly low rate than if it is firing at a high rate (Starke, 1987). This kind of effect can be observed in Fig. 4.

These experiments therefore illustrate that there are a number of types of voltage sensitive Ca^{2+} channels in nerve cells and that one type, the N-type Ca^{2+} channel, is frequently associated with the release of neurotransmitters. Modulation of this kind of Ca^{2+} channel can be achieved presynaptically through a G-protein mediated transduction mechanism and this can lead to changes in the $[Ca^{2+}]_i$ transient and the secretion of the neurotransmitter.

Presynaptic inhibition in the central nervous system

The idea that presynaptic inhibition may be due to a blockade of voltage dependent Ca^{2+} channels in the presynaptic neuron can also be shown to have validity in the central

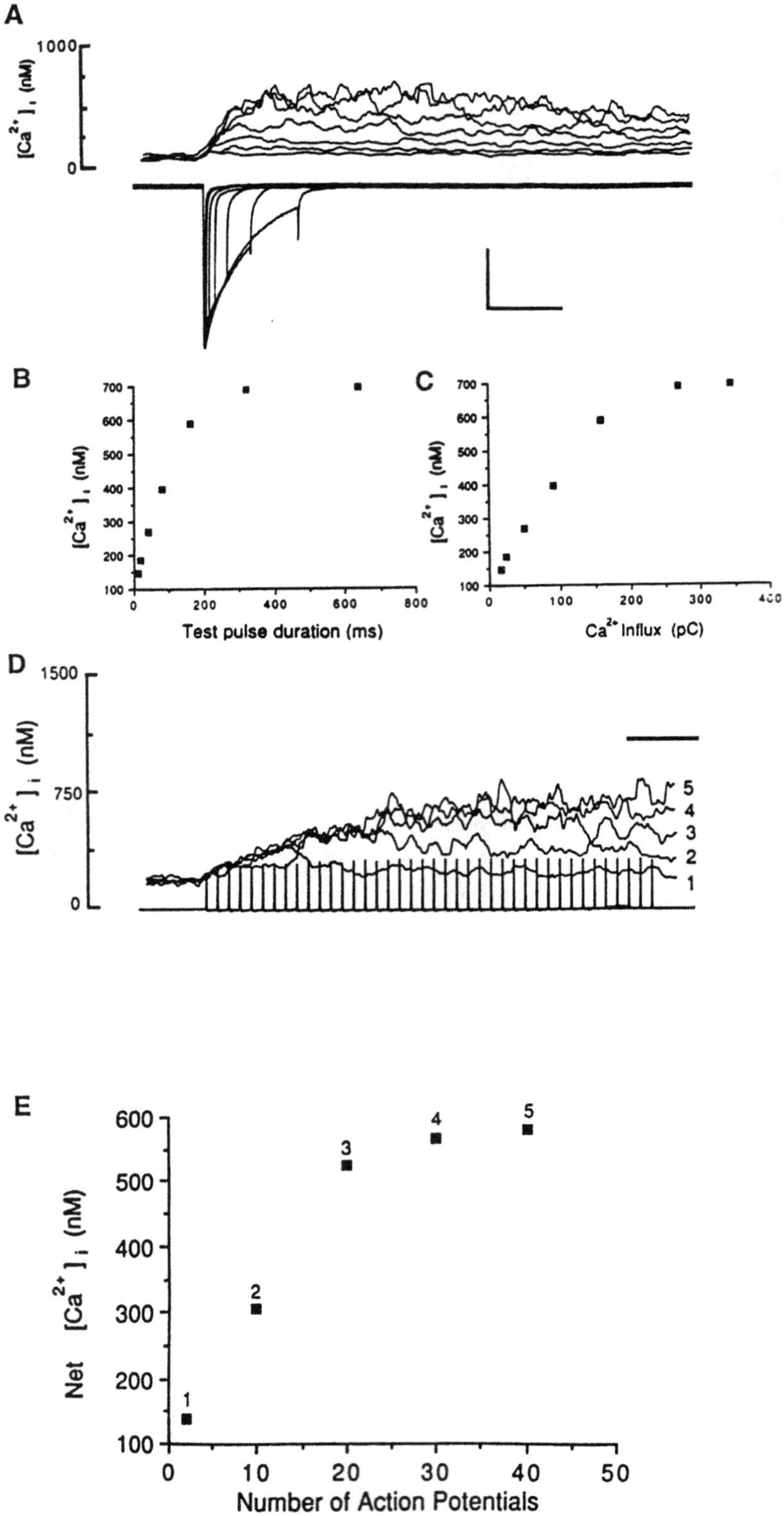
A
1000
0
[Ca2+]i (nM)
B
[Ca2+]i (nM)
Test pulse duration (ms)
C
[Ca2+]i (nM)
Ca2+ Influx (pC)
D
1500
750
0
[Ca2+]i (nM)
5
4
3
2
1
E
Net [Ca2+]i (nM)
Number of Action Potentials

Fig. 3. (A) the increase in the amplitude of the $[Ca^{2+}]_i$ transient in a rat DRG cells in response to step depolarizations from a holding potential of -80 mV to 0 mV for durations varying between 10 and 640 ms using the whole cell voltage clamp technique and simultaneous measurement of fura-2 (100 μM) fluorescent intensities at 340 and 380 nm. The scale bars denote 500 pA (vertical) and 500 ms (horizontal). the lower figures show plots of data in (A), showing (B) The relationship between the net $[Ca^{2+}]_i$ obtained (peak minus basal $[Ca^{2+}]_i$ and test pulse duration and (C) The relationship between the net $[Ca^{2+}]_i$ obtained and the total Ca^{2+} influx (pC, integral of the inward current during the test pulse). (D) Simultaneous current-clamp and $[Ca^{2+}]_i$ recordings from a DRG cell. The rise in $[Ca^{2+}]_i$ is due to action potentials elicited by repeating injections of 890 pA of current for 4 ms. Traces 1-5 are 2, 10, 20, 30 and 40 action potentials in a train respectively with 40 seconds between each series of current injections from the same cell. The horizontal bar denotes 500 ms. (E) Figure showing the asymptote in $[Ca^{2+}]_i$ reached resulting from action potentials evoked by the current injection shown in (D).

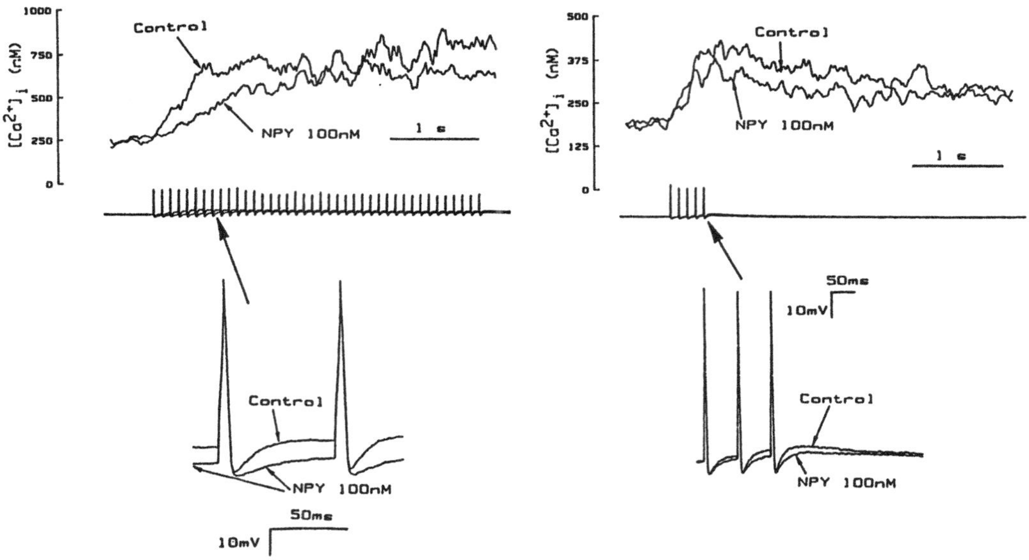

Fig. 4. Effect of NPY on $[Ca^{2+}]_i$ transients in a rat DRG cell. Left hand panel illustrates effects of NPY during an extended spike train in comparison to a short spike train in the right hand panel. Lower traces illustrate effects of NPY on spike afterdepolarization.

nervous system as well. Fig. 5 illustrates some experiments investigating synaptic transmission between hippocampal pyramidal neurons in culture. The neurotransmitter at these synapses is known to be glutamate. It can be seen that under normal ionic conditions cells only fire occasional spikes. However, it is quite clear that under conditions where external Mg^{2+} is removed and extra glycine is added to the medium, a large increase in synaptic activity is observed, including the appearance of large plateau depolarizations with frequent spike trains superimposed on top of them (Abele et al., 1990). This activity is due to the increased stimulation of NMDA receptors by glutamate released at these synapses.

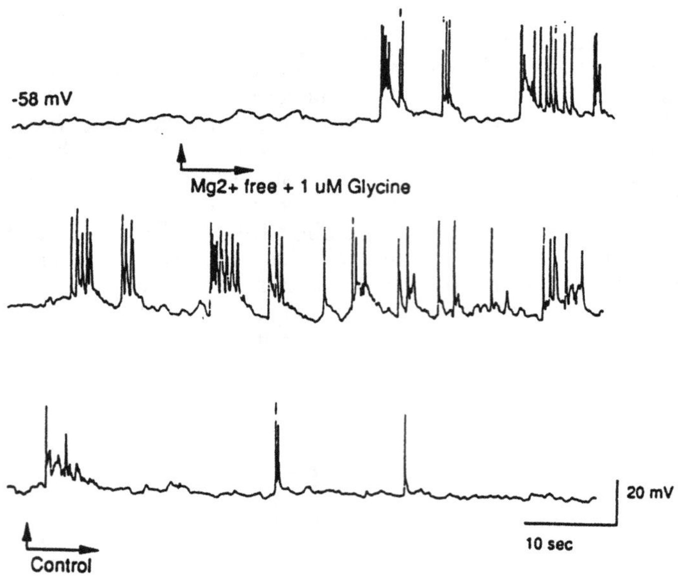

Fig. 5. Whole-cell patch recording of spontaneous activity in a hippocampal pyramidal neuron in culture. The three traces are continuous in time. At the indicated time, bath perfusion of nominally Mg^{2+}-free, glycine (1 μM) supplemented solution was initiated. This cell had been in culture for 12 days.

Synaptic transmission at these synapses can also be assessed in another way which takes advantage of the ability of NMDA to open a Ca^{2+} permeable channel in the postsynaptic membrane. It can be seen that under low Mg^{2+}, glycine supplemented conditions one can also observe large transient increases in the $[Ca^{2+}]_i$ of individual pyramidal neurons. Cells observed under these conditions show either frequent $[Ca^{2+}]_i$ spikes or large maintained increases in $[Ca^{2+}]_i$. Here again it is clear from Fig. 6 that these are due to stimulation postsynaptic NMDA receptors as they can be blocked by drugs that inhibit such receptors. It is also interesting to note that agents that produce presynaptic inhibition in the hippocampus, such as adenosine, also reduce synaptic activity in these cultures presumably by reducing the synaptic release of glutamate. The mechanism by which substances like adenosine reduce the release of glutamate again appears to involve inhibition of voltage sensitive Ca^{2+} current in these cells. Thus, under voltage clamp conditions the Ca^{2+} current recorded in hippocampal pyramidal neurons can be inhibited by adenosine (Scholz et al., 1989). Once again this inhibition is sensitive to pertussis toxin.

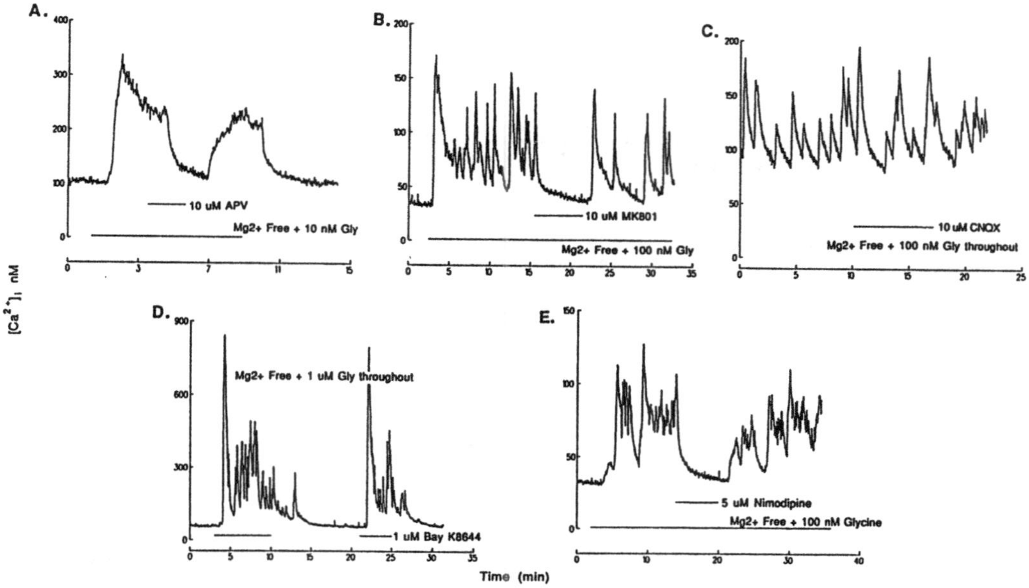

Fig. 6. Pharmacology of $[Ca^{2+}]_i$ fluctuations, observed in single hippocampal pyramidal neurons during perfusion with Mg^{2+} free, glycine supplemented medium. Fluctuations were blocked by inhibitors of NMDA receptors such as APV (A, n=6) or MK-801 (B, n=6) but not by an inhibitor of kainate/quisqualate receptors such as CNQX (C, n=2). In some cells $[Ca^{2+}]_i$ flucutations were not observed but could be triggered by the dihydropyridine Ca^{2+} channel agonist BAY K8644 (n=14 out of 35 cells, D). Fluctuations normally observed be blocked by dihydropyridine Ca^{2+} channel antagonists such as nimodipine (E, n=4). All drugs were effective in every case in which they were tested.

The stimulation of NMDA receptors on these neurons has several consequences. It can be seen in Fig. 7 that when cells are treated with low Mg^{2+}, high glycine conditions for a period of 5 min and then observed 24 hrs later, a considerable number of cells in the culture have died. This is interesting as we know that excessive release of glutamate in the brain under conditions such as stroke or epilepsy leads to the death of nerves. Thus, the observations in cell culture appear to correlate nicely with the situation in vitro. One might therefore predict that anything that interferes with this increased stimulation of NMDA receptors in culture would reduce the observed death of hippocampal neurons. This indeed is the case. Agents that cause presynaptic inhibition such as adenosine, baclofen or NPY or inhibit the actions of NMDA postsynaptically also prevent the death of hippocampal neurons in these cultures.

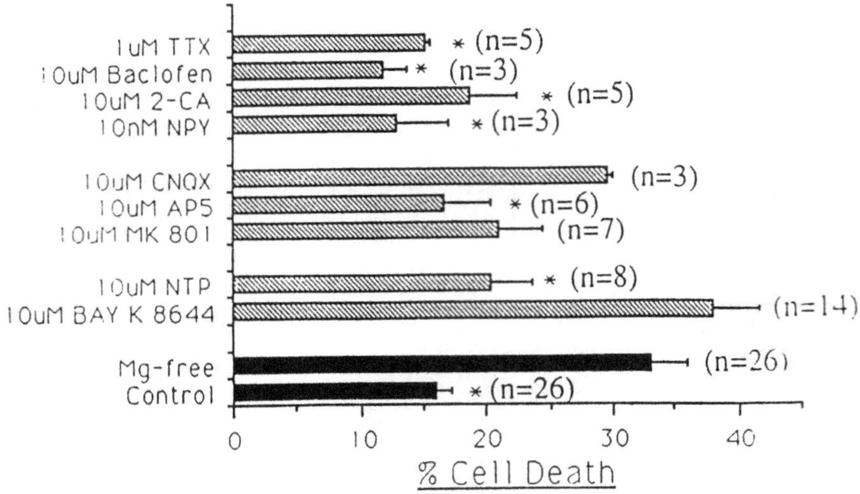

Fig. 7. Hippocampal pyramidal cell death determined 24 hours after cells were treated for 15 min in Mg^{2+}-free solutions with indicated treatments. The percentage of dead cells was determined by counting representative fields from 3 independent coverslips for each experiment. Numbers indicate number of times each treatment was performed. Overall significance was assessed by an analysis of variance ($p < 0.05$). Pairwise comparisons were performed by a Tukey test for multiple comaprisons. Those significantly different from the Mg^{2+} free conditions ($p < 0.05$) are indicated by (*).

Presynaptic inhibition: Some conclusions

The examples I provided above illustrate the fact that inhibition of presynaptic voltage sensitive Ca^{2+} channels, particularly those of the N-type, appear to be responsible for producing presynaptic inhibition throughout the nervous system. There are now a very large number of observations illustrating the ability of neurotransmitters to inhibit the Ca^{2+} current in a variety of vertebrate and invertebrate neurons. However, it should be pointed out that this is not the only way that presynaptic inhibition could be potentially achieved. One should also consider the possibility that activation of presynaptic K^+ or Cl^- conductances could achieve the same ends. Indeed, it is well known that in many cases presynaptic inhibition produced by the inhibitory transmitter GABA proceeds through activation of presynaptic GABA-A receptors and the activation of Cl^- conductances. Furthermore, many of the neurotransmitters that inhibit voltage sensitive Ca^{2+} currents in the way described also have been shown to activate neuronal K^+ conductances presynaptically (Nicoll, 1988; North,

1989). This process is also mediated through pertussis toxin sensitive mechanisms. Such events also seem to occur in invertebrate neurons where a number of neurotransmitters have been shown to both inhibit Ca^{2+} conductances and activate K^+ conductances through G-protein mediated mechanisms (Belardetti and Siegelbaum, 1988). Some attempts have been made to try and determine which of these presynaptic effects is the most important for producing presynaptic inhibition. For example, in one study conducted by Shapiro et al. (1980) on the L-10 neuron of the mollusc Aplysia, it was found that presynaptic inhibition of this neuron was still maintained even when the neuron was stimulated under voltage clamp solely through the production of an inward Ca^{2+} current. Indeed under these conditions presynaptic inhibition appeared to be about 70% as great as in a non-voltage clamped cell. The neurotransmitter responsible for presynaptic inhibition in this case is probably histamine. Exogenously applied histamine clearly does inhibit the Ca^{2+} current in these cells as well as activating a K^+ conductance. Presumably activation of this K^+ conductance is responsible for about 30% of the presynaptic inhibition under normal conditions. At any rate it is possible to see from such an example that these two effects presumably act in a synergistic fashion in many instances.

The discussion in this chapter has concentrated on the regulation of a particular type of neuronal Ca^{2+} conductance and its relationship to the release of neurotransmitters. This is of course an extremely important function of nerve cells. Nevertheless it should be remembered that other forms of voltage sensitive Ca^{2+} channels also exist in neurons and indeed these appear to be frequently modulated as well. Influx of Ca^{2+} through T-channels for example may provide important control of nerve cell excitability and spike pattern generation (Llinas, 1988). Influx of Ca^{2+} through L-channels may also have a number of other functions including control of neurotransmitter release in some cases. Furthermore, the influx of Ca^{2+} through neuronal voltage sensitive Ca^{2+} channels is also important for the regulation of a number of other important conductances in neurons such as Ca^{2+} activated K^+ conductances and Ca^{2+} activated Cl^- conductances. As can be seen in Fig. 4, inhibition of Ca^{2+} through voltage sensitive Ca^{2+} channels during action potentials also leads to a decrease in the activation of a Ca^{2+} activated Cl^- conductance in this particular cell which gives rise to a spike afterdepolarization. This is an example of how inhibition of Ca^{2+} influx may lead to further changes in other ion channels and neuronal excitability. Indeed considering the number of potential functions of Ca^{2+} in nerve cells the ability to regulate its influx is obviously of critical importance in many circumstances.

ACKNOWLEDGEMENTS

I would like to acknowledge the contributions of my colleagues Ken Scholz, Wendy Scholz, April Abele, David Bleakman, Aaron Fox and Lane Hirning all of whom provided data for this paper. This work was supported by DA-02121, DA-02575 and MH-40165.

REFERENCES

Abele, A.E., Scholz, K.P., Scholz, W.K. and Miller, R.J. (1990). Excitotoxicity induced by enhanced excitatory neurotransmission in cultured hippocampal pyramidal neurons. Neuron (in press).

Bean, B.P. (1989a). Classes of calcium channels in invertebrate cells. Ann. Rev. Physiol., 51: 367-385.

Bean, B.P. (1989b). Neurotransmitters inhibit neuronal calcium currents by changes in channel voltage dependence. Nature, 340: 153-156.

Belardetti, F. and Siegelbaum, S.A. (1988). Up and down regulation of single K^+ channel function by distinct second messengers. Trends in Neurosci., 11: 232-238.

Dunlap, K., Holz, G.G. and Rane, S.G. (1987). G-proteins as regulators of ion channel function. Trends in Neurosci.: 10, 241-244.

Ewald, D.A., Sternweis, P.C. and Miller, R.J. (1988). G_o induced coupling of NPY receptors to calcium channels in sensory neurons. Proc. Natl. Acad. Sci. (USA), 85: 3633-3637.

Forscher, P., Oxford, G.S. and Schultz, D. (1988). Noradrenaline modulates calcium channels through tight receptor/channel coupling. J. Physiol., 379: 131-144.

Green, M.A. and Cottrell, G.A. (1988). Actions of baclofen on components of the Ca^{2+} current in rat and mouse dorsal root ganglion neurons in culture. Brit. J. Pharmacol., 94: 235-245.

Hirning, L.D., Fox, A.P., McCleskey, E.W., Olivera, B.M., Thayer, S.A., Miller, R.J. and Tsien, R.W. (1988). Dominant role of N-type calcium channels in evoked release of norepinephrine from rat sympathetic neurons. Science, 239: 57-61.

Hollman, M.M., O'Shea-Greenfield, A., Rogers, S.W. and Heinemann, S. (1989). Cloning by functional expression of a member of the glutamate receptor family. Nature, 342: 643-648.

Holz, G.G., Kream, R.A., Spiegel, A. and Dunlap, K. (1989). G-proteins couple α-adrenergic and GABA-B receptors to inhibiton of peptide secretion from peripheral sensory neurons. J. Neurosci., 9: 657-666.

Kandel, E.R. and Schwartz, J.H. (1982). Molecular biology of learning: modulation of neurotransmitter release. Science, 218: 433-443.

Lemos, J.R. and Nowycky, M.C. (1989). Two types of calcium channels coexist in peptide releasing invertebrate nerve terminals. Neuron, 2: 1419-1426.

Lindgren, C.A. and Moore, J.W. (1989). Identification of ionic currents at presynaptic nerve endings of the lizard. J. Physiol., 414: 201-222.

Lipscombe, D., Kongsamut, S. and Tsien, R.W. (1989). α-adrenergic inhibition of sympathetic neurotransmitter release mediated by modulation of N-type calcium channel gating. Nature, 340: 639-642.

Llinas, R.R. (1988). The intrinsic electrophysiological properties of mammalian neurons: insight into central neuron system function. Science, 242: 1654-1664.

Mayer, M.L. and Miller, R.J. (1990). Excitatory amino acid receptors: regulation of $[Ca^{2+}]_i$ and other second messengers. Trends in Pharmacol. Sci., (in press).

Miller, R.J. (1987). Multiple calcium channels and neuronal function. Science, 235: 46-52.

Mudge, A.W., Leeman, S.E. and Fischbach, G.D. (1979). Enkephalin inhibits release of substance P from sensory neurons in culture and decreases action potential duration. Proc. Natl. Acad. Sci. (USA), 76: 527-532.

Nicoll, R.A. (1988). The coupling of neurotransmitter receptors to ion channels in brain. Science, 241: 545-551.

North, R.A. (1989). Drug receptors and the inhibition of nerve cells. Brit. J. Pharmacol., 98: 13-28.

Nowycky, M.C., Fox, A.P. and Tsien, R.W. (1985). Three types of neuronal calcium channels with different calcium agonist sensitivity. Nature, 316: 440-443.

Perney, T.M., Hirning, L.D., Leeman, S.E. and Miller, R.J. (1986). Multiple calcium channels mediate neurotransmitter release from peripheral neurons. Proc. Natl. Acad. Sci. (USA), 83: 6651-6659.

Rane, S.G., Walsh, M.P., McDonald, J.R. and Dunlap, K. (1989). Specific blockers of protein kinase C block neurotransmitter induced modulation of sensory neuron calcium current. Neuron, 3: 239-245.

Scholz, K.P., Scholz, W.K. and Miller, R.J. (1989). Effects of 2-Cl-adenosine on membrane currents and synaptic transmission in cultured rat hippocampal neurons. Soc. for Neurosci., 15: 177 (abs.).

Shapiro, E., Castelucci, V.F. and Kandel, E.R. (1980). Presynaptic inhibition in Aplysia involves a decrease in the Ca^{2+} current of the presynaptic neuron. Proc. Natl. Acad. Sci. (USA), 77: 1185-1189.

Starke, K.C. (1987). Presynaptic α-autoreceptors. Rev. Physiol. Biochem. Pharmacol., 107: 73-146.

Sternweis, P.C. and Robishaw, D. (1984). Isolation of two proteins with high affinity for guanine nucleotides from membranes of bovine brain. J. Biol. Chem., 259: 13806-13813.

Thayer, S.A. and Miller, R.J. (1990). Regulation of the intracellular free calcium concentrations in single rat dorsal root ganglion neurons in vitro. J. Physiol. (in press).

Tsien, R.W., Lipscombe, D., Madison, D.V., Bley, K.R. and Fox, A.P. (1988). Multiple types of neuronal calcium channels and their selective modulation. Trends in Neurosci., 11: 431-438.

Walker, M.W., Ewald, D.A., Perney, T.M. and Miller, R.J. (1988). Neuropeptide Y modulates neurotransmitter release and Ca^{2+} currents in rat neurons. J. Neurosci., 8: 2438-2446.

CONTROL OF CALCIUM INFLUX AND INTERNAL CALCIUM RELEASE IN ELECTRICALLY NON-EXCITABLE CELLS

O.H. Petersen
MRC Secretory Control Research Group
University of Liverpool
P.O. Box 147
Liverpool L69 3BX
U.K.

In electrically non-excitable cells Ca^{2+} signals evoked by hormones or neurotransmitters are mediated by inositol (1,4,5) trisphosphate (Ins (1,4,5) P_3) evoking Ca^{2+} release from endoplasmic reticulum and this is followed by Ca^{2+} influx from the extracellular fluid through voltage-insensitive pathways controlled by both Ins (1,4,5) P_3 and inositol (1,3,4,5) tetrakisphosphate (Ins (1,3,4,5) P_4 (Petersen, 1989, 1990, Berridge & Irvine, 1989). Ins (1,3,4,5) P_4 is not an independent acute controller of Ca^{2+} movement, but modulates the Ins (1,4,5) P_3-evoked Ca^{2+} mobilization by an unknown mechanism which is only slowly reversible (Petersen, 1989, 1990).

In many different cell types, hormones and neurotransmitters applied in submaximal concentrations evoke repetitive pulses of internal Ca^{2+} release (Berridge & Irvine, 1989). It was recently shown that internal perfusion of single pancreatic acinar cells with Ins (1,4,5) P_3 or its non-metabolizable phosphorothioate derivative evoked repetitive pulses of internal Ca^{2+} release as did stimulation of the muscarinic receptors with acetylcholine (ACh) (Wakui et al., 1989). It therefore now seems unlikely that the Ca^{2+} oscillations are due to pulsatile Ins (1,4,5) P_3 formation (Wakui et al., 1989; Berridge & Irvine, 1989). The currently most attractive hypothesis explaining stimulant-evoked Ca^{2+} oscillations is based on the existence of at least two separate intracellular non-mitochondrial Ca^{2+} pools, one sensitive and the other insensitive to Ins (1,4,5) P_3. An initial release of Ca^{2+} from the Ins (1,4,5) P_3-sensitive pool could act as a primer for further Ca^{2+} release from Ins (1,4,5) P_3-insensitive pools producing a spike that could spread throughout a cell but although there is evidence for a Ca^{2+}-induced Ca^{2+} release in the electrically excitable muscle, nerve and chromaffin cells the position has until recently been unclear in electrically non-excitable cells, such as, for example, exocrine acinar cells.

By combined and simultaneous recordings of changes in $[Ca^{2+}]_i$ near the plasma membrane (assessed by measurement of Ca^{2+}-dependent Cl^- current in the patch-clamp whole-cell recording configuration) as well as in the cytoplasma as a whole

NATO ASI Series, Vol. H 48
Calcium Transport and
Intracellular Calcium Homeostasis
Edited by D. Pansu and F. Bronner

(assessed by measuring fluorescence of fura-2 by photon counting over the surface of a single cell) the stimulant-evoked Ca^{2+} oscillations in pancreatic acinar cells have now been further characterized. The surface membrane receptors for ACh and CCK have been stimulated or direct activation of G-proteins has been achieved through internal application of the non-hydrolysable GTP analogue GTP-γ-S. We have also infused Ins (1,4,5) P_3, inositol (1,4,5) trisphosphorothioate (Ins (1,4,5) PS_3) or Ca^{2+} into the cells. All these procedures evoke pulsatile Ca^{2+} mobilization, but the pattern of Ca^{2+} spike generation varies considerably. The Ca^{2+} spikes observed during both Ins (1,4,5) P_3 and Ca^{2+} infusion are short and more easily detected near the plasma membrane (Ca^{2+}-dependent Cl^- current) than in the cell at large (microfluorimetry). The ACh-evoked Ca^{2+} pulses can in some cases be equally brief, but are mostly broader. The Ca^{2+} spikes induced by CCK or GTP-γ-S are generally of somewhat longer duration. Caffeine, a drug long known to enhance Ca^{2+}-induced Ca^{2+} release potentiates the action of stimulants so that, for example, brief Ca^{2+} pulses evoked by Ins (1,4,5) P_3 become longer and larger and are seen also in the microfluorimetric traces (Osipchuk et al., 1990). Our data indicate that Ca^{2+}-induced Ca^{2+} release plays an important role in the generation of pulsatile internal Ca^{2+} release, but also suggest that surface receptor activation may influence the pattern of Ca^{2+} pulse generation by factors additional to Ins (1,4,5) P_3 formation.

Ca^{2+} Oscillations

A Ca^{2+} signal is an increase in the cytoplasmic Ca^{2+} concentration ($[Ca^{2+}]_i$) and can be generated either by release of Ca^{2+} from intracellular stores or by opening of Ca^{2+} channels in the plasma membrane. Here I am concerned solely with Ca^{2+} signals evoked by hormone or neurotransmitter activation of receptors linked via GTP-binding proteins (G proteins) to the enzyme phosphoinositidase C (also often referred to as phospholipase C). Binding of neurotransmitters or hormones to such receptor types causes breakdown of the membrane-bound inositol lipid phosphatidyl-inositol (4,5) bisphosphate (PIP_2) producing initially two separate messengers namely the lipid soluble diacylglycerol (DAG) and the water soluble inositol (1,4,5) trisphosphate (Ins(1,4,5)P_3) (Berridge, 1987). In 1983 it was shown for the first time that Ins (1,4,5) P_3 releases Ca^{2+} from an intracellular non-mitochondrial store (Streb, Irvine, Berridge & Schulz, 1983).

The Ca^{2+} signals often oscillate and since receptor-mediated repetitive Ca^{2+} spikes can be observed in the absence of extracellular Ca^{2+} it would appear that pulsatile release of Ca^{2+} from intracellular stores is primarily responsible for the oscillations.

From the simplest possible model concept it seems that pulsatile Ins (1,4,5) P_3 formation could provide a straightforward explanation for the oscillating Ca^{2+} signal. At this point in time it is not possible to measure the Ins (1,4,5) P_3 concentration in single cells with a high time resolution and we therefore do not know whether Ins (1,4,5) P_3 levels oscillate during receptor activation. It is, however, possible to address the question in a different manner by asking whether a constant concentration of Ins (1,4,5) P_3 can evoke pulsatile Ca^{2+} release.

Intracellular injections of Ins (1,4,5) P_3 have been shown to evoke repetitive membrane current changes or depolarizations due to Ca^{2+}-activation of Cl^- channels. In such experiments a constant intracellular Ins (1,4,5) P_3 concentration cannot be achieved. Similar findings have been obtained when perfusing mouse pancreatic acinar cells or golden hamster egg cells internally with a constant Ins (1,4,5) P_3 concentration in patch-clamp or microelectrode experiments where the cytoplasmic Ca^{2+} fluctuations have been assessed by measuring the Ca^{2+}-dependent Cl^- current under voltage-clamp conditions (Wakui, Potter & Petersen, 1989) or the Ca^{2+}-dependent hyperpolarizing responses (Swann, Igusa & Miyazaki, 1989). Even such experiments can be criticised since it cannot be ruled out that metabolism and particularly Ca^{2+}-dependent phosphorylation of Ins (1,4,5) P_3 to Ins (1,3,4,5) P_4 occurs. To solve this problem it was therefore necessary to use a non-metabolizable Ins (1,4,5) P_3 analogue with Ca^{2+}-releasing activity. Inositol (1,4,5) trisphosphorothioate (Ins (1,4,5) PS_3) is an effective releaser of Ca^{2+} from intracellular stores, but is not metabolized by phosphatase or kinase pathways. Ins (1,4,5) PS_3 perfused continuously via a patch pipette into single pancreatic acinar cells, evokes regular and repetitive spikes of Ca^{2+}-dependent Cl^- current showing that a constant messenger concentration can cause pulsatile intracellular Ca^{2+} release (Wakui, Potter & Petersen, 1989). While these experiments cannot exclude that receptor activation results in pulsatile Ins (1,4,5) P_3 formation it is an unnecessary complication to postulate that such a phenomenon occurs. The most economical hypothesis is that a constantly elevated Ins (1,4,5) P_3 concentration elicits pulsatile Ca^{2+} release in normal intact cells when stimulated by appropriate concentrations of hormones or neurotransmitters.

Ca^{2+}-Induced Ca^{2+} Release

Since Ins (1,4,5) P_3 at a constant concentration can evoke repetitive spikes of intracellular Ca^{2+} release (Wakui, Potter & Petersen, 1989) and since the simplest suggestion would be that Ins (1,4,5) P_3 primarily evokes a constant movement of Ca^{2+} from the Ins (1,4,5) P_3-sensitive Ca^{2+} pool, it is pertinent to ask whether a small constant flow of Ca^{2+} into the cytosol can evoke repetitive transport of Ca^{2+} from another pool.

Ca^{2+}-induced Ca^{2+} release was originally discovered in skinned cardiac muscle cells where a small increase in the Ca^{2+} concentration of the fluid in contact with the sarcoplasmic reticulum causes a large Ca^{2+} release (Endo, 1977). If Ca^{2+}-induced Ca^{2+} release is directly responsible for the cytoplasmic Ca^{2+} oscillations evoked by stimuli generating Ins (1,4,5) P_3 then intracellular Ca^{2+} infusion should be able to mimick the effect of Ins (1,4,5) P_3. This can be tested in experiments in which Ca^{2+} is infused into a single cell while $[Ca^{2+}]_i$ is assessed both by microfluorimetry using fura-2 and by measurement of Ca^{2+}-dependent Cl^- current (Osipchuk, Wakui, Yule, Gallacher & Petersen, 1990). In this type of experiment the electrical current trace monitors $[Ca^{2+}]_i$ in the immediate vicinity of the inner surface of the plasma membrane whereas the microfluorimetrical recording reports the average $[Ca^{2+}]_i$ in the cell. Intracellular Ca^{2+} infusion results in a gradual rise in $[Ca^{2+}]_i$ and this increase is reversed when the Ca^{2+} infusion is stopped and a Ca^{2+} chelator applied. In the cytoplasmic space near the plasma membrane additional short-lasting repetitive Ca^{2+} spikes which are particularly pronounced just before the major sustained $[Ca^{2+}]_i$ rise and in the phase after discontinuation of the Ca^{2+} infusion are observed. These results indicate that Ca^{2+} can induce pulses of Ca^{2+} release primarily from pools very close to the cell membrane. In the internally perfused mouse pancreatic acinar cells Ins (1,4,5) P_3 infusion also evokes Ca^{2+} spikes near the cell membrane which are mostly not reflected in the average $[Ca^{2+}]_i$ recording, although in small cells it is possible to detect Ins (1,4,5) P_3 evoked Ca^{2+} oscillations synchronously with both microfluorimetry and electrophysiology (Osipchuk, Wakui, Yule, Gallacher & Petersen, 1990). Low doses of ACh also predominantly evoke Ca^{2+} spikes seen only at the cell membrane whereas larger doses cause slightly broader Ca^{2+} signals seen throughout the cell (Osipchuk, Wakui, Yule, Gallacher & Petersen, 1990). The conclusion from these experiments is therefore that Ca^{2+} infusion can mimick the action of Ins (1,4,5) P_3.

In the course of normal signal transduction, Ins (1,4,5) P_3 most likely evokes a small steady flow of Ca^{2+} into the cytosol that subsequently causes repetitive pulses of Ca^{2+} release from separate stores and a minimal quantitative model based on this concept has recently been developed (Goldbeter, Dupont & Berridge, 1990). According to this model two different types of Ca^{2+} channels must exist in intracellular Ca^{2+} storing organelles, namely Ins (1,4,5) P_3-activated and Ca^{2+}-activated Ca^{2+} release channels.

It could be argued that the results demonstrating Ca^{2+}-induced Ca^{2+} oscillations could also be explained by Ca^{2+}-induced Ins (1,4,5) P_3 formation due to the reported Ca^{2+}-sensitivity of phosphoinositidase C (Taylor, Merritt, Putney & Rubin, 1986. It is therefore possible that intracellular Ca^{2+} infusion mimicks the action of intracellular Ins (1,4,5) P_3 not because of direct Ca^{2+}-activation of Ins (1,4,5) P_3-insensitive Ca^{2+} release channels, but simply because it generates Ins (1,4,5) P_3.

The Ins (1,4,5) P_3 receptor antagonist Heparin is a useful tool for testing this point since the prediction is that this substance should inhibit receptor-activated and Ins (1,4,5) P_3-induced, but not Ca^{2+}-induced Ca^{2+} oscillations. If, on the other hand, Ca^{2+} infusion worked via Ins (1,4,5) P_3 generation then heparin should also block the Ca^{2+}-induced Ca^{2+} spikes. Using internally perfused single pancreatic acinar cells my group has recently shown that pulsatile Ca^{2+}-dependent Cl^- current responses to external acetylcholine or internal Ins (1,4,5) P_3 application are blocked by intracellular infusion of heparin whereas in the same experiments the Ca^{2+} oscillations evoked by Ca^{2+} infusion are unaffected (Wakui, Osipchuk & Petersen, unpublished observations). This is direct evidence showing that Ca^{2+} infusion evokes Ca^{2+} release that is not mediated by the Ins (1,4,5) P_3-activated Ca^{2+} channels.

In order to test the hypothesis that the Ca^{2+}-induced Ca^{2+} release in electrically non-excitable cells is mediated by channels of the type found to be responsible for this effect in muscle cells it is useful to employ caffeine, a well established potentiator of Ca^{2+}-induced Ca^{2+} release in muscle (Endo, 1977). The prediction is that caffeine should enhance responses not only to submaximal receptor activation and Ins (1,4,5) P_3 application, but also to submaximal Ca^{2+} infusion.

We have recently shown that 1 mM caffeine can evoke cytoplasmic Ca^{2+} oscillations when applied in the presence of a sub-threshold dose of acetylcholine. This effect of caffeine was rapid and rapidly reversible. Caffeine also markedly enhanced the Ca^{2+} signals evoked by direct G-protein activation (using internal application of GTP-γ-S) or by infusion of Ins (1,4,5) P_3 (Osipchuk, Wakui, Yule, Gallacher & Petersen, 1990). Caffeine (1 mM) evoked regular Ca^{2+} spikes when applied in the presence of a subthreshold intracellular Ca^{2+} infusion, but in the absence of stimulation caffeine did not evoke any effect (Osipchuk, Wakui, Yule, Gallacher & Petersen, 1990). This does not necessarily mean that a normal intact cell is unresponsive to caffeine. In internal cell perfusion studies it is normal to use a Ca^{2+}-free pipette solution with a low concentration (about 0.2 to 0.5) of the Ca^{2+} chelator EGTA. In addition the fluorescent dye fura-2, which also has the properties of a Ca^{2+} chelator, is often present. Thus the Ca^{2+} buffering capacity inside the cell is enhanced under such experimental conditions. If an internal perfusion solution not containing EGTA is used and electrophysiological recordings alone are employed, obviating the need for fura-2, then caffeine (1 mM) does evoke Ca^{2+} spikes when added alone without any other stimulants (Wakui & Petersen, 1990). It is therefore now apparent that caffeine, a drug that lowers the threshold for Ca^{2+}-induced Ca^{2+} release in muscle (Endo, 1977) can under the right circumstances evoke regular Ca^{2+} oscillations also in electrically non-excitable cells. This indicates a functional role for Ca^{2+}-induced Ca^{2+} release channels in the generation of pulsatile Ca^{2+} outflow from intracellular Ca^{2+} storage organelles.

Marty & Tan (1989) have shown in patch-clamp experiments on rat lacrimal acinar cells that immediately after the transition from the cell-attached to the whole-cell configuration there is a transient increase in Ca^{2+}-dependent Cl^- current if the pipette solution contains 0.2 or 1 mM Ca^{2+}. Unlike the situation in pancreatic acinar cells (Osipchuk, Wakui, Yule, Gallacher & Petersen, 1990) regular Ca^{2+} oscillations were not seen with this Ca^{2+} infusion protocol, but the initial Ca^{2+} transient represents Ca^{2+}-induced Ca^{2+} release and this phenomenon was abolished if 50 μM ruthenium red had been introduced into the cell prior to the Ca^{2+} challenge. These experiments (Marty & Tan, 1989) therefore represent further support for our model since ruthenium red is an inhibitor of the Ca^{2+} induced Ca^{2+} release channel.

Negative Feedback Exists

Although the minimal quantitative two-pool model to account for cytoplasmic Ca^{2+} oscillations (Goldbeter, Dupont & Berridge, 1990) proposes that individual Ca^{2+} spikes are terminated simply due to emptying of the relevant Ca^{2+} pool and that a new spike can only occur when the pool is full again it is nevertheless useful to examine an alternative possibility involving Ca^{2+} inhibition of Ca^{2+} release.

Evidence for an inhibitory effect by Ca^{2+} on Ca^{2+} release has been obtained in pancreatic acinar cells. In patch-clamp whole-cell recording studies, Ca^{2+} oscillations initiated by caffeine can be inhibited transiently by intracellular Ca^{2+} infusion and Ca^{2+} oscillations initiated by intracellular Ca^{2+} infusion can be inhibited by the Ca^{2+} ionophore ionomycin. Ionomycin also inhibits Ca^{2+} oscillations evoked by acetylcholine and Ins (1,4,5) P_3 (Wakui & Petersen, 1990). The important point is that Ca^{2+}-induced oscillations are inhibited by further Ca^{2+} mobilization indicating that in this system the negative feed-back occurs at the level of the Ca^{2+} release from the Ins (1,4,5) P_3-insensitive pool. Although it cannot be excluded that in the experiments employing the Ca^{2+} ionophore ionomycin the inhibitory effects were due either to some direct interference by this lipophilic compound with the Ca^{2+} release mechanism or to ionophore-mediated emptying of the relevant Ca^{2+} pools, these explanations cannot apply to the inhibitory effects of Ca^{2+} infusion on, for example, caffeine-evoked oscillations (Wakui & Petersen, 1990).

Ca^{2+} Influx

Up to now the receptor-activated cytoplasmic Ca^{2+} oscillations have been dealt with as if they were completely independent of extracellular Ca^{2+}. This is reasonable in the sense that oscillations in $[Ca^{2+}]_i$ have been observed in many systems in the absence of external Ca^{2+} but in other studies receptor-activated Ca^{2+} spikes disappear when external Ca^{2+} is removed and rapidly reappear after Ca^{2+} readmission.

The mouse pancreatic acinar cells are particularly interesting since extracellular Ca^{2+} dependency is present or absent depending on the precise experimental situation. In intact single cells where the Ca^{2+} oscillations are studied by microfluorimetry after loading with fura-2 acetoxymethyl ester (Yule & Gallacher, 1988) acetylcholine evokes oscillations in $[Ca^{2+}]_i$ which in the initial phase (1-2 minutes) are independent of external Ca^{2+}. In the sustained phase of stimulation, however, the removal of Ca^{2+} immediately causes cessation of the oscillations and these reappear after external Ca^{2+} readmission. Surprisingly the same cells isolated in exactly the same way in the same laboratory behave in a somewhat different manner when investigated in the patch-clamp whole-cell recording configuration. In such experiments both acetylcholine- and Ins (1,4,5) P_3-evoked Ca^{2+} oscillations are completely unaffected by removal and readmission of external Ca^{2+} for about 7-8 min after start of stimulation (Wakui, Potter & Petersen, 1990), but external Ca^{2+} dependency thereafter gradually develops and finally becomes absolute. These experiments show that the oscillation mechanism is not primarily linked to control of Ca^{2+} influx through the surface cell membrane, but indicate that during prolonged periods of stimulation intracellular Ca^{2+} deprivation occurs and reloading from the extracellular Ca^{2+} compartment becomes necessary. The regulation of Ca^{2+} influx and the possible role of inositol (1,3,4,5) tetrakisphosphate in this process is the subject of much discussion, but has been reviewed in detail elsewhere (Petersen, 1989; 1990).

References

Berridge MJ (1987) Inositol trisphosphate and diacylglycerol: two interacting second messengers. Ann Rev Biochem 56:159-193

Berridge MJ, Irvine RF (1989) Inositol phosphates and cell signalling. Nature 341:197-205

Endo M (1977) Calcium release from the sarcoplasmic reticulum. Physiol Rev 57:71-108

Goldbeter A, Dupont G, Berridge MJ (1990) Minimal model for signal-induced Ca^{2+} oscillations and for their frequency encoding through protein phosphorylation. Proc Natl Acad Sci USA 87:1461-1465

Marty A, Tan YP (1989) The initiation of calcium release following muscarinic stimulation in rat lacrimal glands. J Physiol (Lond) 419:665-687

Osipchuk YV, Wakui M, Yulé DI, Gallacher DV, Petersen OH (1990) Cytoplasmic Ca^{2+} oscillations evoked by receptor stimulation, G-protein activation, internal application of inositol trisphosphate or Ca^{2+}: simultaneous microfluorimetry and Ca^{2+}-dependent Cl^- current recording in single pancreatic acinar cells. EMBO J 9:697-704

Petersen OH (1989) Does inositol tetrakisphosphate play a role in the receptor-mediated control of calcium mobilization? Cell Calcium 10:375-383

Petersen OH (1990) Regulation of calcium entry in cells that do not fire action potentials. In: Bronner F (ed) Intracellular Calcium Regulation. Alan Liss, New York, p 77-96

Streb H, Irvine RF, Berridge MJ, Schulz I (1983) Release of Ca^{2+} from a non-mitochondrial intracellular store of pancreatic acinar cells by inositol 1,4,5-trisphosphate. Nature 306:67-69

Swann K, Igusa Y, Miyazaki S (1989) Evidence for an inhibitory effect of protein kinase C on G-protein-mediated repetitive calcium transients in hamster eggs. EMBO J 8:3711-3718

Taylor CW, Merritt JE, Putney JW, Rubin RP (1986) Effects of Ca^{2+} on phosphoinositide breakdown in exocrine pancreas. Biochem J 238:765-772

Wakui M, Petersen OH (1990) Cytoplasmic Ca^{2+} oscillations evoked by acetylcholine or intracellular infusion of inositol trisphosphate or Ca^{2+} can be inhibited by internal Ca^{2+}. FEBS Letts 263:206-208

Wakui M, Potter BVL, Petersen OH (1989) Pulsatile intracellular calcium release does not depend on fluctuations in inositol trisphosphate concentration. Nature 339:317-320

Yule DI, Gallacher DV (1988) Oscillations of cytosolic calcium in single pancreatic acinar cells stimulated by acetylcholine. FEBS Letts 239:358-362

PATCH CLAMP STUDY OF Ca CHANNELS IN ISOLATED RENAL TUBULE SEGMENTS

J.C.J. Saunders and L.C. Isaacson,
Department of Physiology,
Medical School, University of Cape Town,
South Africa

Renal tubular reabsorption of filtered calcium (Ca) is believed to be mediated by both active and passive transport mechanisms, operative to varying degrees in each segment of the nephron (6). While both transcellular and paracellular reabsorptive pathways have been posited, the means by which Ca might cross the apical membrane to enter the tubular cells remains unknown; conceivably, this could be by cation exchange, cotransport, a carrier mediated process, or via Ca channels. We have sought the presence of Ca channels in the apical membranes of isolated tubule segments of the rabbit kidney, using the patch clamp technique (9) with barium (Ba) as a charge carrier.

METHODS

Rabbits (either sex, 1-2 kg bw) were sacrificed by exsanguination, and the left kidney removed. Renal tubule segments were isolated (2) in the cold (6-8°C), and transferred to the patch clamp chamber, where they were bathed in Ringer at room temperature (20-24°C). The segments comprised proximal straight tubules (PST), cortical thick ascending limbs (TAL), distal convoluted tubules (DCT), and cortical collecting ducts (CCD). The apical surfaces of the cells lining the individual tubule segments were then accessed either by perfusing the tubule (with Ringer), and inserting the patch pipette through the open distal end (7), or more frequently, by tearing open the tubule. Formation of an inside-out patch from an apical cell membrane was then attempted.

The Ringer contained (mM): NaCl 143, KCl 4.7, $CaCl_2$ 1.3, $MgCl_2$ 1, glucose 5, HEPES 10, and Na acetate 3; the pH was brought to 7.4 by addition of NaOH. The patch pipette contained (mM): $BaCl_2$ 90, HEPES 10, and amiloride .01; the pH was adjusted to 7.4 by addition of $Ba(OH)_2$. In some experi-

NATO ASI Series, Vol. H 48
Calcium Transport and
Intracellular Calcium Homeostasis
Edited by D. Pansu and F. Bronner

ments, in which the channel's Cl permeability was assessed, the Ringer was diluted with an equal volume of isosmotic mannitol (8). In still other experiments, a Ca channel blocker (5 uM Verapamil or 100 uM Nifedipine), or the agonist Bay K-8644 (100 uM), was added to the bath; in these studies both the bath and the patch pipette contained 70 mmol CaCl2. Channel currents were detected by an Axopatch 1B patch clamp amplifier. The signal was filtered with a 6 pole Bessel filter at cut-off frequencies (-3db) of 200 Hz or sometimes 500 Hz, sampled at 1 or 2 kHz, logged on an Archimedes 440 computer, and stored on disc. Recordings were of up to 2 min duration. Consecutive recordings were obtained with the patch voltage clamped at 30 mV to -60 mV, the actual range for any one patch being determined by the 'tightness' of the seal. Amplitude, open- and closed -time histograms were generated by computer. The fractional open time was calculated as the total time spent in the open state, divided by the total time of the recording. In the following, the polarity of the clamp potential is given as that of the pipette interior relative to the bath. Currents corresponding to movement of cations from the pipette into the bath are presented as downward deflections in the Figures.

RESULTS

Apical patches with 'tight' seals (1-10 gigohms) were obtained from fewer than 10% of the kidneys dissected. Channel activity was observed in about 75% of patches. The data reported here were obtained on patches from 10 PSTs, 3 TALs, 5 DCTs, and 9 CCDs. Channel activity was essentially similar in all the tubule segments studied. Three temporal patterns of channel activity were noted. Typically, channel activity manifested as current spikes of short duration (1-10 ms) and variable amplitude, interspersed with episodes of bursting and flickering. In 13 of the 27 patches, channel events of longer duration (10-50 ms) and of relatively constant amplitude were also present, but were of very infrequent occurrence. Finally, channel events of 250-500 ms duration, of relatively constant current amplitude and of frequent occurrence, were seen in three patches (one PST, two CCDs).

Successive channel events were frequently at different multiples of a unit current amplitude. While the filter settings and the sampling rate used in this study preclude precise analysis of variations in current amplitude

In channel events of less than a couple of ms duration, most of the patches displayed bursts of activity at relatively low

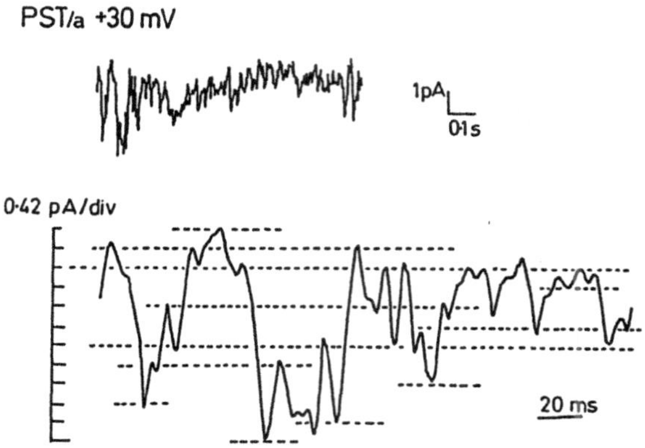

Fig 1

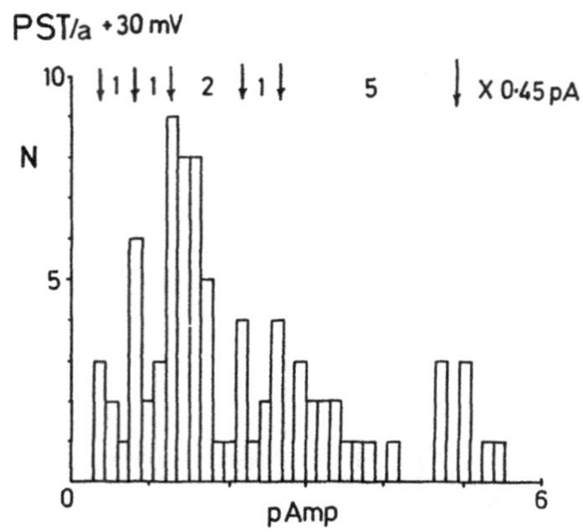

Fig 2

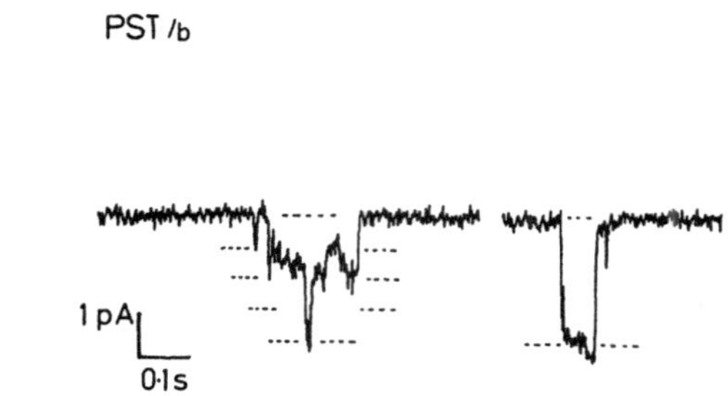

Fig 3

frequencies. Fig. 1 depicts (top) 1 s of such a burst in a PST patch, clamped at +30 mV, and (bottom) an expanded view of the first 256 ms. The current fluctated between different levels (dashed lines) of unit amplitude, with minor inconsistencies presumably attributable to background noise, to interspersed channel events of very short duration, or to intrinsic variability in single channel conductance. In a number of instances successive current peaks differed by several levels of unit amplitude. These appearances were confirmed by the computer-generated amplitude histogram of the entire 1 s burst which revealed current peaks at multiples of 0.45 pA (Fig. 2). Similar observations were made at other clamp voltages, both in this and in most of the other patches. In addition, and despite the bandwidth limitation noted above, amplitude histograms of prolonged recordings (0.3 to 2 min) of predominantly or only spike activity frequently also displayed current peaks at multiples of a unit amplitude. Fig. 3 depicts two channel events, of relatively constant amplitude and longer duration, as observed in another PST patch; the events occurred within a few sec of each other, at the same clamp voltage. The dashed lines denote unit levels of current amplitude. In the second event, the amplitude of the channel current rose abruptly to a level 4 times that of the unit current level in the first, before falling equally abruptly to zero.

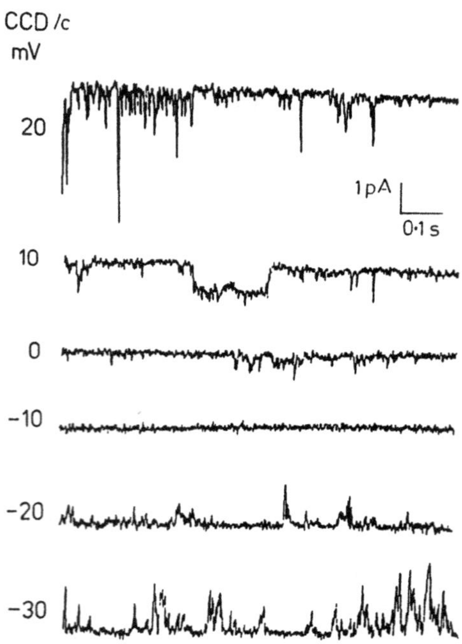

Fig 4

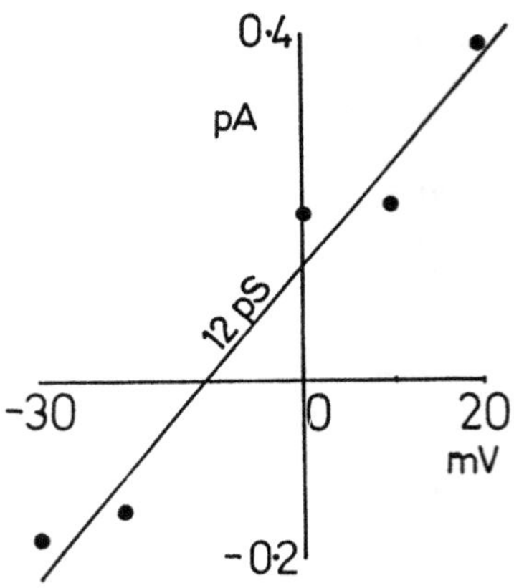

Fig 5

Fig 4 depicts 1 s segment of 1-2 min recordings of channel events in a single CCD patch, clamped at a number of voltages between 20 mV and -30 mV. Both the amplitude and the frequency of channel events increased as the clamp voltage departed from the reversal potential (-10 mV). The fractional open-time increased symmetrically about the reversal potential. The unit current amplitude (derived as above) was linearly related to clamp voltage; the slope conductance, as calculated by linear

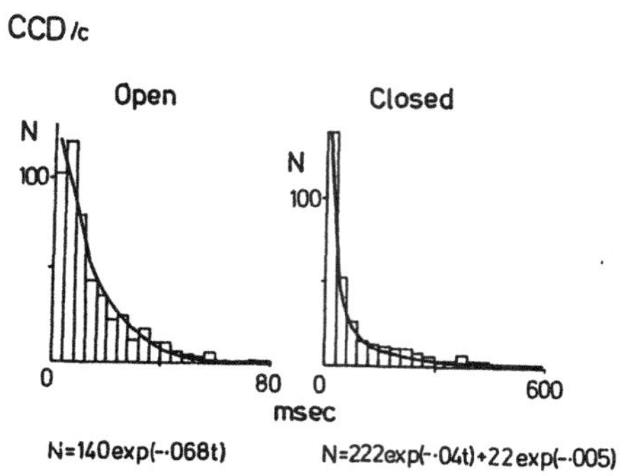

Fig 6

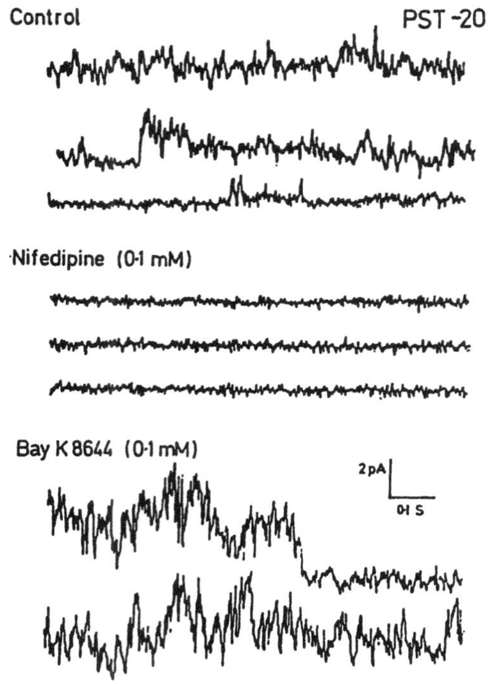

Fig 7

regression, was 12 pS (Fig. 5). Fig. 6 depicts the open- and closed-time histograms of the 2 min recording at +10 mV; these, as also the open- and closed time histograms at the other clamp voltages, could be fitted by single and double exponentials, respectively, Similar observations were made in the remaining patches.

Reversal potentials were usually close to zero mV (mean -1.2 mV; s.d. 4.3. mV). The channels were voltage-sensitive, the fractional open time increasing symmetrically with displacement of the clamp voltage from the reversal potential. There were no apparent differences in the slope conductances of the channels in the various tubule segments; overall, these ranged between 6 pS and 68 pS; mean 27.3 pS, s.e.m. 3.2 pS. With few exceptions, the open- and closed-time histograms could be fitted by one and two exponentials, respectively, so suggesting (5) that the channels possessed a single open state and two closed states. On replacing the bath fluid with isosmotic diluted Ringer (n=2), the reversal potential became more negative, implying channel impermeability to Cl. Addition of Verapamil to the bath reduced the fractional open time at negative but not at positive clamp potentials. Channel activity previously totally abolished by 100 uM Nifedipine was near maximal after exposure to the agonist Bay K-8644 (Fig. 7).

DISCUSSION

The occurrence of currents at multiples of some unit amplitude suggest either the simultaneous activity of many identical channels, or the presence of multiple subconductance levels in a single channel. In terms of binomial probability theory, however, the probability of several discrete and identical channels opening or closing simultaneously (Fig. 1, 3) is vanishingly small. The data are thus consistent with the presence of multiple subconductance levels in single channels, rather than with the presence of multiple identical channels. Reversal potentials close to zero mV reveal that the channels were permeable to more than one ion species. They were not permeable to Cl. Thus current flow at positive clamp voltages reflected predominantly the trans-channel movement of Ba from the pipette into the bath, while that at negative clamp voltages, reflected predominantly the movement of Na in the opposite direction (the concentration of K in the bath fluid being relatively negligible). A number of non-specific ion channels have been reported in renal tissues.

Non-specific cation channels have been described in primary cultures of the rat inner medullary collecting duct (12), as also in the baso-lateral membrane of the thick ascending limb of Henle's loop in the mouse (16). The former, which was found in the apical membrane, was inhibited by amiloride; neither have been reported as occurring elsewhere in the nephron, and neither was Ba permeable. Another non-selective ion channel has been reported in the basolateral membrane of the late proximal tubule of the rabbit (8); this, however, was Cl permeable and Ba impermeable. It is clear that the channel described here is unlike any of these. Voltage sensitive Ca channels have been described in a number of tissues other than renal, with characteristics similar to those found here; multiple temporal patterns of channel activity (10), multiple subconductance levels (3,11,13), permeability to both Ba and Na (11,13,14), kinetics suggestive of a single open and two closed states (1,3,4), comparable magnitude of slope conductance (11,14,15), and appropriate response to Ca channel blockers and agonists.

Thus the data presented here suggest that Ca channels are ubiquitously distributed in the apical membrane of the renal tubule, a suggestion in accord with the concept of Ca reabsorption along the whole length of the nephron (6).

Acknowledgements: The authors are indebted to the Medical Research Council and the University of Cape Town Staff Research Fund for financial support.

REFERENCES

1. Almers, W., and E.W. McCleskey. Non-selective conductance in calcium channels in frog muscle: calcium selectivity in a single file pore. J.Physiol. 353, 586-608, 1984.

2. Burg, M., J. Granthan, M. Abramow and J. Orloff. Preparation and study of fragments of single rabbit nephrons. Am.J.Physiol. 210, 1293-1298, 1966.

3. Cull-Candy, S.G. and M.M. Usowicz. Multiple conductance channels activated by excitatory amino acids in cerebellar neurones. Nature. 325, 525-528, 1987.

4. Fenwick, E.M., A.Marty and E. Neher. Sodium and calcium channels in bovine chromaffin cells. J.Physiol. 332, 599-635, 1982.

5. Franciolini, F., and A. Petris. Single channel recording and gating function of ionic channels. Experientia 44(3), 183-280, 1988.

6. Friedman, A. Renal Calcium Transport: Sites and Insights. News in Physiological Sciences, 3, 17-21, 1988.

7. Gogelein, H. and R. Greger. Single channel recordings from basolateral and apical membranes of renal proximal tubules. Pflugers Arch. 401, 424-426, 1984.

8. Gogelein, H., and R. Greger. A voltage dependent ionic channel in the basolateral membrane of late proximal tubules of the rabbit kidney. Pflugers Archiv. 407, S142-S145, 1986.

9. Hamill, O.P., A. Marty, E. Neher, B. Sakman and F.S. Sigworth. Improved patch clamp techniques for high resolution current recording from cells and cell-free membrane patches. Pflugers Archiv. 391, 85-100, 1981.

10. Hess, P., J.B. Lansman and R.W. Tsien. Different modes of Ca channel gating behaviour favoured by dihydropyridine Ca agonists and antagonists. Nature. 311, 538-544, 1984.

11. Jahr, C.E. and C.F. Stevens. Glutamate activates multiple single channel conductances in hippocampal neurons. Nature. 325, 522-525, 1987.

12. Light, D.B., F.V.McCann, T.M. Keller and B.A. Stanton. Amiloride-sensitive cation channel in apical membrane of inner medullary collecting duct. Am.J.Physiol. 255 (Renal Fluid Electrolyte Physiol. 24), F278-F286, 1988.

13. Qi.Yi, L., F.A.Lai, E. Rousseau, R.V. Jones and G. Meissner. Multiple conductance states of the purified calcium release channel complex from skeletal sarcoplasmic reticulum. Biophys.J. 55, 415-424, 1989.

14. Reuter, H. Calcium channel modulation by neurotransmitters, enzymes and drugs. Nature. 301, 569-574, 1983.

15. Rorsman, P., F.M. Ashcroft and G. Trube. Single Ca channel currents in mouse pancreatic beta-cells. Pflugers Arch. 412, 597-603, 1988.

16. Teulon, J., M. Paulais and M. Bouthier. A Ca++ activated cation-selective channel in the basolateral membrane of the cortical thick ascending limb of Henle's loop of the mouse. Biochimica et biophysica acta. 905, 125-132, 1987.

RECENT DEVELOPMENTS IN THE MOLECULAR CHARACTERIZATION OF THE L-TYPE Ca^{2+}-CHANNEL AND MITOCHONDRIAL Ca^{2+}-ANTAGONIST SITES

F. Scheffauer, M. Canonaco-Friedrich, H.-G. Knaus, J. Striessnig, C. Zech, G. Zernig and H. Glossmann

Institute for Biochemical Pharmacology
Peter-Mayr-Straße 1
A-6020 Innsbruck
Austria

Heparin is an allosteric modulator of specific L-type calcium channel labelling

Several putative endogenous modulators of L-type calcium channels have been described. These modulators are peptides or less defined organic substances which have been isolated from rat or bovine brain. Hanbauer et al., 1989 e.g. described a putative endogenous ligand which inhibited nitrendipine binding and modulated cardiac and neuronal Ca^{++}-channel function (Callewaert et al., 1989). Evidence exists that heparin may serve important functions (e.g. on endothelial cells) in addition to its role in blood coagulation. In this study we have investigated the effects of heparin on L-type calcium channel labelling and function.

The left panel of Fig.1 shows the effect of heparin on (+)-[^{3}H]PN200-110 equilibrium binding parameters of skeletal muscle T-tubule membranes. The heparin inhibition is explained by a decrease of B_{max} (control 53.6 pM; 1 μg/ml heparin 12.1 pM) and by an increase of the apparent K_D (control 131.8 pM; 1 μg/ml heparin 553.3 pM). Preincubation of rabbit skeletal muscle T-tubule membranes with 1 μM (+)-cis-diltiazem protected partially against the inhibitory effect of heparin (K_D 340 pM, B_{max} 26.2 pM). The right panel of Fig.1 exemplifies the effect of heparin on (+)[^{3}H]PN200-110 labelling in the absence and presence of the diastereoisomers of cis-diltiazem. Under control conditions heparin inhibited (+)-[^{3}H]PN200-110 labelling of rabbit skeletal muscle T-tubule membranes with an IC_{50} value of 0.215 ± 0.02 μg/ml (maximal inhibition 94.8%), addition of 3 μM (+)-cis-diltiazem increased the heparin inhibition constant to 0.96 ± 0.23 μg/ml and decreased the maximal inhibition (12.6 ± 3.45 % of control binding remained). The diastereoisomer (-)-cis-diltiazem was without effect. The apparent Hill slopes were always greater than unity. Similar heparin effects were observed when L-type calcium channels were probed with tritiated phenylalkyamines or benzothiazepines (not shown).

NATO ASI Series, Vol. H 48
Calcium Transport and
Intracellular Calcium Homeostasis
Edited by D. Pansu and F. Bronner

The Heparin interaction with skeletal muscle T-tubule calcium channels does not require Ca^{++} and occurs at pH 7.0 to 9.5

It is well established, that heparin interacts with "specific" binding domains (e.g. apolipoprotein B and E, (Weisgraber et al., 1987; Cardin et al., 1986) antithrombin III (Lane et al., 1987), platelet factor IV (Walz et al., 1977)) that are clusters of basic amino acids. These relevant binding domains can be distinguished from non-specific interactions of the highly charged polyanion heparin by chelating divalent cations (e.g. by EDTA) or at different pH values to exclude ion-exchange or cross-linking effects by Ca^{++}.

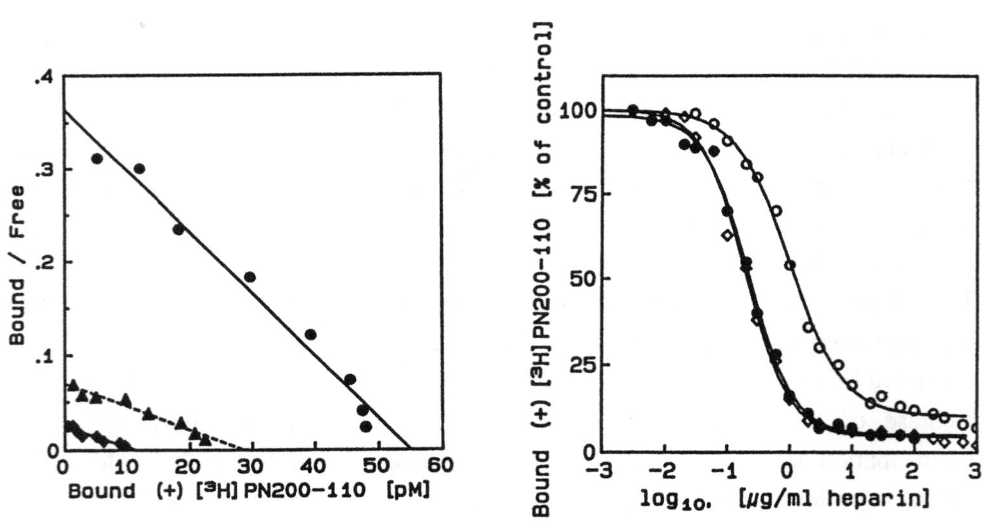

Figure 1
Effect of heparin on 1,4 dihydropyridine Ca^{2+} channel labelling
Left panel: Saturation isotherm (Scatchard transformation) for the (+)-[^{3}H]PN200-110 labelled 1,4 dihydropyridine receptor. Rabbit skeletal muscle T-tubule membrane protein (0.006 mg/ml) was incubated with (+)-[^{3}H]PN200-110 (28.9 - 2931 pM) in the absence (●) and presence (▲,◆) of 1 µg/ml heparin. The filled triangles (▲) indicated identical conditions as (◆) except that 1 µM (+)-cis-diltiazem is present. The corresponding K_D and B_{max} values are given in the text. **Right panel:** Rabbit skeletal muscle T-tubule membranes (0.014 mg/ml) were incubated with 0.346 nM (+)-[^{3}H]PN200-110 in the presence of increasing concentrations of heparin. (○) indicates the presence 3 µM (+)-cis-diltiazem, (◇) the presence of 3 µM (-)-cis-diltiazem.

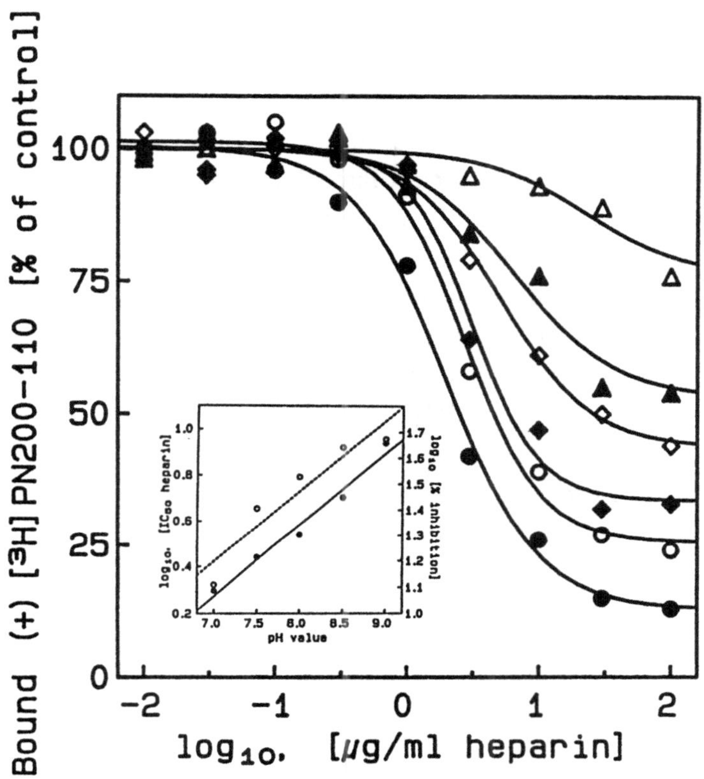

Figure 2
pH effects on heparin inhibition of (+)-[^{3}H]PN200-110 labelling of rabbit skeletal muscle T-tuble membranes
Skeletal muscle T-tubule membranes were washed twice in pH value adjusted 100 mM Bis-Tris-Propane-buffer. 0.012 mg/ml skeletal muscle T-tubule membrane protein (in buffer with the corresponding pH value) was incubated with (+)-[^{3}H]PN200-110 (0.267 nM) in the presence of increasing concentrations of heparin. The symbols used are: (●), pH 7.0; (○), pH 7.5; (◆), pH 8.0; (◇), pH 8.5; (▲), pH 9.0; (△), pH 9.5. The inset shows a semilogarithmic transformation of the pH values versus apparent IC_{50} values calculated from inhibition experiments (left Y axis, solid circles) or % of heparin inhibition of (+)-[^{3}H]PN200-110 labelling (right Y axis, open circles).

Fig.2 shows the effect of heparin on specific (+)-[^{3}H]PN200-110 labelling at different pH values. Increasing the pH value from 7.0 up to 9.5 decreased the inhibitory potency of heparin 5-fold and reduced the maximal inhibition from 91.2% to 27.4% of specific (+)-[^{3}H]PN200-110 binding. The presence of 10 mM EDTA or 3 mM $CaCl_2$ exhibited almost no effect (IC_{50} values for heparin were: control: 1.39 µg/ml, 10 mM EDTA present 1.34 µg/ml, 3 mM $CaCl_2$ present 0.66 µg/ml). 500 mM NaCl significantly decreased (but not eliminated) the inhibitory potency of heparin. With 0.5 M NaCl the IC_{50} value was 5.45 µg/ml and and maximal inhibition was 55.5 % (Fig.3).

Evidence for an extracelluar binding domain for heparin

Heparin effects on L-type calcium channels were investigated by clamping isolated guinea-pig cardiac myocytes at a membrane potential of -35 mV. Heparin, when applied extracellulary, enhanced the ion current through the voltage-gated calcium channel by 40% and shifted the maximum for $I_{Ca}2+$ to the hyperpolarizing direction (not shown). These results provided evidence for a binding site located on an extracellular domain. To investigate the proposed binding domain in more detail, the adsorbtion behavior of purified and reconstituted calcium channel preparations on WGA and heparin Sepharose^R affinity columns was determined. The purified L-type calcium channel is extremely sensitive to heparin (IC_{50} value 0.05 μg/ml) and can be adsorbed to Sepharose^R-immobilized WGA or heparin. Reconstitution into SBL-cholesterol vesicles resulted in a complete loss of the ability of heparin to inhibit (+)-[^{3}H]PN200-

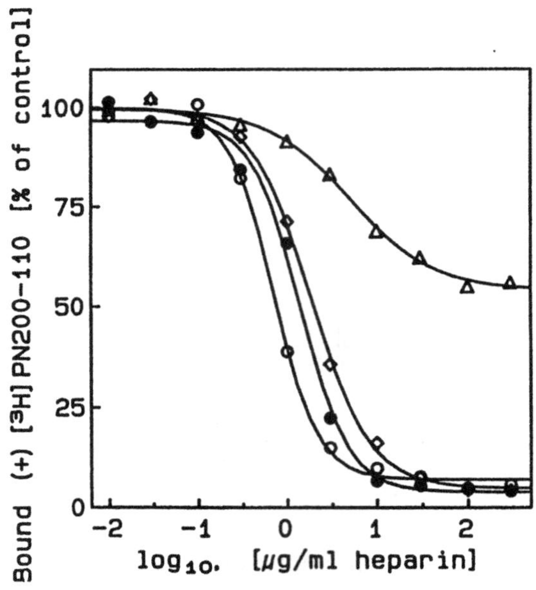

Figure 3
Effects of EDTA, $CaCl_2$ and NaCl on heparin inhibition of Ca^{2+}-antagonist labelling
Rabbit skeletal muscle T-tubule membranes (0.012 mg of protein /ml) were incubated with 0.367 nM (+)-[^{3}H]PN200-110 in the of increasing concentrations of heparin. The following symbols indicate the presence of 500 mM NaCl (Δ), 3 mM $CaCl_2$ (◇) and 10 mM EDTA (○). (●) shows control conditions without added salts or chelators. Data are normalized with respect to binding in the absence of heparin.

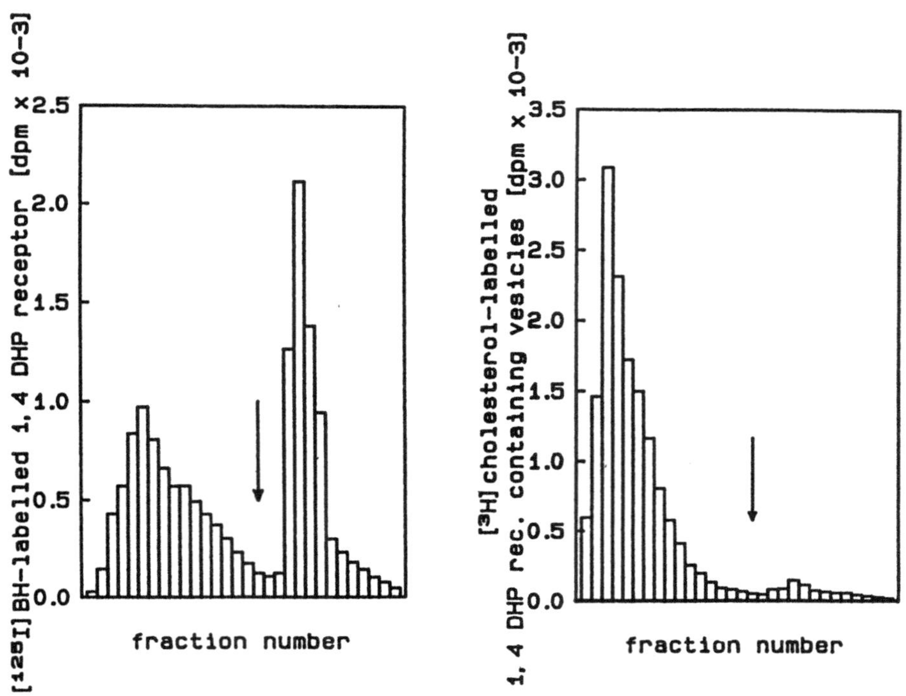

Figure 4
Elution profiles of purified and reconstituted Ca^{2+} channels
Purified rabbit skeletal muscle L-type calcium channels were iodinated with [^{125}I]Bolton-Hunter protein labelling reagent. An aliquot was reconstituted into SBL-cholesterol according to Nunoki et al., 1989 and the adsorbtion behavior of both preparations onto a WGA-sepharose Cl4B (20 x 8 mm) was investigated. Part of the cholesterol was [^{3}H]-labelled. The left panel shows the elution profile prior to reconstitution, the right panel after reconstitution into SBL-cholesterol vesicles. The arrow indicates the start of the biospecific elution with 5% N-acetylglucosamine.

110 binding even up to a final concentration of 300 µg/ml and reconstituted channels were not retained on heparin Sepharose^R. In contrast reconstituted channels are sensitive to EDTA inhibition of 1,4 dihydropyridine labelling (not shown). When purified and reconstituted L-type calcium channels were probed for adsorbtion onto WGA-sepharose, the reconstituted calcium channels failed to bind in contrast to the purified preparation (Fig.4). This suggests that our reconstituted calcium channels were mainly in an outside-in orientation - with an intravesicular location of the carbohydrate side chains. Therefore the lack of heparin effects on reconstituted calcium channel preparations can be explained by the orientation of the extracellular located channel portions unaccessible for both, WGA lectin and heparin.

Effects of heparin on the labelling of non-L-type calcium channel receptors

To prove the selectivity of our heparin probe for voltage-dependent L-type calcium channels, its effects on labelling of other membrane spanning receptors was investigated. Heparin failed to alter the specific binding of radioligands that are markers for i.e. voltage-dependent sodium channels, alpha and beta adrenergic, serotonin, dopamine, opiate and benzodiazepine receptors (not shown). Most notable was the lack of effects of heparin on [^{3}H]batrachotoxinin and [^{3}H]saxitoxin labelling of the rat cortex sodium channel, which is known being structurally very close related to L-type calcium channels (Noda et al., 1984). Therefore heparin seems to act exclusively on L-type calcium channels in comparison with other membrane spanning receptors.

Immuno-photoaffinity labelling of calcium channel drug receptors

Photoaffinity ligands are valuable tools in the characterization of receptors. These ligands often show selective binding and incorporate covalently at or near their site of action by UV-irradiation. It is therefore possible to locate and identify even amino acids which form the binding domain. The irreversible fixation of the incorporated drug allows the use of electrophoretic, chromatographic and immunological methods.

1,4-Dihydropyridine and phenylalkylamine binding-sites have been shown to be localized on the alpha$_1$-subunit of the voltage-dependent L-type calcium channel in different tissues by photoaffinity labelling with the arylazide drugs [^{3}H]azidopine (Ferry et al., 1984; Striessnig et al., 1987; Ferry et al., 1987, Vaghy et al., 1987; Takahashi et al., 1987; Sieber et al., 1987; Sharp et al., 1987) and [N-methyl-^{3}H]LU49888 (Striessnig et al., 1987; Vaghy et al., 1987; Sieber et al., 1987). Recently a benzothiazepine photoaffinity ligand, (+)-cis-azidodiltiazem [(+)-cis-(2S,3S)-5-[2-(4-azido-benzoyl)aminoethyl]-2,3,4,5-tetrahydro-3-hydroxy-2-(4-methoxyphenyl)-4-oxo-1,5-benzothiazepine) and the respective tritiated derivative (+)-cis-[^{3}H]azidodiltiazem were synthesized (Striessnig et al, 1990). It was shown that the radioactive benzothiazepine specifically photo-incorporates into the alpha$_1$-subunit of the skeletal muscle channels.

In all of the conventional photoaffinity-labelling experiments a radioactive photoaffinity-ligand is essential for detection. We decided to raise polyclonal antibodies against the label for immunological detection of the non-radioactive azidodiltiazem. For the production of benzothiazepine-specific antisera, rabbits were immunized with a ligand-carrier conjugate. The antigen was prepared by incubating (+)-cis-azidodiltiazem (AZD) with human serum albumin (HSA) in a molar ratio of about 7:1

including low amounts of (+)-cis-[^{3}H]azidodiltiazem (final specific activity 5-11 dpm/pmol)) and subsequent UV-irradiation.

New Zealand white rabbits were immunized by subcutaneous injection of the antigen emulsified in complete Freund's adjuvant in monthly intervalls and blood was taken before each injection. Titers of the antisera were determined by RIA as well as by a solid phase immunoassay (ELISA, Engvall and Perlman, 1972) using the AZD-HSA-conjugate as a solid phase immobilized antigen. Anti-HSA directed antibodies were

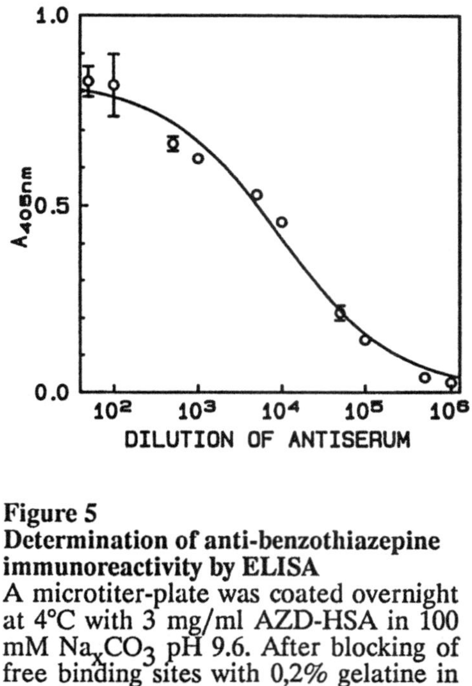

Figure 5
Determination of anti-benzothiazepine immunoreactivity by ELISA
A microtiter-plate was coated overnight at 4°C with 3 mg/ml AZD-HSA in 100 mM Na_xCO_3 pH 9.6. After blocking of free binding sites with 0,2% gelatine in Tris-buffered saline (TBS) for 1 h at 20°C the plate was incubated with serial dilutions of serum 101/9 in TBS (2 h, 20°C). Immunoreactivity was measured by the standard procedure of Engvall and Perlman (1972) using anti-rabbit alkaline phosphatase conjugate as an immunolabel and p-nitrophenyl-phosphate as the chromogenic substrate. Absorbance at 405 nm was read by a micro-ELISA-reader. Each incubation step was followed by a three times washing with TBS-0,05%Tween 20.

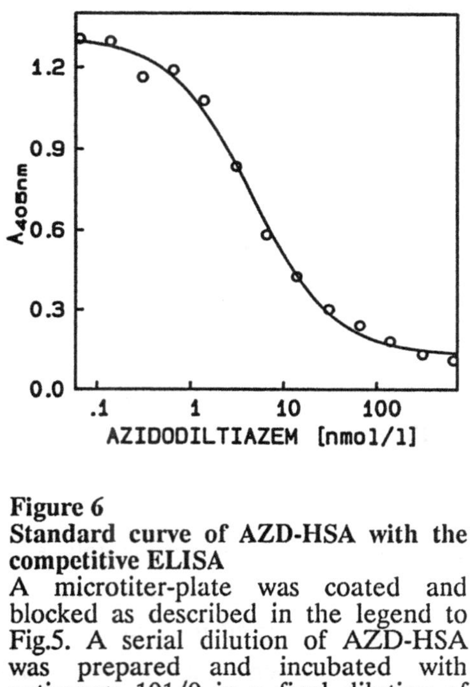

Figure 6
Standard curve of AZD-HSA with the competitive ELISA
A microtiter-plate was coated and blocked as described in the legend to Fig.5. A serial dilution of AZD-HSA was prepared and incubated with antiserum 101/9 in a final dilution of 1:20000 for 1 h at 20°C. To block HSA-reactive antibodies the serum was preblocked with 0.5 mg HSA/ml final assay volume. The samples were allowed to compete with the solid-phase fixed antigen for 2 h at 37°C. Bound antibodies were detected as described in the legend to Fig.5 and values of absorbance were plotted against the log of concentration of protein-bound azidodiltiazem.

TABLE 1.
Crossreactivity of the anti-azidodiltiazem antiserum with calcium channel blockers as determined by competitive ELISA

Drug	Concentration	Inhibition of antibody binding (% of control)
HSA-AZD	$5x10^{-7}M$	100 %
(+)-cis-Azidodiltiazem	$10^{-6}M$	94 %
(+)-cis-Diltiazem	$10^{-6}M$	75 %
(±)-Verapamil	$10^{-6}M$	11 %
(±)-Nitrendipine	$10^{-6}M$	17 %

neutralized by preincubating the sera with 0.5 mg HSA/ml final assay volume (Fig.5). Specificity of the sera was tested by a competitive ELISA. Microtiter-plates coated with AZD-HSA were incubated with diluted antiserum (working dilution 1:20000) supplemented with serial dilutions of (+)-cis-diltiazem, (±)-nitrendipine and (±)-verapamil, using (+)-cis-azidodiltiazem and AZD-HSA as positive controls (Table 1). Due to the cross-reactivity of the antibodies with other benzothiazepines the assay can be also used for quantitative determination of these drugs.

For the detection of protein-fixed non-radioactive (+)-cis-azidodiltiazem, a competitive ELISA was established and standard curves for AZD incorporated into HSA were calculated (Fig.6). The ELISA was then tested for its suitability in detecting receptor-incorporated (+)-cis-azidodiltiazem. For this purpose the purified calcium channel from skeletal muscle was photolabelled with (+)-cis)-azidodiltiazem and separated under denaturing conditions on a size exclusion HPLC (TSK3000SW). The collected fractions were tested for immunoreactivity by ELISA. The inhibition peak co-eluted with the non-resolved alpha-subunits of the channel complex and corresponded to the radioactivity peak of a sample photolabelled with [N-methyl-^{3}H]LU49888 separated in the same manner (not shown). To discriminate alpha$_1$ and alpha$_2$-subunits, the photolabelled receptor complex was separated by reducing SDS polyacrylamide gel electrophoresis (Laemmli, 1970). The gels were cut into 4 mm slices and eluted into ELISA buffer. The elution peak with maximal inhibition of antibody binding corresponded exactly to the location of the alpha$_1$-subunit (Fig.7). The sensitivity of the assay was determined by log-logit transformation of standard curves in the range of 3% to 97% inhibition of antibody binding. Repetitive analysis of serial dilutions of AZD-HSA as a competitor showed IC_{50}-values of 3.7 ± 1.7 nM (mean of duplicate determinations on 10 independent plates ± standard deviation).

The (+)-cis-azidodiltiazem-content of unknown samples is calculated by extrapolation of percent-inhibition values of the probes versus the x-axis (log of concentration) of the standard curve. With this method IC_{50}-values of (+)-cis-azidodiltiazem-labelled receptor in the range of 0.4 to 5 nM were determined.

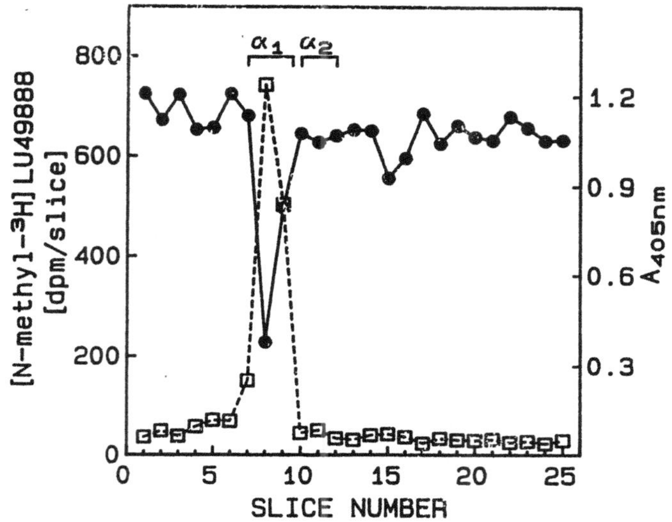

Figure 7
Immunoreactivity and radioactivity distribution of photolabelled calcium channels separated by reducing SDS-gel electophoresis:
Purified calcium channels from skeletal muscle were photolabelled with (+)-cis-azidodiltiazem or [N-methyl-^{3}H]LU49888 and seperated on reducing SDS-gels. Gels were sliced, eluted into TBS-Gelatine and fractions were tested in the competitive ELISA. The figure shows the immunoreactivity profile (circles, representing the absorbance at 405 nm), the [N-methyl-^{3}H]LU49888-associated radioactivity (squares) and the location of $alpha_1$ and $alpha_2$-subunits as detected by staining of the gel with Coomassie-blue.

Thus with the help of ligand-directed antibodies we are able to specifically detect non-radioactive (+)-cis-azidodiltiazem irreversibly incorporated into HSA and the $alpha_1$-subunit with equal sensitivity. This immunological approach should also provide a tool for immunoaffinity-chromatographic purification of the benzothiazepine-binding domain of the L-type calcium channel.

Ca^{++}-antagonists define a novel receptor on mitochondrial membranes

Beside their interaction with the alpha$_1$-subunit of the L-type Ca^{2+} channel, more than a dozen sytems have been reported to be modulated by certain Ca^{2+} antagonists (for a review see Zernig, 1990). Of special interest is their interaction with mitochondrial sites. Several animal studies have shown that Ca^{2+} antagonists protect mitochondria in (ischemically) compromized tissues against Ca^{2+} overload and preserve their structure and function (for reviews see Fleckenstein, 1983; Kloner and Braunwald, 1987). These beneficial effects cannot be solely explained by Ca^{2+} channel block (Zsoter and Church 1983). The mitochondrial inner membrane possesses low-affinity high-capacity Ca^{2+} antagonist binding sites for 1,4-dihydropyridines (Zernig and Glossmann 1988), phenylalkylamines (Glossmann et al., 1989) (and possibly benzothiazepines; Zernig et al., 1990). The 1,4 dihydropyridine and phenylalkylamine sites are allosterically coupled to each other and are regulated by nucleotides, e.g. ATP (Zernig et al., 1989). The partially purified mitochondrial 1,4-dihydropyridine binding site consists of four polypeptides with molecular masses around 50 kDa, 34 kDa, and 25 kDa and retains full allosteric control by phenylalkylamines and ATP (Glossmann et al., 1989).

Ca^{2+} antagonist binding to the mitochondrial sites strongly depends on anions (Zernig and Glossmann, 1988). The mitochondrial inner membrane also contains a structure which transports a variety of anions (Beavis and Garlid, 1987) and is different from any of the other anion shuttles (e.g. the dicarboxylate-, phosphate-, or citrate transporter or the ADP/ATP translocase). By use of patch-clamp experiments in cuprizone-induced giant mitochondria, single anion channel activity has been demonstrated in the inner mitochondrial membrane (Sorgato et al., 1987). Thus, the existence of an inner mitochondrial anion channel (IMAC), which had been only tentatively characterized by several groups on the basis of indirect swelling (Garlid and Beavis 1986; Beavis and Garlid, 1987; Beavis, 1989) or $^{36}Cl^-$-uptake experiments alone (Comerford et al., 1986), was finally confirmed at the molecular electrophysiological level. Employing mitochondrial swelling conditions according to Garlid and Beavis (1986), we found that Ca^{2+} antagonists of all three chemical classes dose-dependently inhibited the inner mitochondrial anion channel as did amiodarone, a known inhibitor of this structure (Beavis, 1989). Representative concentration-inhibition curves are shown in Figure 8 (top row). Mitochondrial swelling was only inhibited by Ca^{2+} antagonists when it was induced by the activation of the IMAC; swelling induced by activation of the phosphate- and dicarboxylate transporters was not inhibited by the Ca^{2+} antagonists (data not shown). Therefore, nonspecific membrane interaction as well as modulation of the phosphate- or dicarboxylate transporter can be ruled out as the basis of Ca^{2+}

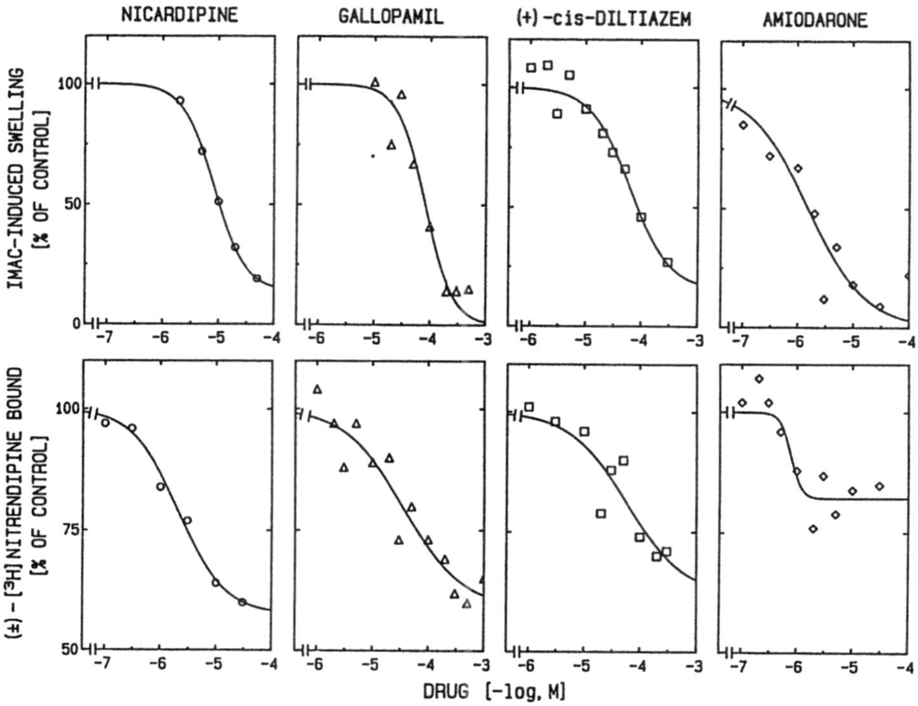

Figure 8
Inhibition of inner mitochondrial membrane anion channel (IMAC) - induced swelling and of mitochondrial (±)-[^{3}H]nitrendipine binding by Ca^{2+} antagonists and amiodarone
Top row: Inhibition of IMAC-induced swelling: Mitochondria (equivalent to 80 - 362 μg of protein determined according to Lowry et al., 1951) were incubated in 1.35 ml of buffer KB (55 mM KCl, 5 mM Tris-HCl (pH 7.4), 0.2 mM Tris-EGTA, 0.2 mM Tris-EDTA) supplemented with 0.8 μg/ml rotenone either alone (control, blank) or in presence of various drug concentrations for 7 min at 25°C. After transferring the suspension to a Hitachi U2000 photometer, activation of the IMAC was measured at 25°C as the swelling-associated absorption decrease at 520 nm according to Garlid and Beavis 1986. Respective pIC_{50} values (means ± S.E. of n determinations were: (±)-Nicardipine 5.077 ± 0.022 (10), (±)-gallopamil 4.158 ± 0.032 (4), (+)-cis-diltiazem 4.212 ± 0.216 (6), and amiodarone 5.931 ± 0.119 (8).
Bottom row: Inhibition of reversible (±)-[^{3}H]nitrendipine binding to mitochondrial membranes under the conditions of the swelling experiment: Binding was measured essentially as described (Glossmann and Ferry, 1985; Zernig and Glossmann, 1988) with the following modifications: To simulate the conditions of the mitochondrial swelling experiments, mitochondrial membranes (80 - 268 μg of protein/ml) were incubated with 1.8 - 4.3 nM (±)-[^{3}H]nitrendipine in 250 or 500 μl of the KB buffer (as used in the swelling experiments) at 25°C for 120 min either alone or in the presence of various drug concentrations. Under the above conditions, total (±)-[^{3}H]nitrendipine binding to mitochondrial membranes ranged from 0.09 and 0.6 pmol/mg protein depending on radioligand and membrane concentration. Nonspecific binding was determined either in presence of 30 μM nicardipine (58 % of total binding). Respective pIC_{50} values (means ± S.E. of n determinations were: (±)-Nicardipine 5.678 ± 0.289 (22), (±)-gallopamil 4.438 ± 0.273 (10), (+)-cis-diltiazem 4.222 ± 0.736 (7), and amiodarone 6.109 ± 0.116 (6).

antagonist-mediated swelling inhibition. Under the conditions of the swelling experiments (see legend to Figure 8 for details), (±)-[^{3}H]nitrendipine reversibly and saturably bound to mitochondrial membranes, displaying a K_D value of 7.2 ± 2.0 μM and a B_{max} value of 1.03 ± 0.37 nmol/mg protein. Examples of (±)-[^{3}H]nitrendipine binding-inhibition by various Ca^{2+} antagonists and amiodarone are shown in Figure 8 (bottom row). Comparison of pIC_{50} values for IMAC-inhibition and pIC_{50} values for inhibition of mitochondrial (±)-[^{3}H]nitrendipine binding yielded the following results: For all 1,4-dihydropyridines tested (nitrendipine, nicardipine, (+)- and (-)-isradipine, Bay K 8644, niludipine, Bay M 5579, nisoldipine, nimodipine, nifedipine) and amiodarone, the correlation coefficient was 0.89 ($p<0.001$), the relationship being: $pIC_{50,IMAC} = -0.37 + 0.95 \times pIC_{50,binding}$. For all tested 1,4-dihydropyridine-, phenyalkylamine-, and benzothiazepine Ca^{2+} antagonists (those listed above plus verapamil, gallopamil, devapamil, LU 47781, (+)- and (-)-cis-diltiazem) and amiodarone, the correlation coefficient was 0.69 ($p<0.01$), the relationship being: $pIC_{50,IMAC} = 1.32 + 0.56 \times pIC_{50,binding}$. The data strongly suggest the association of the mitochondrial Ca^{2+} antagonist binding sites with the IMAC. As permeation of anions through this channel could remove any restraint for mitochondrial Ca^{2+} overload (Fiskum et al., 1980; Akerman and Nicholls, 1983), inhibition of the IMAC by Ca^{2+} antagonists might prevent Ca^{2+} overload and resulting structural and functional impairment of mitochondria in (ischemically) compromized tissue.

Acknowledgement

Research of the authors is funded by FWF and by Dr. Legerlotz foundation.

References

Akerman KEO, Nicholls DG (1983) Rev Physiol Biochem Pharmacol 95: 149-201

Beavis AD (1989) On the inhibition of the mitochondrial inner membrane anion uniporter by cationic amphiphiles and other drugs. J Biol Chem 264: 1508-1515

Beavis AD, Garlid KD (1987) The mitochondrial inner membrane anion channel. Regulation by divalent cations and protons. J Biol Chem 262: 15085-15093

Becker GL, Fiskum G, Lehninger AL (1980) Regulation of free Ca^{2+} by liver mitochondria and endoplasmic reticulum. J Biol Chem 255: 9009-9012

Callewaert G, Hanbauer I, and Morad M (1989) Modulation of calcium channels in cardiac and neuronal cells by an endogenous peptide. Science 243: 663-666

Cardin AD, Hirose N, Blankenship DT, Jackson RL, and Harmony JAK (1986) Binding of a high reactive heparin to human apolipoprotein E: identification of two heparin-binding domains. Biochem Biophys Res Commun 134: 783-789

Comerford JG, Dawson AP, Selwyn MJ (1986) Anion transport in sub-mitochondrial particles prepared from rat liver mitochondria. Trans Biochem Soc 14: 1044-1045

Ferry DR, Rombusch M, Goll A, and Glossmann H (1984) Photoaffinity labelling of Ca2+ channels with [3H]azidopine. FEBS Lett 169: 112-118

Ferry DR, Goll A, and Glossmann H (1987) Photoaffinity labelling of the cardiac calcium channel. (-)-[3H]azidopine labels a 165 kDa polypeptide, and evidence against a [3H]-1,4-dihydropyridine-isothiocyanate being a calcium-channel-specific affinity ligand. Biochem J 243: 127-35

Fleckenstein A (1983) History of calcium antagonists. Circ Res 52: I3-I16

Garlid KD, Beavis AD (1986) Evidence for the existence of an inner membrane anion channel in mitochondria. Biochim Biophys Acta 853: 187-204

Glossmann H, Ferry DR (1985) Assay for calcium channels. Meth Enzymol 109: 513-550

Glossmann H, Zernig G, Graziadei I, and Moshammer T (1989) Non L-type Ca^{2+} channel linked receptors for 1,4-dihydropyridines and phenylalkylamines. In: Gispen WH, Traber J (eds) Nimodipine and central nervous system function. Schattauer, Stuttgart, pp 51-67

Hanbauer I, and Sanna E (1989) Presence in brain of an endogenous ligand for nitrendipine-binding sites that modulates Ca^{2+} channel activity. Ann N Y Acad Sci 560: 96-104

Kloner RA, Braunwald E (1987) Effects of calcium antagonists on infarcting myocardium. Am J Cardiol 59: 84B-94B

Laemmli EK (1970) Cleavage of structural proteins during the assembly of the head of bacteriophage T4. Nature 227: 680-685

Lane DA, Flynn AM, Pejler G, Lindahl U, Choay J, and Preissner K (1987) Structural requirements for te neutralisation of heparin-like saccharides by complement S protein/ vitronectin. J Biol Chem 262: 16343-16348

Lowry DH, Rosebrough NH, Farr A, Randall FJ (1951) Protein measurement with the Folin phenol reagent. J Biol Chem 193: 265-275

Noda M, Shimizu S, Tanabe T, Takai T, Kayano T, Ikeda T, Takahashi H, Nakayama H, Kanaoka Y, Minamino N, Kangawa K, Matsuo H, Raferty MA, Hirose T, Inayama S, Hayashida H, Miyata T, and Numa S (1984) Primary structure of electrophorus electricus sodium channel deduced from cDNA sequence. Nature 312: 121-126

Nunoki K, Florio V, and Catterall WA (1989) Activation of purified calcium channels by stoichiometric protein phosphorylation. Proc Natl Acad Sci USA 86: 6816-6820

Sharp AH, Imagawa T, Leung AT, and Campbell KP (1987) Identification and characterization of the dihydropyridine binding subunit of the skeletal muscle dihydropyridine receptor. J Biol Chem 262: 12309-12315

Sieber M, Nastainczyk W, Zubor V, Wernet W, and Hofmann F (1987) The 165-kDa peptide of the purified skeletal muscle dihydropyridine receptor contains the known regulatory sites of the calcium channel. Eur J Biochem 167: 117-122

Sorgato MC, Keller BU, Stuehmer W (1987) Patch-clamping of the inner mitochondrial membrane reveals a voltage-dependent ion channel. Nature 330: 498-500

Striessnig J, Knaus HG, Grabner M, Moosburger K, Seitz W, Lietz H, and Glossmann H (1987) Photoaffinity labelling of the phenylalkylamine receptor of the skeletal muscle transverse-tubule calcium channel. FEBS Lett 212: 247-253

Striessnig J, Knaus HG, and Glossmann H (1988) Photoaffinity-labelling of the calcium-channel-associated 1,4-dihydropyridine and phenylalkylamine receptor in guinea-pig hippocampus. Biochem J 253, 39-47

Striessnig J, Scheffauer F, Mitterdorfer J, Schirmer M and Glossmann H (1990) Identification of the benzothiazepine-binding polypetide of skeletal muscle calcium channels with (+)-cis-azidodiltiazem and anti-ligand antibodies. J Biol Chem 265: 363-370 (1990)

Takahashi M, Seagar MJ, Jones JF, Reber BF, and Catterall WA (1987) Subunit structure of dihydropyridine-sensitive calcium channels from skeletal muscle. Proc Natl Acad Sci USA 84: 5478-5482

Vaghy PL, Striessnig J, Miwa K, Knaus HG, Itagaki K, McKenna E, Glossmann H, and Schwartz A (1987) Identification of a novel 1,4-dihydropyridine- and phenylalkylamine-binding polypetide in calcium channel preparations. J Biol Chem 262: 14337-14342

Walz DA, Wu V-Y, de Lamo R, Dene H, and McCoy LE (1977) Primary structure of human platelet factor 4. Thromb Res 11: 893-898

Weisgraber KH, and Rall Jr SC, (1987) Human apolipoprotein B-100 heparin-binding sites. J Biol Chem 262: 11097-11103

Zernig G (1990) Widening potential for Ca^{2+} antagonists: non-L-type Ca^{2+} channel interaction. Trends Pharmacol Sci 11: 38-44

Zernig G, Glossmann H (1988) A novel 1,4-dihydropyridine-binding site on mitochondrial membranes from guinea-pig heart, liver, and kidney. Biochem J 253: 49-58

Zernig G, Graziadei I, Moshammer T, Kandler D, Peschina W, Reider N, and Glossmann H (1990) Association of the mitochondrial Ca^{2+} antagonist binding sites with an inner mitochondrial membrane anion channel. Naunyn-Schmiedeberg's Arch Pharmacol (in press)

Zernig G, Moshammer T, Graziadei I, Glossmann H (1989) The mitochondrial high-capacity low-affinity (±)-[^{3}H]nitrendipine binding site is regulated by nucleotides. Eur J Pharmacol 157: 67-73

Zsoter TT, Church JG (1983) Calcium antagonists. Pharmacodynamic effects and mechanism of action. Drugs 25: 93-112.

EVIDENCE FOR AN INTRACELLULAR RYANODINE- AND CAFFEINE-SENSITIVE CALCIUM ION-CONDUCTING CHANNEL IN EXCITABLE TISSUES

G. Meissner, A. Herrmann-Frank, F.A. Lai, and L. Xu
Departments of Biochemistry and Physiology
University of North Carolina
Chapel Hill, NC 27599-7260
USA

In skeletal and cardiac muscle, an action potential triggers the release of calcium ions from the intracellular membrane compartment, sarcoplasmic reticulum (SR), via a ligand-gated, high-conductance "Ca^{2+} release" channel. This channel has been identified as the receptor for the plant alkaloid ryanodine and also shown to be morphologically identical with the large protein structures ("feet") that span the gap between junctional SR and the muscle surface membrane and T-tubule. The detergent-solubilized Ca^{2+} release channel has been purified from skeletal and cardiac muscle as a tetrameric 30 S protein complex and reconstituted into planar lipid bilayers. The skeletal and cardiac channels conduct monovalent cations in addition to Ca^{2+}, are activated by Ca^{2+}, and modulated by ATP, Mg^{2+}, calmodulin, and the two Ca^{2+} releasing drugs caffeine and ryanodine. Ruthenium red, a polyvalent cation, inhibits the channels at micromolar concentrations. Recently, our laboratory has also obtained evidence for the presence of a 30 S mono- and divalent cation-conducting channel complex in canine and porcine aorta and rat and bovine brain. A similar high molecular weight Ca^{2+} release channel complex may be present therefore, in most, if not all, mammalian excitable tissues.

INTRODUCTION

In excitable tissues, release of Ca^{2+} ions from intracellular membrane compartments is triggered either by a change in surface membrane potential, or via a chain of voltage-independent steps that involve the agonist-induced formation of inositol 1,4,5-trisphosphate (IP_3) and its subsequent binding to an intracellular membrane receptor/channel complex, the IP_3 receptor (van Breemen and Saida, 1989). The voltage-dependent mechanism, commonly referred to as excitation-contraction (E-C) coupling, has been extensively studied in striated muscle. In skeletal and cardiac muscle, rapid release of calcium ions from

NATO ASI Series, Vol. H 48
Calcium Transport and
Intracellular Calcium Homeostasis
Edited by D. Pansu and F. Bronner

the intracellular membrane compartment, sarcoplasmic reticulum (SR), is triggered by an action potential which is propagated along the fiber surface membrane and into the cell interior through invaginations, the transverse tubule system (T-tubule) (Endo, 1977). The electrical excitatory signal is communicated to the SR at specialized areas where large protein structures ("feet") span the gap between junctional SR and the muscle surface membrane and T-tubule (Peachey and Franzini-Armstrong, 1983). Although the mechanism of *in vivo* SR Ca^{2+} release has not yet been fully defined (Fabiato, 1983; Rios and Pizzaro, 1988), there is now substantial evidence that SR Ca^{2+} release in skeletal and cardiac muscle occurs via a ligand-gated "Ca^{2+} release" channel, which serves as the receptor for the plant alkaloid ryanodine and is identical with the feet (Lai and Meissner, 1989). The primary structure of the rabbit and human skeletal ryanodine receptors has been determined by cDNA cloning and sequencing and shown to comprise a single polypeptide of ~5032 amino acids (Takeshima *et al.*, 1989; Zorzato *et al.*, 1990). Much progress has also been made recently in the identification of the intracellular IP_3 receptor. Cloning and expression of a membrane glycoprotein from cerebellar Purkinje neurons has identified an IP_3-binding protein of 2,749 amino acids (Furuichi *et al.*, 1989). The neuronal IP_3 receptor displays partial sequence homology with the skeletal muscle ryanodine receptor and, like the ryanodine receptor, is thought to assemble as a tetramer with the four subunits possibly surrounding a central pore that forms the Ca^{2+} channel pathway. Thus, current evidence suggests the existence of at least two structurally distinct, intracellular Ca^{2+} release pathways. Since smooth (Walker *et al.*, 1987; Ehrlich and Watras, 1988; van Breemen and Saida, 1989) and skeletal (Donaldson *et al.*, 1988; but see also Walker *et al.*, 1987) muscle as well as neurons (Thayer *et al.*, 1988) contain IP_3- and surface membrane potential (and ryanodine)-sensitive Ca^{2+} release pathways, it would appear that two intracellular Ca^{2+} release structures can co-exist in excitable tissue.

In our studies, we have prepared junctionally-derived SR membrane fractions to purify and characterize a Ca^{2+}-gated, ryanodine- and caffeine-sensitive 30 S Ca^{2+} release channel complex from skeletal (Lai *et al.*, 1988, 1989) and cardiac (Anderson *et al.*, 1989) muscle. More recently, we have also obtained from smooth muscle (Herrmann-Frank et al, 1990) and brain (Lai *et al.*, 1990) a partially purified 30 S protein fraction, which upon addition to planar lipid bilayers, showed single channel currents with properties in several respects similar to those observed for the skeletal and cardiac muscle ryanodine receptor-Ca^{2+} release channel complexes.

RESULTS AND DISCUSSION

Table 1 summarizes several of the presently known properties of the skeletal and cardiac SR ryanodine receptor-Ca^{2+} release channel complexes as determined in our laboratory. The two channel complexes have been purified from junctionally-derived SR membranes in essentially one step by sucrose gradient centrifugation, following their solubilization in the presence of the zwitterionic detergent Chaps (3-[3-cholamidopropyl) dimethylammonio]-1-propanesulfonate) and radiolabeling with the channel-specific ligand [^{3}H]ryanodine. The skeletal and cardiac ryanodine receptors migrate on the gradients as tetrameric 30 S complexes comprised of polypeptides of apparent relative molecular mass (M_r) of ~400,000 on SDS-PAGE (Lai *et al.*, 1988; Anderson *et al.*, 1989) and with a calculated molecular weight of ~564,000 from cDNA sequence analysis (Takeshima *et al.*, 1989; Zorzato *et al.*, 1990). Negative stain electron microscopy of the purified receptors has revealed four-leaf clover (quatrefoil) structures similar to those described for the feet spanning the T-tubule-SR junctional gap. The purified skeletal and cardiac channel complexes bind, in a Ca^{2+}-dependent manner with nanomolar affinity, approximately 500 pmol ryanodine per mg protein, which, taking into account a calculated molecular weight of 564,000 for the skeletal receptor, corresponds to one high-affinity binding site per 30 S tetramer. Recent, more extensive [^{3}H]ryanodine binding studies with the membrane-bound skeletal muscle receptor have revealed the presence of both

Table 1

Properties of skeletal and cardiac muscle SR ryanodine-sensitive Ca^{2+} release channel complexes

1. **Composition: 30 S homotetramer of M_r ~2.3 x 10^6 (four ~565,000 subunits)**
2. **Structure: Four-leaf clover (quatrefoil)**
3. **Ryanodine binding: One high-affinity and three low-affinity [^{3}H]ryanodine binding sites per 30S tetramer**
4. **Conductance:**

50 mM Ca^{2+} trans	**100 pS**	**(skeletal)**
	80 pS	**(cardiac)**
500 mM Na^+ trans	**600 pS**	**(skeletal)**
	550 pS	**(cardiac)**

5. **Regulation: Activation by µM Ca^{2+}, and modulation by mM ATP, mM Mg^{2+} and µM calmodulin**
6. **Exogenous ligands: Activation by nM ryanodine and mM caffeine**
   Inhibition by mM ryanodine and µM ruthenium red

high and low affinity sites with a ratio of 1:3 (Lai *et al.*, 1989).

The Chaps-solubilized purified 30 S Ca^{2+} release channel complex, isolated in the absence of [^{3}H]ryanodine, has been reconstituted into Mueller-Rudin type lipid bilayers (Lai *et al.*, 1988). The skeletal and cardiac muscle Ca^{2+} release channel complexes conduct monovalent cations more efficiently than Ca^{2+} ions (Table 1). Our planar lipid bilayer - single channel (Smith *et al.*, 1986, 1989; Rousseau *et al.*, 1986, 1988, 1989; Lai et al., 1988; Liu et al., 1989; Anderson et al., 1989) and macroscopic vesicle ion flux (Meissner *et al.*, 1986; Meissner, 1986; Meissner and Henderson, 1987) measurements have further indicated that the skeletal and cardiac SR Ca^{2+} release channels are activated by cytoplasmic Ca^{2+} in a similar, although not identical manner, and modulated by Mg^{2+}, ATP, calmodulin, and the two Ca^{2+} releasing drugs caffeine and ryanodine.

Figs. 1 and 2 show the current traces of two Ca^{2+}-activated rabbit skeletal muscle channels, and further contrast the action of the two plant alkaloids caffeine and ryanodine on the SR Ca^{2+} release channel. In Fig. 1, a single purified skeletal muscle Ca^{2+} release channel was recorded in the presence of a symmetrical NaCl medium. Single channel conductance was 570 pS with 500 mM Na^+ as the current carrier.

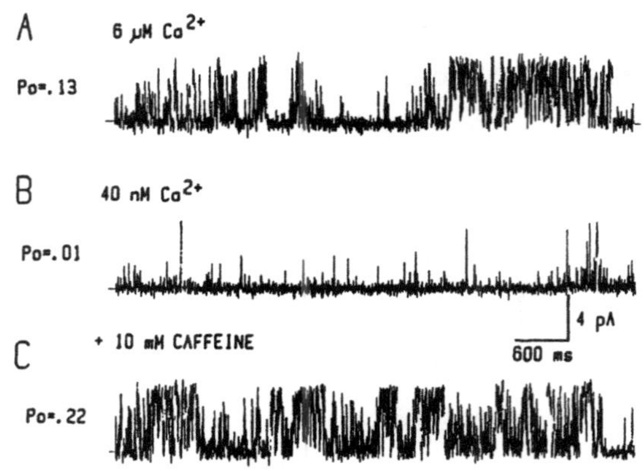

Figure 1. Activation of skeletal muscle Ca^{2+} release channel by Ca^{2+} and caffeine. The 30 S purified channel complex was reconstituted into a planar lipid bilayer. Single channel currents, shown as upward deflections, were recorded in symmetric 500 mM NaCl, 6 μM free Ca^{2+} (A), after cis free Ca^{2+} was decreased to 40 nM by the addition of EGTA (B), and after the addition 10 mM caffeine to the 40 nM free Ca^{2+} cis medium (C). Holding potential = -10 mV. The unitary conductance of the Ca^{2+}- and caffeine-activated channel was 570 pS (with permission from Rousseau *et al.*, 1988).

In the upper trace of Fig. 1, the channel was partially activated by 6 μM free Ca^{2+} in the cis chamber (the cytoplasmic side of the SR). Channel activity was decreased by lowering the free Ca^{2+} concentration to 40 nM in the cis chamber (trace B) and again increased by the addition of 10 mM caffeine to the low cis Ca^{2+} medium (trace C of Fig. 1). Caffeine activated the channel by increasing channel open time without affecting single channel conductance. The activity of the skeletal and cardiac channels is increased by caffeine in a dose-dependent manner, without appearing to reach a maximal value at 20 mM caffeine, which has been the highest concentration tested (Rousseau *et al.*, 1988, 1989). However, the caffeine-activated channel can be nearly maximally activated by the addition of millimolar ATP to a 20 mM caffeine medium. Further, the caffeine-activated channel, like the Ca^{2+}-activated channel, is inhibited by Mg^{2+} and ruthenium red, two effective inhibitors of the SR Ca^{2+} release channel (Meissner *et al.*, 1986).

Ryanodine modifies both the conductance and the gating behavior of the skeletal and cardiac channels in a characteristic manner. In Fig. 2, a single Ca^{2+}-activated, K^+-conducting channel is shown which entered into a subconductance state upon the addition of μM ryanodine cis. The ryanodine-

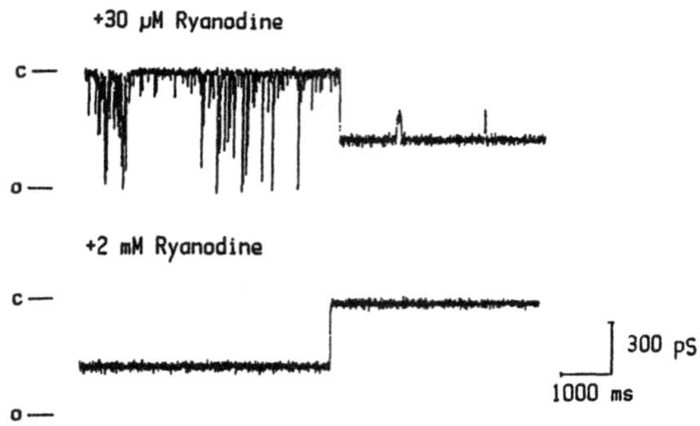

Figure 2. Effect of ryanodine on a single reconstituted, purified skeletal muscle Ca^{2+} release channel. Single channel currents were recorded in symmetric 250 mM KCl, 50 μM free Ca^{2+}. The upper trace shows the appearance of a subconducting state with an open probability of ~1, several minutes after the addition of 30 μM ryanodine to the cis chamber. The lower trace illustrates the sudden transition from the subconductance state to a fully closed state within 1 min after the cis addition of 2 mM ryanodine. Bars on the left represent the open (o) and closed (c) state of the channel (with permission from Lai *et al.*, 1989).

modified channel had a conductance about one half of that of the unmodified channel and an open probability close to unity. In Fig. 2, a relatively high ryanodine concentration of 30 μM was used to reduce the time required to observe the otherwise very slow interaction of ryanodine with the channel. Upon addition of millimolar cis ryanodine, the channel's subconductance state abruptly disappeared and the channel entered into a fully closed state. The effects of ryanodine were long-lasting and on the time scale (~30 min) of the bilayer measurements essentially irreversible in that removal of ryanodine did not return the channel to its original state. Further, both the subconducting and the fully closed channel became insensitive to other ligands such as Ca^{2+}, ATP or Mg^{2+}, which otherwise greatly affect the gating behavior of the unmodified channel. We have suggested that the appearance of the subconductance state in the presence of micromolar Ca^{2+} is the consequence of an interaction of ryanodine with a high affinity site, which has a stoichiometry of 1 per 30 S tetramer (Lai *et al.*, 1989). The subsequent complete closing of the channel in the presence of millimolar ryanodine can be interpreted as being due to the occupation of one or more low affinity sites, which are thought to have a stoichiometry of 3 per 30 S tetramer (Table I).

In preliminary studies, we have also obtained evidence for the presence of a 30 S cation conducting channel complex in canine and porcine aorta (Herrmann-Frank *et al.*, 1990) and rat and bovine brain (Lai *et al.*, 1990) (Table II). Microsomal proteins were solubilized in Chaps and centrifuged through linear sucrose gradients. As has been observed for the skeletal muscle ryanodine receptor, a single [^{3}H]ryanodine receptor peak with apparent sedimentation coefficient of ~30 S was obtained. Incorporation of the solubilized brain 30 S peak fractions into planar bilayers induced single channel currents with properties similar to those observed for the skeletal muscle ryanodine receptor-Ca^{2+} release channel complex. The vascular 30 S protein fraction also induced the formation of a monovalent ion and Ca^{2+} conducting channel whose activity was regulated by Ca^{2+}, Mg^{2+}, caffeine and ruthenium red. However, in our preliminary studies we were not able to observe the formation of a subconducting state upon the addition of µM ryanodine, although mM ryanodine fully closed the channel. A second noticeable difference has been that the maximal conductance observed in 250 mM K^+ buffer possessed half the conductance value of the main conductance state of the skeletal muscle and brain channels. The appearance of intrinsic

Table 2

Properties of K^+- and Ca^{2+}-conducting channel complexes from skeletal and vascular muscle and brain

	Skeletal	Vascular	Brain
Apparent sedimentation coefficient	30 S	30 S	30S
Main conductance			
In 50 mM Ca^{2+}	100 pS	~100 pS	N.D.
In 250 mM K^+	750 pS	360 pS	820 pS
Regulation			
Activation by μM Ca	Yes	Yes	Yes
Activation by mM caffeine	Yes	Yes	Yes
Inhibition by mM Mg^{2+}	Yes	Yes	Yes
Inhibition by μM ruthenium red	Yes	Yes	Yes
Modification by ryanodine to			
subconducting state	Yes	Not observed	Yes
fully closed state	Yes	Yes	N.D.

N.D. not determined

subconductances present within the channel tetramer is a frequently observed phenomenon upon reconstitution of the detergent solubilized, purified channel (Liu *et al.*, 1989). It may, therefore, be that additional studies will also reveal a conductance level of 700-800 pS for the vascular channel in 0.25 M K^+ medium, as observed for the skeletal muscle and brain 30 S channel complexes. The observation of a high molecular weight Ca^{2+}-gated, caffeine- and ryanodine-sensitive channel complex in skeletal, cardiac and vascular muscle and brain suggests this channel's presence in most, if not all, mammalian excitable tissues.

ACKNOWLEDGEMENTS

Supported by USPHS grants and fellowships from Deutsche Forschungsgemeinschaft (AHF) and Muscular Dystrophy Association (FAL).

REFERENCES

Anderson K, Lai FA, Liu QY, Rousseau E, Erickson HP, Meissner G (1989) Structural and functional characterization of the purified cardiac ryanodine receptor - Ca^{2+} release channel complex. J Biol Chem 264:1329-1335.

Donaldson SK, Goldberg ND, Walseth TF, Huetteman DA (1988) Voltage dependence of inositol 1,4,5-trisphosphate-induced Ca^{2+} release in peeled skeletal muscle fibers. Proc Natl Acad Sci, USA 85:5749-5753.

Endo M (1977) Calcium release from the sarcoplasmic reticulum. Physiol Rev 57:71-108.

Ehrlich BE, Watras J (1988) Inositol 1,4,5-trisphosphate activates a channel from smooth muscle sarcoplasmic reticulum. Nature 336:583-586.

Fabiato A (1983) Calcium-induced release of calcium from the cardiac sarcoplasmic reticulum. Am J Physiol 245:C1-C14.

Furuichi T, Yoshikawa S, Miyawaki A, Wada K, Maeda N, Mikoshiba K (1989) Primary structure and functional expression of the inositol 1,4,5-trisphosphate-binding protein P_{400}. Nature 342:32-38.

Herrmann-Frank A, Darling E, Meissner G (1990) Single channel measurements of the Ca^{2+}-gated ryanodine-sensitive Ca^{2+} release channel of vascular smooth muscle. Biophys J 57: 156a.

Lai FA, Erickson HP, Rousseau E, Liu QY, Meissner G (1988) Purification and reconstitution of the calcium release channel from skeletal muscle. Nature 331:315-319.

Lai FA, Meissner G (1989) The muscle ryanodine receptor and its intrinsic Ca^{2+} channel activity. J Bioenerg Biomembr 21:227-246.

Lai FA, Misra M, Xu L, Smith HA, Meissner G (1989) The ryanodine receptor-Ca^{2+} release channel complex of skeletal muscle sarcoplasmic reticulum. Evidence for a cooperatively coupled, negatively charged homotetramer. J Biol Chem 264:16776-16785.

Lai FA, Xu L, Meissner G (1990) Identification of a ryanodine receptor in rat and bovine brain. Biophys J 57:529a.

Liu QY, Lai FA, Rousseau E, Jones RV, Meissner G (1989) Multiple conductance states of the purified calcium release channel complex from skeletal sarcoplasmic reticulum. Biophys J 55:415-424.

Meissner G (1986) Evidence of a role for calmodulin in the regulation of calcium release from skeletal muscle sarcoplasmic reticulum. Biochemistry 25:244-251.

Meissner G, Darling E, Eveleth J (1986) Kinetics of rapid Ca^{2+} release by sarcoplasmic reticulum. Effects of Ca^{2+}, Mg^{2+} and adenine nucleotides. Biochemistry 25:236-244.

Meissner G, Henderson JS (1987) Rapid calcium release from cardiac muscle sarcoplasmic reticulum vesicles is dependent on Ca^{2+} and is modulated by Mg^{2+}, adenine nucleotide and calmodulin. J Biol Chem 262:3065-3073.

Peachey LD, Franzini-Armstrong C (1983) Structure and function of membrane systems of skeletal muscle cells. In Skeletal Muscle (Peachey LD, Adrian RH, Geiger SR, eds) Handbook of Physiology. American Physiological Society, Bethesda, MD pp 23-71.

Rios E, Pizarro G (1988) Voltage sensors and calcium channels of excitation-contraction coupling. NIPS 3:223-227.

Rousseau E, Smith JS, Henderson JS, Meissner G (1986) Single channel and $^{45}Ca^{2+}$ flux measurements of the cardiac sarcoplasmic reticulum calcium channel. Biophys J 50:1009-1014.

Rousseau E, LaDine J, Liu QY, Meissner G (1988) Activation of the Ca^{2+} release channel of skeletal muscle sarcoplasmic reticulum by caffeine and related compounds. Arch Biochem Biophys 267:75-86.

Rousseau E, Meissner G (1989) Single cardiac sarcoplasmic reticulum Ca^{2+} release channel: activation by caffeine. Am J Physiol 256:H328-H333.

Smith JS, Coronado R, Meissner G (1986) Single channel measurements of the calcium release channel from skeletal muscle sarcoplasmic reticulum. Activation by Ca^{2+} and ATP and modulation by Mg^{2+}. J Gen Physiol 88:573-588.

Smith JS, Rousseau E, Meissner G (1989) Calmodulin modulation of single sarcoplasmic reticulum Ca^{2+}-release channels from cardiac and skeletal muscle. Circ Research 64:352-359.

Takeshima H, Nishimura S, Matsumoto T, Ishida H, Kangawa K, Minamino N, Matsuo H, Ueda M, Hanaoka M, Hirose T, Numa S (1989) Primary structure and expression from complementary DNA of skeletal muscle ryanodine receptor. Nature 339:439-445.

Thayer SA, Hirning LD, Miller RJ (1988) The role of caffeine-sensitive calcium stores in the regulation of the intracellular free calcium concentration in rat sympathetic neurons *in vitro*. Mol. Pharmacol. 34:664-673.

Van Breemen C, Saida K (1989) Cellular mechanisms regulating $[Ca^{2+}]_i$ in smooth muscle. Annu Rev Physiol 51:315-329.

Walker JW, Somlyo AV, Goldman YE, Somlyo AP, Trentham DR (1987) Kinetics of smooth and skeletal muscle activation by laser pulse photolysis of caged inositol 1,4,5-trisphosphate. Nature 327:249-252.

Zorzato F, Fujii J, Otsu K, Phillips M, Green NM, Lai FA, Meissner G, MacLennan DH (1990) Molecular cloning of cDNA encoding human and rabbit forms of the Ca^{2+} release channel (ryanodine receptor) of skeletal muscle sarcoplasmic reticulum. J Biol Chem 265:2244-2256.

CHARACTERIZATION OF NOVEL PROBES OF VOLTAGE-DEPENDENT CALCIUM CHANNELS

Gregory J. Kaczorowski, John P. Felix, Maria L. Garcia, V. Frank King, Judith L. Shevell, and Robert S. Slaughter
Merck Institute for Therapeutic Research
Rahway, New Jersey 07065

Voltage-dependent Ca^{2+} channels have been identified in both electrically excitable and non-excitable cells. Several distinct types of Ca^{2+} channels have been detected in these preparations. Perhaps the best characterized channel, in terms of its pharmacological properties, is the L-type, or voltage-dependent, slowly inactivating, Ca^{2+} channel ($P_{Ca,L}$). This channel protein is a multi-drug receptor. Several different structural classes of molecules have been identified which bind in a potent fashion to unique receptor sites that are part of the channel. These agents modulate channel gating behavior. Originally, 3 high affinity drug binding sites were shown to be associated with $P_{Ca,L}$ that is present in cardiac sarcolemmal membranes. These sites recognize dihydropyridine, aralkylamine, and benzothiazepine classes of organic Ca^{2+} entry blockers (for a review see Glossmann and Striessnig, 1988). By analyzing the binding of ligands which interact at these sites (e.g., through studies with $[^3H]$nitrendipine, $[^3H]$D-600, and $[^3H]$diltiazem, respectively), it has been possible to demonstrate that 3 distinct receptors exist in a complex on $P_{Ca,L}$, and that these sites are allosterically coupled. From these data, a model was developed which describes the allosteric interactions between individual sites in the cardiac Ca^{2+} entry blocker receptor complex (Garcia et. al., 1986). Recently, a new structural class of organic Ca^{2+} channel inhibitor (i.e., certain substituted diphenylbutylpiperidines) has been identified and shown to bind to another unique site on the cardiac Ca^{2+} channel (King et. al., 1989). Using these recent results, the model of allosteric coupling between drug binding sites has been further refined, and this series of interactions is presented in Figure 1. By applying this scheme, it is possible to determine the site of action of

NATO ASI Series, Vol. H 48
Calcium Transport and
Intracellular Calcium Homeostasis
Edited by D. Pansu and F. Bronner

Allosteric Coupling Between High Affinity Drug Binding Sites on the Cardiac L-Type Calcium Channel

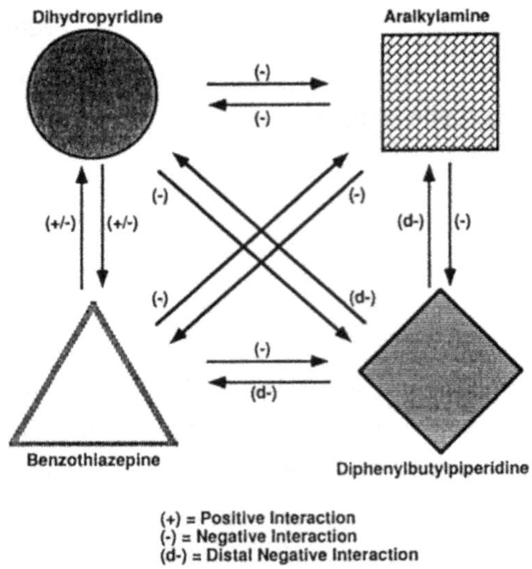

Figure 1.

previously uncharacterized $P_{Ca,L}$ effectors, and identify new modulators of channel activity.

Using ligand binding protocols and this model, a number of novel $P_{Ca,L}$ modulators have been investigated. It is a widely held belief that agents which cause stimulation of dihydropyridine binding must interact exclusively at the benzothiazepine site of the Ca^{2+} entry blocker receptor complex. However, only a few structural types besides diltiazem have been identifed which possess this property. In searching for diltiazem-like compounds, it was noted that the Chinese medicinal herb isolate, tetrandrine, an α-β bis-benzylisoquinoline alkaloid, was a likely candidate for such an agent (King et. al., 1988). Tetrandrine is a natural product Ca^{2+} entry blocker. The Chinese pharmacological literature reveals that tetrandrine possesses negative inotropic activity in isolated cardiac muscle preparations, shortens cardiac action potentials, possesses antiarrhythmic activity in cardiac tissues where intracellular Na^+ is elevated, blocks contraction of K^+-depolarized smooth muscle, possesses hypotensive activity in normotensive and different hypertensive rat models, and lowers blood pressure in dogs with a mild bradycardic effect. These properties are consistent with the postulate that tetrandrine possesses Ca^{2+} entry blocker activity, and this compound has been used in the clinic in China for the treatment of angina and hypertension. Upon investigating the effects of tetrandrine on ligand binding in cardiac sarcolemmal membranes, it was noted that tetrandrine produces concentration-dependent inhibition of diltiazem and D-600 binding, but that it stimulates binding of nitrendipine. It is an effective inhibitor of diltiazem binding, displaying a K_i value of 100 nM. Scatchard analyses of the effects of tetrandrine under equilibrium binding conditions reveal that this agent decreases the affinity of both diltiazem and D-600 for their respective receptors, while it increases the affinity of nitrendipine. In the first two cases, the increase in K_d values could be due to either a competitive or allosteric effect. To resolve this issue, the effect of tetrandrine was monitored on dissociation kinetics of diltiazem, D-600, and nitrendipine from cardiac sarcolemmal vesicles. While tetrandrine increases the dissociation rate of D-600, and slows the dissociation rate of nitrendipine, it has absolutely no effect whatsoever on the dissociation kinetics of diltiazem. These data suggest that tetrandrine competes at the benzothiazepine site of the cardiac Ca^{2+} entry blocker receptor complex. The postulate that tetrandrine possesses Ca^{2+} entry blocker activity was verified by demonstrating that it blocks inward Ca^{2+} currents through $P_{Ca,L}$ in GH_3 pituitary cells under whole cell voltage clamp conditions. In summary, tetrandrine interacts at the benzothiazepine site, dispite it being structurally dissimilar from diltiazem, and it stimulates binding of nitrendipine, as might be predicted for a diltiazem-like agent.

Recently, 23 different bis-benzylisoquinoline alkaloids, representing α-β, β-α, β-β, and α-α conformational classes, have been analyzed for their ability to modulate $P_{Ca,L}$ activity (Kaczorowski, et. al., 1989). As expected, all of these compounds inhibit binding of diltiazem and D-600. However, their effects on nitrendipine binding are different than that predicted by the tetrandrine study. Strikingly, some compounds stimulate nitrendipine binding, while others inhibit binding of this ligand, and some agents have no effect at all. This pattern of activity was noted throughout the different conformational groups, and could not easily be correlated with any given structural type, although most alkaloids representing the α-β class were stimulators of nitrendipine binding. Scatchard analyses indicate that inhibition of diltiazem and D-600 binding, as well as either stimulation or inhibition of nitrendipine binding, occur through K_d effects. Evidence that all different conformational classes of bis-benzylisoquinolines interact at the benzothiazepine site was obtained from experiments showing that none of these compounds affect dissociation kinetics of diltiazem, whereas they do modify the kinetics of aralkylamine and dihydropyridine dissociation. These results suggest that all tetrandrine analogs are competitive at the benzothiazepine receptor. To confirm this idea, a correlation was made between the ability of various compounds to block binding of diltiazem in cardiac membranes and inhibit Ca^{2+} fluxes through $P_{Ca,L}$ in GH_3 cells.

Importantly, a linear relationship is observed with 22 out of 23 alkaloids in their ability to block diltiazem binding *vs.* their ability to inhibit $P_{Ca,L}$ activity. Data obtained with diltiazem also support this correlation. These findings suggest that all conformational classes of bis-benzylisoquinoline alkaloids compete at the benzothiazepine site of the Ca^{2+} entry blocker receptor complex. These are the first data indicating that binding at the diltiazem receptor can result in either positive or negative allosteric interactions with the dihydropyridine receptor.

Other structurally unique Ca^{2+} entry blockers have previously been identified as being diltiazem-like agents because of their ability to stimulate dihydropyridine binding. Two such compounds, KB 944 (Holck et. al., 1984), and MDL 12330A (Rampe et. al., 1987), are inhibitors of $P_{Ca,L}$. They also completely block binding of diltiazem, are partial inhibitors of D-600 binding, and produce both a concentration-dependent, and temperature-dependent, stimulation of nitrendipine binding in cardiac sarcolemmal membranes (Garcia et. al., 1987). A Scatchard analysis reveals that dilitazem binding is inhibited by increasing ligand K_d values, while inhibition of D-600 binding results from either a lowering in affinity (KB 944), or a decrease in B_{max} (MDL 12330A). Nitrendipine binding is stimulated in each case by an increase in ligand affinity. Interestingly, these agents alter dissociation kinetics of diltiazem, D-600, and nitrendipine, indicating that they cannot be competitive at any of the 3 well defined sites in the Ca^{2+} entry blocker receptor complex. They must recognize a unique site on $P_{Ca,L}$. Therefore, stimulation of dihydropyridine binding is not diagnostic of a strictly diltiazem-like agent. The most critical test for a true competitive agent at this site is whether or not the compound affects dissociation kinetics of diltiazem.

In addition to benzothiazepine, aralkylamine, and dihydropyridine structural classes of organic Ca^{2+} entry blockers, other drugs interact with $P_{Ca,L}$ in a high affinity fashion. [^{3}H]fluspirilene, a substituted diphenylbutylpiperidine, binds with nM affinity to a unique site that is coupled by allosteric mechanisms to the other sites in the cardiac Ca^{2+} entry blocker receptor complex (King et. al., 1989). Fluspirilene binds with a K_d of 0.6 nM and a density which is in a stoichiometry of 1:1:1 with the other 3 receptors for Ca^{2+} entry blockers in heart. A unique feature of the fluspirilene site is that it is also coupled to the pore structure of $P_{Ca,L}$. For example, metal ions which inhibit $P_{Ca,L}$ by binding at the mouth of the channel (e.g., Cd^{2+}, La^{3+}, Ni^{2+}, Co^{2+}) stimulate binding of fluspirilene, while substrates for $P_{Ca,L}$ (e.g., Ca^{2+}, Sr^{2+}, Ba^{2+}) inhibit fluspirilene binding. In the case of Cd^{2+}, this metal ion stimulates fluspirilene binding by increasing ligand affinity to 0.1 nM, as well as by increasing slightly the site density of the receptor. Therefore, the effect of Cd^{2+} on fluspirilene binding can be used as a diagnostic test as to whether or not an unknown agent competes at the substituted diphenylbutylpiperidine receptor site. It has been shown that dihydropyridines, aralkylamines, and benzothiazepines block binding of fluspirilene to its receptor with K_i values that mirror the affinity of each of these respective ligands for their own sites. When this relationship is examined in the presence of Cd^{2+}, members from each Ca^{2+} entry blocker class still block binding of fluspirilene, but their ability to do so is shifted to lower affinity. On the other hand, molecules of the substituted diphenylbutylpiperidine class, such as pimozide and penfluridol, inhibit fluspirilene binding with well defined K_i values in the absence of Cd^{2+}, but in the presence of this metal ion, their inhibitory potencies are enhanced. Thus, fluspirilene binding can not only be used to identify agents which interact at or near the pore structure of $P_{Ca,L}$, but, in the presence of Cd^{2+}, these protocols can also determine whether a previously uncharacterized agent interacts competitively at the substituted diphenylbutylpiperidine receptor. It is interesting that compounds binding at this site are potent inhibitors of $P_{Ca,L}$, but they do not display the typical cardiovascular pharmacology profile of a Ca^{2+} entry blocker (King et. al., 1989). These latter observations suggest that some drugs can modulate the activity of $P_{Ca,L}$ in a unique fashion, and this might be exploited in the development of drugs for novel therapeutic applications.

A series of derivatives of the pyrazine diuretic amiloride have been shown to have potent Ca^{2+} entry blocker activity in patch clamp experiments that monitored the activity of $P_{Ca,L}$ in GH_3 cells (Suarez-Kurtz and Kaczorowski, 1989). Upon investigating the mechanistic basis for the action of amiloride analogs, it was found that these agents block binding of [^{3}H]ligands at all 4 sites in the cardiac Ca^{2+} entry blocker receptor complex (Garcia et. al., 1990). In this fashion, binding of nitrendipine, D-600, diltiazem, and fluspirilene are inhibited with the same rank order of potency as that in which these compounds block $P_{Ca,L}$ activity. Scatchard analyses indicate that the action of the amiloride analogs is primarily to decrease the affinity of Ca^{2+} entry blocker ligands for their receptors. That the amiloride analogs do not interact at any of the well defined sites in the Ca^{2+} entry blocker receptor complex is demonstrated by data showing that these agents modify the dissociation kinetics of nitrendipine, D-600, diltiazem, and fluspirilene from their respective receptor sites. However, the pattern by which amiloride analogs modulate various Ca^{2+} entry blocker binding reactions does give some indication as to a possible mechanism of interaction with $P_{Ca,L}$. As noted above, Cd^{2+} stimulates binding of [^{3}H]fluspirilene in cardiac sarcolemmal membranes. When this experiment is repeated in the presence of fixed concentrations of amiloride analogs, it was observed that the concentration dependency of Cd^{2+} was shifted to higher concentrations, but that the pattern of the stimulatory effect remained the same. A Scatchard analysis indicates that the amiloride analogs inhibit modulation of fluspirilene binding by Cd^{2+} in an apparent competitive fashion. This suggests that amiloride derivatives may block $P_{Ca,L}$ by interacting at the pore of the channel. This is consistent with independent electrophysiological findings regarding the mechanism of channel inhibition (Suarez-Kurtz and Kaczorowski, 1989). To confirm that amiloride analogs interact with $P_{Ca,L}$ in a specific manner, rather than by simply affecting the protein through perturbation of the membrane environment, two different photoreactive derivatives of amiloride have been investigated. Bromobenzamil and 2-methoxy-5-nitrobenzamil were employed to determine whether the photolysis of these molecules would produce irreversible inhibition of Ca^{2+} entry blocker binding in cardiac membranes. Photoreaction conditions can be controlled so that protein nucleophiles will displace either the halogen in the 6-position of the pyrazine ring or the 2-methoxy group on the nitrophenyl ether moiety to cause covalent incorporation of photoprobe into the channel protein. Both compounds cause concentration and time-dependent irreversible inactivation of Ca^{2+} entry blocker binding. With either photolabel, protection from photoinactivation occurs if the reaction is carried out in the presence of either excess D-600 or excess diltiazem. These data strongly suggest that amiloride analogs interact in a specific fashion with $P_{Ca,L}$, and that their binding site is coupled allosterically to receptors for other Ca^{2+} entry blockers. Unfortunately, amiloride analogs are not selective inhibitors of $P_{Ca,L}$. They also affect other types of voltage-dependent Ca^{2+} channels with nearly the same rank order of potency as they inhibit $P_{Ca,L}$. It has been suggested that amiloride itself is a selective blocker of T-type Ca^{2+} channels (Tang et. al., 1988). This is clearly not the case as demonstrated in patch clamp experiments with GH_3 cells where the inhibition by amiloride and amiloride derivatives was characterized for both L-type and T-type Ca^{2+} channels (Suarez-Kurtz and Kaczorowski, 1989; Garcia et. al., 1990). Despite their lack of selectively as Ca^{2+} channel blockers, amiloride analogs may be probes at the external face of $P_{Ca,L}$, and, in this fashion, the amiloride photoaffinity labels might be used to map the pore structure of a purified Ca^{2+} channel preparation.

Several other organic Ca^{2+} entry blockers have been described in the literature. Of particular note is a series of substituted benzhydrylpiperazines such as cinnarizine, flunarizine, and DPI 201-106. The latter agent in this series is also a Na^+ channel agonist with cardiotonic properties. The mechanism by which these compounds modulate $P_{Ca,L}$ has been investigated (Siegl et. al., 1988). Both cinnarizine and DPI 201-106 block binding of nitrendipine, D-600 and diltiazem to cardiac sarcolemmal membrane vesicles. It is known that the interacton of DPI 201-106 with cardiac Na^+ channels is stereospecific. However, there is no stereospecificity displayed for the interaction of DPI 201-106 with cardiac Ca^{2+} channels, indicating that there are dissimilar binding sites for this agent on the two channel types. Scatchard analyses indicate that cinnarizine and DPI 201-106 inhibit binding of nitrendipine, D-600 and diltiazem in a mixed fashion (i.e., both K_d and B_{max} effects are noted). Moreover, when dissociation kinetics of nitrendipine, D-600 and diltiazem are investigated in the presence of cinnarizine or DPI 201-106, both agents increase off-rates of the 3 [^{3}H]ligands. These results, taken together with electrophysiological findings, indicate that substituted benzhydrylpiperazines may interact at a unique, low affinity, local anesthetic-like site on $P_{Ca,L}$.

The cardiac L-type Ca^{2+} channel is multi-drug receptor with binding sites for various channel effectors coupled by defined allosteric interactions. Using the diagnostic capability of the model shown in Figure 1 describing these interactions, it has been possible to discover new agents that can influence the activity of $P_{Ca,L}$. Moreover, it has also been possible to use binding data, along with electrophysiological and flux studies which address function, to elucidate the mechanism of action of novel channel modulators. At least 7 distinct drug binding sites have been identified on $P_{Ca,L}$, and more sites willl probably be discovered. There are many similarities between $P_{Ca,L}$ and voltage-dependent Na^+ channels in that both systems comprise multi-drug receptors which possess a variety of allosterically coupled binding sites. Using these studies as a paradigm, it is possible to develop the molecular pharmacology of ion channels that have not previously been studied. By identifying a single high affinity ligand that binds to an uncharacterized channel, it is possible to detect agents that interact competitively at the first site, as well as find compounds that modulate ligand binding by allosteric effects. In this way, one can discover different structural series of molecules that modulate channel activities by unique mechanisms. Such studies can form the basis for the rationale discovery and design of new therapeutically effective ion channel modulators.

REFERENCES

Garcia ML, King VF, Siegl PKS, Reuben JP, Kaczorowski GJ (1986) Binding of calcium entry blockers to cardiac sarcolemmal membrane vesicles: Characterization of diltiazem binding sites and their interaction with dihydropyridine and aralkylamine receptors. J Biol Chem 261:8146-8159

Garcia ML, King VF, Kaczorowski GJ (1987) Interaction of KB 944 and MDL 12330A with the calcium entry blocker receptor complex in cardiac sarcolemmal membrane vesicles. Fed Proc 46:345a

Garcia ML, King VF, Shevell JL, Slaughter RS, Suarez-Kurtz G, Winquist RJ, Kaczorowski GJ (1990) Amiloride analogs inhibit L-type calcium channels and display calcium entry blocker activity. J Biol Chem 265: in press

Glossmann H, Striessnig J (1988) Structure and pharmacology of voltage-dependent calcium channels. ISI Atlas Sci. Pharmacol., 2:202-210

Holck H, Fischli W, Hengartner U (1984) Effects of temperature and allosteric modulators of [^{3}H]nitrendipine binding: Methods for detecting potential Ca^{2+} channel blockers. J Recept Res 4:557-569

Kaczorowski GJ, King VF, Shevell JL, Felix JP, Slaughter RS, Garcia ML (1989) Drug discoveries through receptor studies. Eur J Clin Pharmacol 36:A21

King VF, Garcia ML, Himmel D, Reuben JP, Lam YK, Pan JX, Han GQ, Kaczorowski GJ (1988) Interaction of tetrandrine with slowly inactivating calcium channels: Characterization of calcium channel modulation by an alkaloid of Chinese medicinal herb origin. J Biol Chem 263:2238-2244

King VF, Garcia ML, Shevell JL, Slaughter RS, Kaczorowski GJ (1989) Substituted diphenylbutylpiperidines bind to a unique high affinity site on the L-type calcium channel: Evidence for a fourth site in the cardiac calcium entry blocker receptor complex. J Biol Chem 264:5633-5641

Rampe D, Triggle DJ, Brown AM (1987) Electrophysiological and biochemical studies on the putative calcium channel blocker MDL 12330A in an endocrine cell. J Pharmacol Exp Ther 243:402-407

Siegl PKS, Garcia ML, King VF, Scott AL, Morgan G, Kaczorowski GJ (1988) Interactions of DPI 201-106, a novel cardiotonic agent, with cardiac calcium channels. Naunyn-Schmiedeberg's Arch Pharmacol 338:684-691

Suarez-Kurtz G, Kaczorowski GJ (1989) Effects of dichlorobenzamil on calcium currents in clonal GH_3 pituitary cells. J Pharmacol Exp Ther 247:248-253

Tang CH, Presser F, Morad M (1988) Amiloride selectively blocks the low threshold (T) calcium channel. Science 240:213-215

SECTION II

CALCIUM EXTRUSION - THE Ca ATPases

THE CALCIUM PUMPING ATPase OF THE PLASMA MEMBRANE STRUCTURE-FUNCTION RELATIOSHIPS

E. CARAFOLI
Laboratory of Biochemistry
Swiss Federal Institute of Technology (ETH)
8092 Zürich, Switzerland.

The plasma membrane Ca ATPase is the largest of the P-type ion pumps. Its primary structure has been elucidated in rat brain and a number of human cells. The sequence of three isoforms has been described in rat brain (Shull and Greeg, 1988) and several human isoforms have been identified and sequenced in erythrocytes, teratoma cells, and other cells (Verma et al., 1988). The human teratoma pump contains 1220 amino acids, corresponding to a M_r of 134,683. Asp 475 forms the acyl phosphate during the reaction cycle, and Lys 601 binds the ATP antagonist fluoroscein isothiocyanate, and is thus part of the ATP binding site. Ten hydrophobic domains, spanning the membrane have been tentatively identified ; 4 are in the N-terminal portion of the pump, 6 in the C-terminal portion. The mid portion of the pump (about 500 residues) contains no transmembrane stretches. The calmodulin (CaM) binding domain has been identified next to the C-terminus (residues 1100-1127) : it ressembles the putative CaM binding regions of other CaM-modulated enzymes in its strongly basic character and its propensity to form an amphiphilic helix. The Ca dependent protease calpain attacks the CaM binding domain, removing it in two steps. In the first step about one third of the CaM binding domain is cut away leaving behind a truncated pump which still binds CaM. In the second step the entire CaM binding domain is removed. Trypsin also attacks the CaM binding domain, gradually removing the CaM sensitivity of the pump. In the absence of CaM, the pump is activated by a number of acidic phospholipids, including the phosphorylated derivatives of phosphatidyl-inositol. Trypsin work has located the site of interaction of acidic phospholipids between transmembrane helices two and three. The phospholipid-sensitive domain is a 44 residue stretch which has a predominance of basic amino acids (Zvaritch et al., 1990). The translated sequence of the APTase contains at the N- and C-side of the CaM binding domain sequences very rich in Asp and Glu which may play a role in regulating

NATO ASI Series, Vol. H 48
Calcium Transport and
Intracellular Calcium Homeostasis
Edited by D. Pansu and F. Bronner

the interaction of Ca and CaM with the pump. However, the high affinity Ca binding site of the catalytic cycle is somewhere else in the pump since these acidic stretches can be removed by trypsin without impairing the high affinity interaction of the pump with Ca. Work with synthetic peptides has shown that two aromatic residues located in the N-terminal portion of the CaM binding domain may be important in the interaction of the domain with CaM. It has also shown that the CaM binding domain is involved in the dimerization of the pump, which is another mechanism for its activation in the absence of CaM. A serine (1178), located on the C-terminal side of the CaM binding domain is phosphorylated by the cAMP-dependent kinase in one of the human isoforms of the pump. This isoform is expressed in erythrocytes together with another isoform which is much more abundant and does not contain the cAMP-responsive site. Phosphorylation increases the Ca-affinity of the pump. Most of the pump isoforms differ in the C-terminal portion which contains the regulatory domains (the CaM binding domain and the cAMP-dependent phosphorylation site), and could thus display different regulation properties.

The sheme shown in Figure 1 offers a view of the molecular architecture of the pump consistent with the information described here.

REFERENCES

Green NM (1989) APT-driven cation pumps, alignment of sequences. Trans Biochem Soc 17 : 970-972.

Shull GE, Greeg J (1988) Molecular cloning of two isoforms of the plasma membrane Ca^{2+}-transporting APTase from rat brain. J Biol Chem 263 : 8646-8657.

Verma AK, Filoteo AG, Stanford DR, Wieben ED, Penniston JT, Strehler EE, Fischer R, Heim R, Vogel G, Mathews S, Strehler-Page MA, James P, Vorherr T, Krebs J, Carafoli E (1988) Complete primary structure of a human plasma membrane Ca^{2+} pump. J Biol Chem 263 : 14152-14159.

Zvaritch E, James P, Vorherr T, Falchetto R, Modyanov N and Carafoli E (1990) Mapping of functional domains in the plasma membrane Ca^{2+} pump. Biochemistry, in press

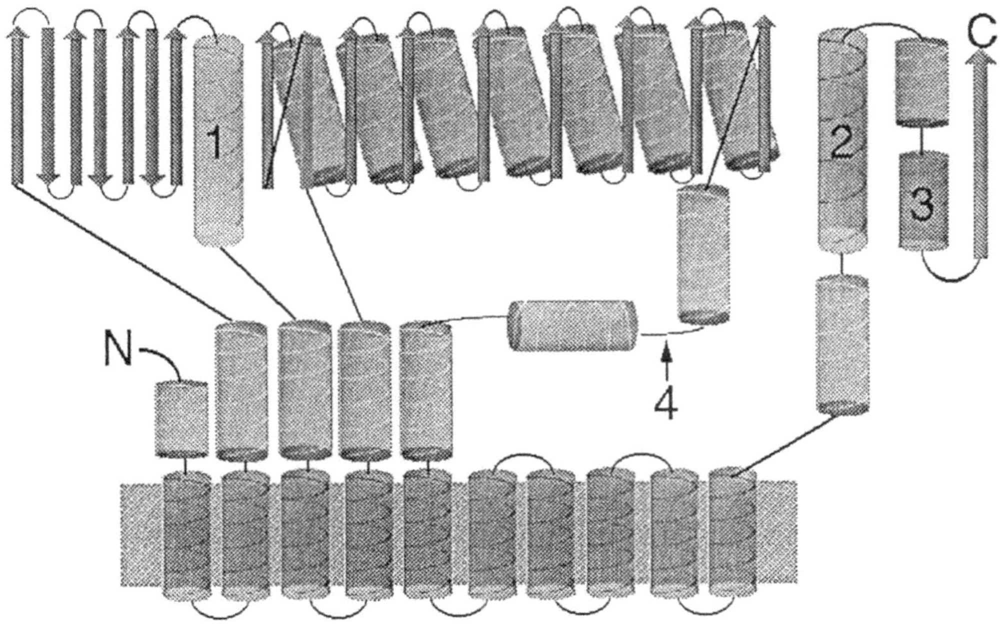

FIGURE 1 :

Predicted secondary structure of the plasma membrane Ca^{2+} pump. The University of Wisconsin Genetic Computer Group sequence analyses software package was used, taking into account the alignements of P-type ATPases proposed by Green 1989 α-helices are represented by cylinders, β-sheets by arrows. 1 = putative domain responsive to acidic phospholipids (see Zvaritch et al. 1990). 2 = calmodulin binding domain. 3 = domain containing the substrate sequence for the cAMP-dependent protein kinase. 4 = flexible hinge, which permits the movement of the aspartylphosphate and of the ATP-binding lysine.

TERTIARY STRUCTURE OF THE MUSCLE CALCIUM PUMP

A.E. SHAMOO, T. LOCKWICH, C.J. CAO, AND C. DOUGLASS

Department of Biological Chemistry
University of Maryland at Baltimore
School of Medicine
660 West Redwood Street
Baltimore, Maryland 21201 (USA)

ABSTRACT

The tertiary structure of the (Ca^{2+} + Mg^{2+})-ATPase between its ATP hydrolytic site and its calcium binding regions can undergo changes to effectuate true uncoupling between uptake, occlusion, and ATPase activity. Tryptic digestion at site T_2 (Arg 198) has no effect on the ATPase activity while Ca^{2+}-uptake is abolished. The loss of Ca^{2+}-uptake appears to be due to uncoupling of uptake from ATPase and theoretical modeling of uptake data reveals that the functional enzyme may exist as a dimer. In contrast, T_2 digestion has no influence on ATP-dependent calcium occlusion as measured by Eu^{3+} luminescence spectroscopy. The study of the synthetic peptide analogues and their binding characteristics indicate that Asp 196 and Asp 149 are important for cation binding.

INTRODUCTION

The maintenance of highly regulated intracellular calcium concentrations plays a vital role in the contraction-relaxation mechanism in the skeletal muscle. Sarcoplasmic Reticulum (SR), an intracellular organelle, removes calcium from and releases it into the cytoplasm, initiating relaxation and contraction cycle (Ebashi and Lipman, 1962; Hasselbach and Makinose, 1963). SR membranes are rich in (Ca^{2+} + Mg^{2+})-ATPase, a single polypeptide enzyme solely responsible for the active transport of calcium from the cell cytoplasm into SR lumen (Hasselbach and Makinose, 1963;

NATO ASI Series, Vol. H 48
Calcium Transport and
Intracellular Calcium Homeostasis
Edited by D. Pansu and F. Bronner

MacLennan, 1970; Racker, 1972, for review see Berman, 1982). The primary components of the Ca^{2+}-pump ATPase are two calcium binding sites, an ATP hydrolytic site, and a region spanning the membrane. Energy transduction must occur from ATP hydrolysis to the calcium sites in order to achieve the vectorial movement of the two calciums from the cytoplasm across the membrane and into the SR.

Utilizing europium as a luminescence analogue of calcium, we have found that luminescence data from europium bound to SR-(Ca^{2+} + Mg^{2+})-ATPase indicates that there are two high affinity calcium binding sites. Furthermore, the two calcium ions at the binding sites are highly coordinated by the protein as the number of H_2O molecules remaining surrounding the calcium ions are reduced from eight (for the aquo Ca^{2+}) to 3 and 0.5. When the enzyme is phosphorylated by ATP, the number of water molecules surrounding the ion are lowered further to 2 and zero water molecules, respectively. By measuring energy transfer between the two luminescent lanthanide ions, we are able to estimate the intersite distances. The Ca^{2+}-Ca^{2+} intersite distance is estimated to be 8-9 Å (Scott, 1985; Herrman et. al. 1986, Klemens et. al., 1988), and decreases by 1 Å when the enzyme is phosphorylated (Herrmann and Shamoo, 1988). The average distance from the Ca^{2+} sites to CrATP is about 18 Å (Shamoo and Herrmann, 1988; Herrman and Shamoo, 1988; Joshi and Shamoo, 1988; Gangola and Shamoo, 1987).

One of the key questions in elucidating the molecular mechanism of active transport is the understanding of energy coupling of ATP hydrolysis to the two calcium ions. For years, we have attempted to develop methodologies to study energy transduction. For example, we have used trypsin digestion of SR-ATPase to achieve uncoupling of Ca^{2+}-transport from ATP hydrolysis. Basically, the effects of trypsin digestion of SR are: T_1 digestion (Arg 505) has no effect on any of the measured functions of SR and results in the production of fragments A and B; and T_2 digestion (Arg 198) achieved at 0^oC and in the presence of calcium results in further digestion of fragment A into A_1 and A_2. In recent experiments, we have shown that when SR vesicles or the reconstituted ATPase are subjected to true T_2 (0^oC and mM Ca^{2+}) digestion, it causes a decrease in Ca^{2+}-uptake that correlates with the digestion of the A fragment into A_1 and A_2 without any change in the ATPase activity or amount of phosphorylated intermediate. In addition, in T_2 digested SR there is no change in Ca^{2+} leakage due to tryptic digestion, but the leakage in the digested SR system cannot be blocked by ruthenium red and Mg^{2+}, whereas in the control SR, Mg^{2+} + ruthenium red blocks Ca^{2+}-leakage (Cao et. al., 1989, 1990).

We have studied the issue of uncoupling by utilizing europium luminescence which is not influenced by SR leakage. These experiments were performed by measuring the level of occlusion retained by control and digested (Ca^{2+}+Mg^{2+})-ATPase as measured by Eu^{3+}fluorescent lifetimes which are sensitive to the occlusion process. As we have stated earlier, T_2 digestion decreases Ca^{2+}-uptake, but occlusion due to E-P formation remains unchanged. Further tryptic digestion beyond T_2 in the presence of micromolar ATP diminishes Ca^{2+} occlusion to zero while the ATPase hydrolytic activity remains at 50%.

RESULTS AND DISCUSSIONS

Trypsin digestion and Ca^{2+}-Transport System

	Ca^{2+}-uptake (nmole Ca^{2+}/mg.min)	
Preparation	Trypsin/protein ratio (w/w) 1/8.75 (3 hr.)	
	Average ± SD	%
Control SR	33.4 ± 8 .24	100
Trypsin Digested SR	0	0
Purified (Ca^{2+} + Mg^{2+})-ATPase (R3)	249. 7 ± 50.56	100
Trypsin Digested R3	0	0

Table 1

Calcium uptake was conducted with 20 mM Tris-HCl buffer (pH 7.0) 0.1 mKCl, 5 mM $CaCl_2$ at $0^{o}C$. Liposomes were prepared from asolectin.

Table 1 presents summary data adapted from Cao et. al.,(1989, 1990). The data clearly shows that when SR vesicles or purified (Ca^{2+} + Mg^{2+})-ATPase reconstituted into liposomes are subjected to true T_2 digestion, there is a near complete loss of Ca^{2+}-uptake without any effect on the $(Ca^{2+} + Mg^{2+})$-ATPase activity. This apparent uncoupling of Ca^{2+}-transport from $(Ca^{2+} + Mg^{2+})$-ATPase activity (Scott and Shamoo, 1982, 1984) was criticized by some subsequent reports as the loss of uptake may be due to increased leakage of the vesicles to Ca^{2+} (Shoshan-Barmatz et. al., 1987; Torok et. al., 1988). However, we have found that in the case of purified $(Ca^{2+} + Mg^{2+})$-ATPase reconstituted into liposomes, tryptic digestion does not increase calcium leakage. Therefore, one cannot account for the loss of Ca^{2+}-uptake by increased Ca^{2+} leakage.

Figure 1 is a summary of the data taken from Cao et. al. (1990). The data show Ca^{2+}-uptake as a function of digestion time. As we have mentioned earlier, throughout the time of digestion $(Ca^{2+} + Mg^{2+})$-ATPase activity is practically unaffected.

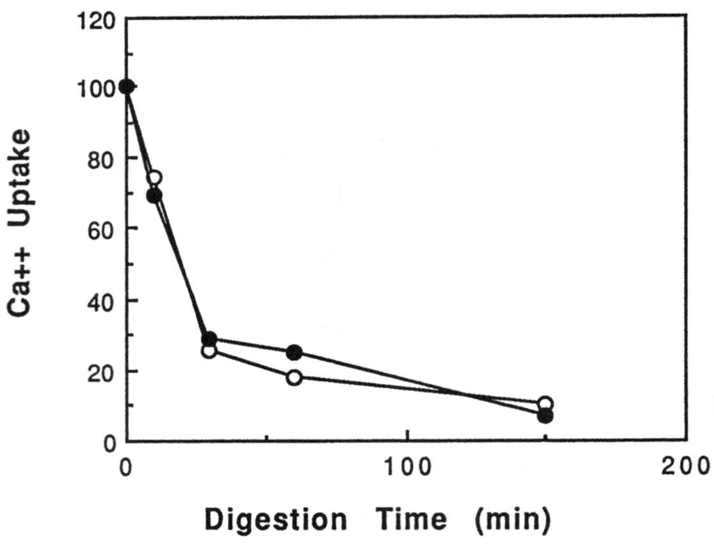

Figure 1

Calcium uptake versus time of digestion. SR membranes at 1 mg/ml were digested at a trypsin/protein ratio of 1/10 (w/w) in 20mM Tris-HCl buffer, pH 7.0, 0.1 M KCl, 5 mM $CaCl_2$ at 0° C. The ATPase was purified and reconstituted into liposomes prepared from asolectin . Open and solid circles represent SR vesicles and purified reconstituted ATPase respectively.

Table 2 data, directly addresses the issue of the effect of tryptic digestion on Ca^{2+}-efflux (passive Ca^{2+}-leakage).

Table 2

	Ca^{2+}-Leakage	
	Control Solution	Mg^{2+} + Ruthenium Red
Control SR	Large	Small
T_2 SR	Large	Large
Purified and Reconstituted (Ca^{2+} + Mg^{2+})-ATPase	Small	Small
T_2 - Purified and Reconstituted (Ca^{2+} + Mg^{2+})-ATPase	Small	Small

Control SR in the presence of Ca^{2+}-channel blockers Mg^{2+} + Ruthenium Red (RR) (Meissner et. al., 1985), has little Ca^{2+} leakage compared to the control solution. However, the presence of Mg^{2+} + RR has no effect on the large rate of Ca^{2+}-leakage of digested SR. Two possible explanations for the channel blockers inability to block Ca2+ after trypsin digestion are: either the SR no longer has these Mg^{2+} + RR regulatory sites or another protein becomes leaky to calcium.

In the purified (Ca^{2+} + Mg^{2+})-ATPase reconstituted into liposomes, tryptic digestion in the presence or absence of Mg^{2+} + RR does not increase Ca^{2+} leakage when compared to control SR. This indicates that tryptic digestion per se of the molecule (Ca^{2+} + Mg^{2+})-ATPase does not cause increased Ca^{2+} leakage. Therefore, the loss of Ca^{2+}-uptake in the purified (Ca^{2+} + Mg^{2+})-ATPase reconstituted into liposomes cannot be due to increased Ca^{2+} leakage, but rather to an uncoupling of the two functions.

Modeling

We have developed theoretical modeling equations to fit the loss of Ca^{2+}-uptakes versus the loss of A fragment when it is degraded by T_2 digestion into A_1 and A_2 (Cao et. al., 1990). In this modeling program, we addressed the issue of whether the functional (Ca^{2+} + Mg^{2+})-ATPase consists of a dimer or a monomer of the single polypeptide of 110 KDa. The loss of Ca^{2+}-uptake due to digestion could fall into three cases: (1) the enzyme is a monomer and the cleavage of monomer results in the loss of Ca^{2+}-uptake (2) the enzyme is a dimer and cleavage of either one of the two polypeptides results in the loss of Ca^{2+}-transport and (3) the enzyme is a dimer and cleavage of both polypeptides are required for loss of Ca^{2+}-uptake. Our data (Cao et al., 1989;1990) and those published by Shoshan-Bamatz et. al. (1987) fit case 2 where the enzyme is a dimer and that tryptic digestion of either polypeptide results in loss of Ca^{2+}-uptake (Cao et al., 1990).

Trypsin Digestion and Ca^{2+}-Occlusion

Ca^{2+} occlusion is a necessary step in achieving Ca^{2+}-transport. Calcium occlusion represents the key step of dehydrating and internalizing the calcium ion into the enzyme (Dupont, 1980; Vilsen and Andersen, 1986). Using Eu^{3+} as an analogue to calcium and utilizing laser-pulsed fluorescence spectroscopy to estimate lifetime of fluorescent decay of Eu^{3+} luminescence, we can estimate the number of water molecules surrounding the ion and the degree of occlusion.

TABLE 3

		%	
	Ca^{2+}-uptake	ATPase Activity	Occlusion
T_2	0	100	100
T_2+ (- ATP)	0	35	100
T_2+ (+ ATP)	0	50	0

T_2 cleavage as discussed earlier causes loss of uptake without concommittant loss of ATPase activity and without an increase in

Ca^{2+}-leakage. T_2 cleavage was also found not to affect occlusion. However, digestion carried out past T_2 w/out ATP while causing no loss of occlusion, resulted in a loss of 65% of the ATPase activity (table 3). Tryptic digestion past T_2 in the presence of mM ATP resulted in the loss of 50% of the ATPase and complete loss of occlusion. The enzyme can still phosphorylate to about 65% of control undigested level whether the digestion is in the presence or absence of ATP. Furthermore, the loss of occlusion due to tryptic digestion of ATPase in the presence of ATP appears to require the phosphorylation step since occlusion is not lost when an ATP analogue (AMP-PNP) is used instead of ATP during the digestion procedure (Lockwich et. al., 1990a).

Synthetic Peptides

The location of the two calcium binding sites within the polypeptide of (Ca^{2+} + Mg^{2+})-ATPase is not yet fully elucidated. We (Scott and Shamoo, 1982,1984), and others (Ludi and Hasselbach, 1984) reported data to indicate that the calcium binding site may be located in the enzymes's N-terminus,also known as the A_2 fragment. However, recent reports suggest that the A_1 fragment (Sumbilla et. al., 1990) or the B-fragment (Clark et al., 1989) could contain the calcium binding sites. We (Scott and Shamoo, 1984) reported that the two calcium sites are heterogeneous in nature with respect to temperature and tryptic digestion at Arg 198 (T_2). For example, T_2 digestion at Arg-198 alters one of the high affinity sites into an intermediate affinity site (Scott and Shamoo, 1984). We then utilized the published primary structure of the enzyme (MacLennan et al., 1985; Brandle et al., 1986) and our data (on the importance of the site of T_2 cleavage) to propose putative Ca^{2+} binding site regions (Gangola and Shamoo, 1986).

Recently, there has been conflicting reports as to the location of the two calcium binding sites. For example, Clark et. al. (1989) reports that the crucial amino acids needed for Ca^{2+}-dependent phosphorylation (and by inference the calcium binding sites) reside almost exclusively in the B fragment (C-terminus). However, Sumbilla et. al. (1990) showed that NCD-4 inhibits Ca^{2+} binding at a location on the A_1 fragment. We (Scott and Shamoo, 1982, 1984) and Ludi and Hasselbach (1984) reported the A_2 fragment as the potential site for calcium binding sites. Therefore, to date, no clear evidence has been reported on the exact location of the Ca^{2+} binding sites.

Table 4

Peptide	No. of H_2O molecule	Quenching
MOPS alone	9.0	---
T2 Site		
Thr-Asp-Pro-Val-Pro-Asp-Pro-Arg	8.0	No
Thr-Thr-Ala-Pro-Val-Pro-Asp-Pro-Arg	8.0	No
Thr-Asp-Pro-Val-Pro-Ala-Pro-Arg	9.0	Yes
Thr-Thr-Ala-Pro-Val-Pro-Ala-Pro-Arg	9.0	Yes
Thr-Asp-Pro-Val-Pro-Asp-Lys-Arg	8.0	No
Thr-Asp-Pro-Pro-Val-Pro-Pro-Asp-Pro-Arg	8.0	No
Thr-Asp-Gly-Gly-Pro-Val-Pro-Gly-Gly-Asp-Pro-Arg	8.0	No
Thr-Asp-Asp-Pro-Val-Pro-Gly-Asp-Pro-Arg	8.0	No
Temp Site		
Lys-Asp-Ile-Val-Pro-Gly-Asp-Ile	8.0	No
Lys-Ala-Ile-Val-Pro-Gly-Asp-Ile	8.0	No
Lys-Asp-Ile-Val-Pro-Gly-Ala-Ile	9.0	Slightly
Lys-Asp-Gly-Ile-Val-Pro-Gly-Gly-Asp-Ile	8.0	No

We synthesized (table 4) an octapeptide nearest to the T_2 digestion site which contained a unique sequence of Thr_{191}-Asp_{192}-Pro_{193}-Val_{194}-Pro_{195}-Asp_{196}-Pro_{197}-Arg_{198} as a potential candidate for the Ca^{2+}-binding site (Gangola and Shamoo, 1986). The synthetic peptide for both cardiac and skeletal muscle ATPase (which differ in that Asp_{190} versus Glu_{192}) exhibited characteristics consistent with the proposed site or at least part of it. We proceeded to study the physical chemistry of these synthetic peptides and their analogues on the assumption that at the least these structures are similar to whatever the calcium binding sites would turn out to be. Therefore, an understanding of the ion-ligand interaction of various peptide side chains and their carbonyl group with calcium or its analogue Eu^{3+}, will assist us in the elucidation of the mechanism and specificity of calcium binding and its eventual role in Ca^{2+}-transport.

Our data from the synthetic peptide approach indicates that the unique position of Asp 196 and Asp 149 are important in cation binding and dehydration (Lockwich et al., 1990). Mutation of Asp 196 reduces the expression of the entire enzyme (Clark et al., 1989). Also, when two peptide sequences near T_2 are altered (Asp_{196} to Ala_{196} and Asp_{192} to Ala_{192}) these two peptides can quench europium fluorescence. The (Ca^{2+} + Mg^{2+})-ATPase also quenches europium fluorescence; however, T_2 digested enzyme does not. Therefore, there appears to be a unique small sequence that interact in a very special way with the cations.

CONCLUSION

Figure 2 summarizes the data presented in this paper. T_2 cleaves the enzymes into fragments A and B. Further digestion in the presence of millimolar calcium and at 0^oC results in the cleavage of the A fragment into A_1 and A_2 fragments. T_2 digestion results in the loss of calcium uptake without any effect on ATPase and occlusion. Further digestion past T_2 in the presence of ATP result in the complete loss of calcium occlusion and about half of the ATPase activity. However, past T_2 digestion in the absence of ATP, has no effect on occlusion while about 35% ATPase activity is retained. It appears that the presence of ATP exposes a digestion site in the B fragment that is important for the process of occlusion. The loss of calcium occlusion with about half of the ATPase activity intact may point out that the functional enzyme to achieve occlusion is a dimer and that both monomers are required for overall calcium transport. This is supported by the modeling fit of the data on the loss of calcium uptake versus remaining A fragment where the data fit that the functional enzyme for calcium transport requires the integrity of both monomers (i.e. dimer) to be intact.

The location of calcium binding/transport sites on the polypeptide chain appears to be unresolved yet. Clark's et. al. data (1989) point that the calcium sites are located in the B fragment embeded in the membrane region. Others (Ludi and Hasselbach, 1984, Sumbilla et. al., 1990) including our reports (Scott and Shamoo, 1982, 1984) indicate that the location of the calcium sites are in the A1-A2 region. Data from differential calorimetry (Lepock et. al., 1990) also indicate that the A1-A2 region is the site of calcium binding and not the membrane bound region. Furthermore, data from

Figure 2

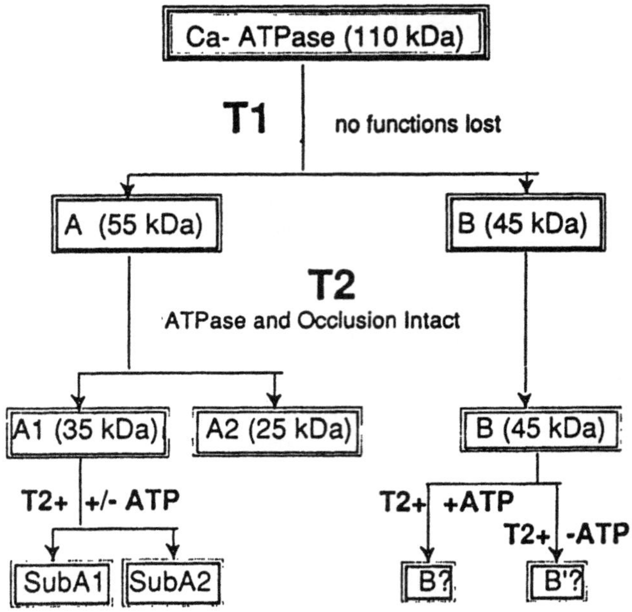

<u>T2+ (+ ATP)</u>---> ATPase 50%, Occlusion Lost

<u>T2+ (- ATP)</u>---> ATPase 35%, Occlusion Intact

the mutagenesis of yeast H^+-ATPase (Serrano and Portillo, 1989) also indicate that the A_1-A_2 region is the site of cation binding. One way to reconcile all of the data to date, we propose that the calcium binding region is in the A_1-A_2 region and that occluded calcium is in the membranous B fragment region. We have shown that calcium occlusion is dependent on phosphorylation. Clark's et. al. (1989) data also showed that several amino acids in membranous B region were critical for the calcium dependent phosphorylation.

ACKNOWLEDGEMENT

This work was supported in part by the Muscular Dystrophy Association of America and the American Heart Association.

REFERENCES

Aubard, J., Levor, P., Denis, A., and Claverie, P. (1987). Direct analysis of chemical relaxation signals by a method based on the combination of laplace transform and pade approximants. Comput.Chem. 11: 163-178.

Berman, M.C. (1982). Energy coupling and uncoupling of active calcium transport by sarcoplasmic reticulum. Biochem. Biophys. Acta 694: 95-121.

Brandle, C.J., Green, N.M., Korczak, B., MacLennan, D.H. (1986). Two Ca^{2+}- ATPase genes: Homologies and mechanistic implications of deduced amino acid sequences. Cell 44: 597-607.

Cao, C.J., Lockwich, T., Shamoo, A.E. (1989). In: The effects of trypsin digestion on sarcoplasmic reticulum (SR) active and passive transport. Biophysical Journal 55(2): 204a.

Cao, C., Lockwich, T., Scott, T.L., Blumenthal, R. and Shamoo, A.E. (1990). Uncoupling of Ca^{2+} transport from ATP hydrolysis activity of sarcoplasmic reticulum (Ca^{2+} + Mg^{2+})-ATPase. Submitted to Biochemistry.

Clarke, D.M., Loo, T.W., Inesi, G., MacLennan, D.H. (1989). Location of high- affinity Ca^{2+}- binding sites within the predicted transmembrane domain of the sarcoplasmic reticulum Ca^{2+}-ATPase. Nature 339: 476-478.

Dupont Y. (1980). Occlusion of divalent cations in the phosphorylated calcium pump of sarcoplasmic reticulum. Eur. J. Biochem. Dec.: 445-449.

Ebashi, S. and Lipmann, F. (1962). Adenosine triphosphate- linked concentration of calcium ions in a particulate function of rabbit muscle. J. Cell. Biol. 14: 389-400.

Gangola, P. and Shamoo, A.E. (1987) Characterization of (Ca^{2+} + Mg^{2+})- ATPase of sarcoplasmic reticulum by laser-excited europium luminescence. Eur. J. Biochemistry 162: 357-363.

Hasselbach, W. and Makinose, M. (1963). The calcium pump of the relaxing granules of muscle and its dependence on ATP splitting. Biochem. Z. 333: 518- 528.

Herrmann, T.R., Gangola, P., Shamoo, A.E. (1986). Estimation of inter- binding-site distances in sarcoplasmic reticulum (Ca^{2+} + Mg^{2+})-ATPase using Eu (III) luminescence. Biochemistry 25: 5834-5838.

Herrmann, T.R., Shamoo, A.E. (1988). Estimation of inter-binding-site distances in sarcoplasmic reticulum (Ca^{2+} + Mg^{2+})-ATPase under occluded and non-occluded conditions. Molecular and Cellular Biochem. 82: 55-58.

Joshi, N.B. and Shamoo, A.E. (1988) Distance between functional sites in cardiac sarcoplasmic reticulum (Ca^{2+} + Mg^{2+})-ATPase. Interlanthanide Energy Transfer, Eur. J. of Biochemistry 178: 483-487.

Lepock, J.R., Rodahl, A.M., Zhang, C., Heynen, B.W., and Cheng,K. (1990). Thermal denaturation of the Ca^{2+}- ATPase of sarcoplasmic reticulum reveals two thermodynamically independent domains. Biochem. 29: 681- 689.

Lockwich, T., Douglass, C. and Shamoo, A.E. (1990). Uncoupling of Ca^{2+} occlusion From ATP hydrolysis activity in sarcoplasmic reticulum (Ca^{2+} + Mg^{2+})-ATPase. Submitted to Biochemistry.

Lockwich, T., Herrmann, T., Douglass, C., and Shamoo, A.E. (1990). Characterization of Ca^{2+} binding to peptide segments of (Ca^{2+} + Mg^{2+})-ATPase - A synthetic approach to determine site requirement. Submitted to International J. of Peptide Chemistry.

Ludi, J., Hasselbach, W. (1984). Separation of the tryptic fragments of sarcoplasmic reticulum ATPase with high performance liquid chromatography. FEBS Lett. 167: 33-36.

MacLennan, D.H. (1970). Purification and properties of an adenosine triphosphate from sarcoplasmic reticulum. J. Biol. Chem. 245: 4508-4518.

MacLennan, D.H., et al. (1980). Ion pathways in proteins of the sarcoplasmic reticulum. Ann. N.Y. Acad. Sci. 358: 138-148.

Meissner, G., Darling, E. and Eveleth, J. (1986). Kinetics of rapid Ca^{2+} release by sarcoplasmic reticulum. Effects of Ca^{2+}, Mg^{2+}, and adenine nucleotides. Biochemistry 25: 236-244.

Racker, E. (1972). Reconstitution of calcium pump with phospholipids and a purified Ca^{++}- Adenosine triphosphatase from sarcoplasmic reticulum. J. Biol. Chem. 247: 8198-8200.

Scott, T.L., Shamoo, A.E. (1982). Disruption of energy transduction in sarcoplasmic reticulum by trypsin cleavage of (Ca^{2+} + Mg^{2+})-ATPase. J. Membr. Biol. 64: 137-144.

Scott, T.L., Shamoo, A.E. (1984). Distinction of the roles of the two high-affinity calcium sites in the functional activities of the Ca^{2+}-ATPase of sarcoplasmic reticulum. Eur. J. Biochem. 109: 231-238.

Scott, T.L. (1985). Distances between the functional sites of the (Ca^{2+} + Mg^{2+})-ATPase of sarcoplasmic reticulum. J. Biol. Chem. 260: 14421-14423.

Serrano, R., Portillo, F. (1989). Active sites of yeast H^{+}- ATPase studied by direct mutagenesis. Biochem. Soc. Trans. 17(6): 973-975.o

Shamoo, A.E., Herrmann, T.R. (1988). Tertiary structure of the (Ca^{2+} + Mg^{2+})-ATPase probed by laser-pulsed lanthanide luminescence spectroscopy. In: R. Verna (ed) Proceedings of International Symposium on "Advances in Biotechnology of Membrane Ion Transport". Aula Maga dell' University, A'Aquila, Italy, Sept. 19-20.

Shamoo, A.E., Lockwich, T., and Cao, C.J. (1990). Tertiary structure and energy coupling in Ca^{2+}-pump system. In Press. Molecular and Cellular Biochemistry.

Shamoo, A.E. and Herrmann, T., In: Proc. of Int. Symp. on "Advances in Biotechnology of Membrane Ion Transport" p. 125-133. P.L. Jergensen and R. Verna, editors, Serona Symposia Vol. 51. Raven Press, 1988.

Shoshan-Barmatz, V., Ouziel, N., Chipman, D.M. (1987). Tryptic digestion of sarcoplasmic reticulum inhibits Ca^{2+} accumulation by action on a membrane component other than the Ca^{2+}-ATPase. J. Biol. Chem. 262: 11559-11564.

Sumbilla, C., Malek, H., Lakowicz, J., and Inesi, G. (1990). Flourescent labeling of the transmembrane segment of the sarcoplasmic reticulum (SR) ATPase is accompanied by specific inhibition of calcium binding. Biophys. J. 57: 204a.

Torok, K., Trinnaman, B.J., Green, M. (1988). Tryptic cleavage inhibits but does not uncouple Ca^{2+}-ATPase of sarcoplasmic reticulum. Eur. J. Biochem. 173: 361-367.

Vilsen, B. and Andersen, J.P. (1986). Occlusion of Ca^{2+} in soluble monomeric sarcoplasmic reticulum Ca^{2+}- ATPase. Biochim. Biophys. Acta. 855: 429-431.

A 19KD C-TERMINAL TRYPTIC FRAGMENT OF THE ALPHA CHAIN OF NA/K-ATPase, ESSENTIAL FOR CATION OCCLUSION AND TRANSPORT

W.D.Stein*, R.Goldshleger**, and S.J.D.Karlish**#

Institute of Life Sciences
Hebrew University
Jerusalem* ISRAEL

Biochemistry Department
Weizmann Institute of Science
Rehovot** ISRAEL

Trypsin, in the presence of Rb but absence of Ca, digests off half the protein of pig renal Na/K-ATPase leaving a 19KD membrane-embedded fragment of the alpha chain with almost normal Rb and Na occlusion, but no ATP-dependent functions. Subsequent digestion, now in the absence of Rb but the presence of Ca, leads to rapid loss of the 19KD peptide and parallel loss of cation occlusion. Thus the peptide is essential for occlusion. Its N-terminal sequence is NPKTDKLVNERLISMA. It begins at residue 830 and extends towards the C-terminus. 19KD-containing membranes can reconstitute into phospholipid vesicles, to sustain a slow Rb-Rb exchange. Cation occlusion sites and the transport pathway, within trans-membrane segments, are thus quite separate from the ATP binding site, located on the cytoplasmic domain of the alpha chain.

INTRODUCTION

The Na/K-ATPase or sodium pump has been purified, cloned and sequenced (Ovchinnikov, 1987; Jorgensen and Anderson, 1988). Much is known about its transport, ion occlusion properties and conformational changes . However, the cardinal question - the manner in which the free energy of hydrolysis of ATP is transduced into the active pumping of cations - remains unanswered. Several ATP-binding residues have been located to the central cytoplasmic loop of the alpha chain, but little is known about the cation binding sites or the pathway for cation movement (for a review see Glynn and Karlish,(1990)). In an attempt to identify cation-binding residues, we have recently studied inactivation by N,N'-dicyclohexylcarbodiimide (DCCD) of Rb and Na occlusion by Na/K-ATPase. We concluded that carboxyl residues are involved and that Rb and Na bind to the same residues (Shani-Sekler et al, 1988).

Footnotes :

(i) # To whom correspondence should be addressed

(ii) This is an abbreviated version of a full paper (Karlish, Goldshleger and Stein, to be published), containing a complete Methods and Materials section and Bibliography.

(iii) This work was supported by the Weizmann Renal Research Fund.

NATO ASI Series, Vol. H 48
Calcium Transport and
Intracellular Calcium Homeostasis
Edited by D. Pansu and F. Bronner

Labelling with ^{14}C-DCCD leads to incorporation of about 2 moles of DCCD per mole of the alpha chain for full inactivation of Rb occlusion. On attempting to locate the label in the primary sequence using selective proteolysis, we were surprised to find a condition in which, in control experiments, occlusion was retained while the alpha chain was largely but incompletely fragmented, yielding a stable 19KD membrane-embedded fragment.

This paper describes conditions for preparation of 19KD-containing membranes, presents the N-terminal sequence of the 19KD fragment (which locates it in the primary sequence of the alpha chain) and characterises functional properties of these membranes. These properties include retention of cation occlusion and transport, but loss of ATP-dependent functions.

RESULTS

Materials and methods used are fully described in Karlish et al (see Footnote (ii)). Fig. 1A depicts the time-course of tryptic digestion of the Na/K-ATPase in a medium containing Rb ions and EDTA, at a roughly equimolar trypsin-to-membrane-protein weight ratio. The gel shows only fragments remaining within the membrane. Within 10s the alpha chain disappeared and, with time, successively smaller fragments progressively appear and disappear. By one hour, the major product is a stable 19KD fragment, with minor fragments of 16 and 14 KD and other smaller fragments. At this stage, 40 to 50 % ofthe membrane protein has been released into the medium. The beta chain is largely resistant to proteolysis . The 19KD fragment is only slowly digested further (t1/2 about 4 h). The striking result is that the capacity to occlude Rb is essentially unchanged, although no fragment of the alpha chain larger than 19KD remains embedded in the membrane. Fig 1B depicts an experiment to compare the effects of the presence and absence of Rb and of Ca. The presence of Rb and the absence of Ca are essential. In contrast, in the absence of Rb, the 19KD fragment was largely destroyed with loss of Rb occlusion. In the presence of Ca, both the 19KD fragment and the beta chain are digested and a 13 KD fragment produced, with loss of Rb occlusion.

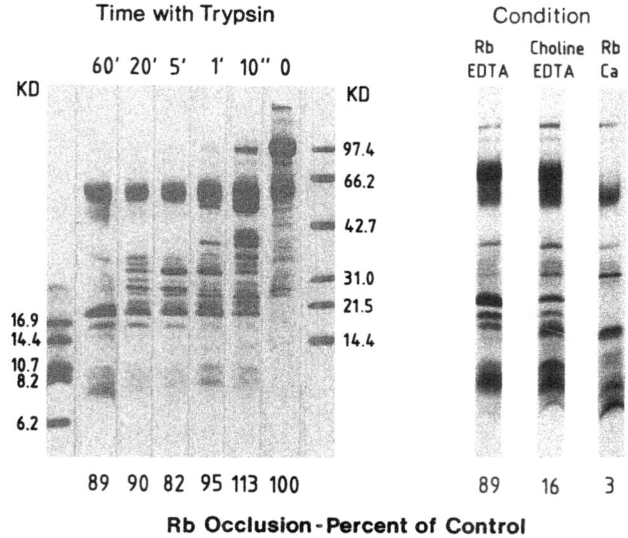

Fig.1. Time-course and Conditions for accumulation of the 19KD fragment. (A)Na/K-ATPase was incubated with equimolar Trypsin in the standard Tris/RbCl/EDTA medium. At times indicated, aliquots were mixed with tryptic inhibitor. Rb occlusion at 5.5mM Rb was measured on samples . The remainder was centrifuged, washed with pH7.5/25mM Imidazole, 1mM EDTA, and the pellets dissolved in the 1:5 diluted gel buffer. Equal volumes of dissolved control enzyme and proteolysed membranes were applied to a 10% Tricine gel. (B). Na/K-ATPase was incubated for 1hr at 37°C with Trypsin in media containing 10mM RbCl/1mM EDTA or 10mM choline chloride/1mM EDTA or 10mMRbCl/1mM CaCl2. Tryptic inhibitor was added. Rb occlusion (at 4.5mM Rb) was measured on aliquots, and the remainder processed as described in Karlish et al (1990). A quantity of proteolysed membranes equivalent to 50 ug of control enzyme was applied to each lane of the gel.

The experiments of Figs 1 suggested that the 19KD fragment is essential for Rb occlusion. Fig 2 tests this suggestion quantitatively. 19KD-membranes were suspended in a medium containing Rb and Ca ions. Trypsin (1:50 w/w) was added and, at various times (10 s to 4 min), digestion was halted by adding trypsin inhibitor. On portions of the suspension, occlusion of Rb was measured and others were applied to 10 to 20 % mini-gels. The gels were stained and extensively destained, the 19KD fragment extracted into 2% SDS, and the absorbance of the Coomassie Blue (proportional to protein concentration in this range) measured at 595 nm. The regression line in fig 2 has slope of 1.06 ± 0.06, indicating that the 19KD fragment is essential for Rb occlusion.

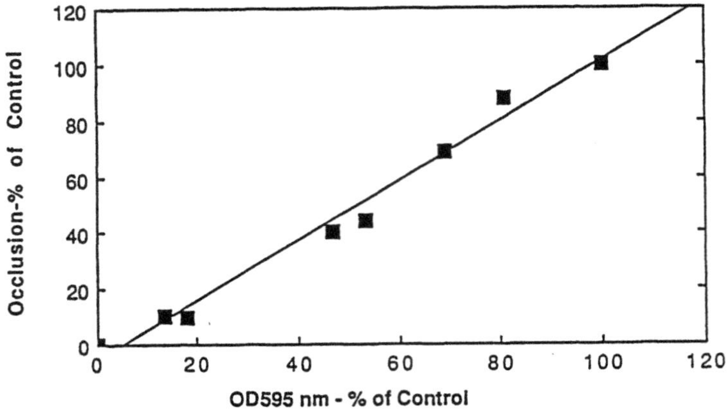

Fig.2 Correlation between Rb Occlusion and the Amount of the 19KD fragment. 19KD- membranes were incubated with Trypsin in a medium containing 10mM Rb/1mM CaCl2. At times from 10" to 4' aliquots were removed to tryptic inhibitor , Rb occlusion was measured, membranes pelleted in the Airfuge, and dissolved in gel buffer. Volumes of control enzyme were applied to a 10 to 20% Mini-Gel. After running, staining and thorough destaining, the band of the 19KD fragment was cut out , homogenised in 2% SDS, and acrylamide removed by centrifugation. The absorbance extracted from each lane was measured at 595nm. In each case the absorbance from a piece of extracted gel outside the lanes was subtracted .

For sequencing from its N-terminus, the 19KD fragment, was transferred to Immobilon paper from a 10 % Tricine gel. In two separate experiments, yields of N-terminal residues in each cycle were sufficient to identify the first sixteen residues as

Asn.Pro.Lys.Thr.Asp.Lys.Leu.Val.Asn.Glu.Arg.Leu.Ile.Ser.Met.Ala

This unambiguously identifies the N-terminus of the 19KD fragment as residue 830 of the alpha chain (see fig 4).

Studies of ion occlusion and ATP-dependent functions in control and 19KD- membranes showed that the latter occlude both Rb and Na. The specific activity for both Na and Rb occlusion is about twice that of control membranes, consistent with the loss of about half of the membrane protein. The ratio of Na to Rb occlusion is not significantly different from 1.5 in both control and proteolysed membranes. The stoichiometry of Na and Rb occlusion is 3 Na or 2 Rb per phosphoenzyme molecule. Thus, the present result indicates that for the 19KD- membranes all three sodium or two Rb , occluding sites are intact.

Since the 19KD- fragment lacks the major hydrophilic loop of the alpha chain, one might expect ATP-binding and ATP-dependent functions to have been lost. This expectation was verified. In particular, the high affinity binding of eosin, an ATP analogue is absent, as is also the inhibitory effect of ATP on Rb occlusion. Evidently, ATP - binding and - dependent activities are totally abolished while ion occlusion is intact.

There are interesting differences in properties of cation occlusion between control and 19KD- membranes even though the maximal capacities are unaltered. For Rb, control and 19KD- membranes are practically indistinguishable, yielding simple hyperboli with best-fit Km values of 47 ± 3 and 38 ±6 microM, respectively. In contrast, while Na occlusion is essentially hyperbolic for control membranes (Km = 1.56 ± 0.30 mM, Hill number n = 0.92) , it is highly co-operative for 19KD- membranes, with a somewhat lower affinity (K1/2 = 2.69 ± 0.13 mM, Hill number of n = 1.90).

An obvious point to examine was whether the entire cation transport pathway was intact in the 19KD- membranes. To test this, 19KD- membranes were reconstituted into phospholipid vesicles. ^{86}Rb uptake was measured into Rb-loaded vesicles from media containing low (0.25 mM) and high concentrations (40 mM) of RbCl, (fig 3A). The difference between the time-courses at low and high Rb represents the saturable (pump-mediated) component of ^{86}Rb uptake. Saturable Rb uptake occurs in two phases, one extremely rapid, the other far slower. The Km for this slow uptake is 0.48 mM (data not shown), very similar to that for Rb/Rb exchange in control vesicles (21), but the rate is 1 to 2 % of that of control. The amount of Rb taken up in the rapid phase (in experiments using 1mM Rb) is 5-7 picomoles/10 microl of vesicles. This is close to the pump density in vesicles made using control Na/K-ATPase, estimated by the amount of phosphoenzyme . A similar biphasic time-course of ^{86}Rb uptake is observed for control vesicles at 0°C (unpublished). We propose (see Discussion) that the fast phase reflects a single turnover of the proteolysed enzyme and is followed by a very slow steady-state Rb-Rb exchange. This proposal assumes that the ^{86}Rb is transported into the vesicles rather than being merely bound or occluded. To test this (fig 3B), the amount of ^{86}Rb associated with the vesicles was measured over four minutes, in the absence or presence of valinomycin, after loading with ^{86}Rb for 2.5 or 10 min (see legend). Fig 3B shows clearly that valinomycin is needed to remove rapidly the ^{86}Rb , showing that isotope associated with the vesicles both at 2.5 and 10 min had been transported into them.

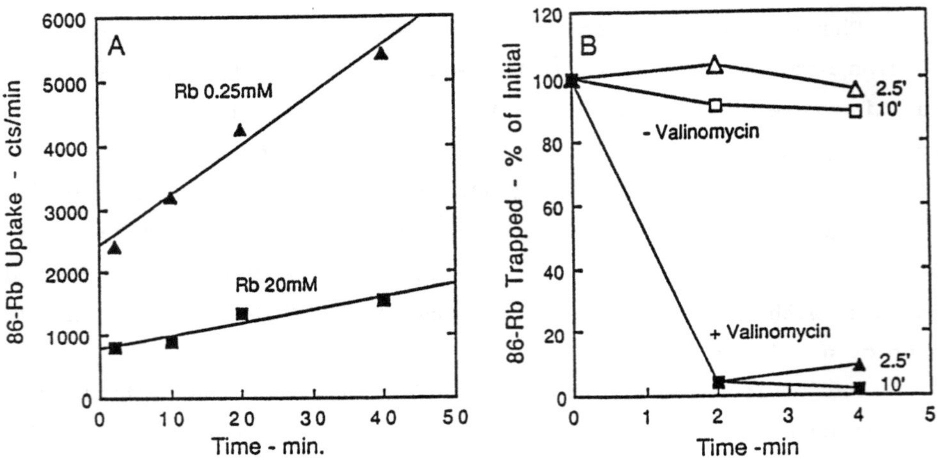

Fig.3. Rb/Rb exchange in Phospholipid Vesicles Reconstituted with 19KD-Membranes , and Release of Accumulated ^{86}Rb by Valinomycin . (A) To calculate the quantity of Rb taken up in the initial phase and steady-state rate of ^{86}Rb flux, the slope and intercept of the regression lines at 20mM Rb were subtracted from those at 0.25mM Rb . (B) Rb-loaded vesicles were incubated for 2.5' or 10' in the standard medium containing 1mM Rb + ^{86}Rb The vesicles with associated ^{86}Rb were eluted on a Dowex-50 column, and unlabelled 150mM RbCl was added. The suspension was divided , 1u.M valinomycin or Ethanol was added, and after incubation at 20°C for 0, 2 or 4 min., aliquots were transferred to a second set of Dowex-50 columns and vesicles eluted.

DISCUSSION

<u>Structure.</u> Fig 4 depicts a hypothetical model for the 19KD- membranes, based on the arrangement of the alpha and beta chains as suggested by Ovchinnikov. We have assumed that the membranes contain, in addition to the 19KD polypeptide and the beta chain, other membrane-spanning segments which cannot be cut further by trypsin. These fragments are shown as being shaved to the tryptic limit peptide. Such might not be exactly the case. We have also assumed that the 19KD fragment extends to the C-terminus, but it might reach only to one of the arginine or lysine residues near the C-terminus.

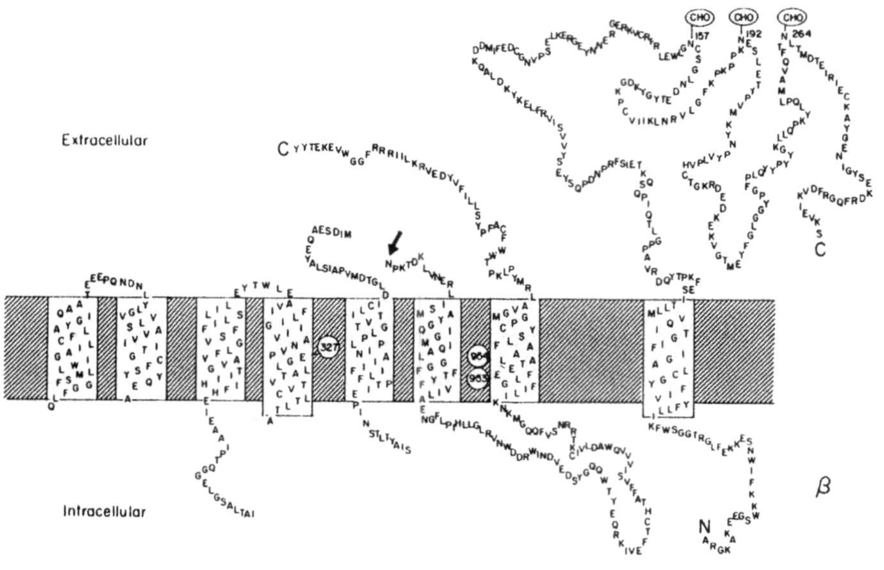

Fig 4. Possible Arrangement of Polypeptides in 19KD- Membranes.

Figs 1 and especially, 2 show that the 19KD fragment is essential for occlusion of Rb but do not by themselves show that the ion-binding residues reside on that segment. This is probably so since recent experiments show ^{14}C-DCCD to label the 19KD fragment almost exclusively, this labelling being highly protected against by potassium . The potassium-binding carboxyls could be glu 953 and glu 954 (fig 4). The 19KD fragment seems necessary for occlusion, but may not be sufficient. Other transmembrane segments and/or the beta chain may donate residues to an ion-binding cage.

Function. ATP-binding and -dependent functions are destroyed in the 19KD-membranes. Cation-binding functions seem largely intact, with interesting differences between control and 19KD-membranes.

The significance of fig. 3 is that it shows that the trans-port pathways for Rb in the 19KD-membranes remain accessible. The observations can be understood in terms of our recent findings on the biphasic release of occluded Rb. The first Rb is released rapidly, the second in minutes. Thus, the intercept in fig.3A reflects entry of one Rb ion in a first turnover, while dissoc-iation of the second , limiting the steady-state rate of Rb/Rb exchange is greatly retarded by unlabelled Rb within the vesicles.

Implications. It seems clear that the cation- and ATP- binding sites are physically separate. ATP-binding residues are located on the large cytoplasmic loop. The present findings and our previous work with DCCD suggest strongly that the cation binding sites are located within transmembrane segments and that the K and two Na sites are the same. Inevitably, coupling between ATP hydrolysis and active cation movements involves indirect interactions between the ATP and cation sites, mediated presumably by conformation changes (the E1 - E2 transitions, see Jorgensen and Andersen, 1988). If we consider the sodium pump as analogous to an automobile, the formation of 19KD-membranes is equivalent to removing the engine, while allowing the wheels to revolve freely and the automobile to move downhill. Pursuing this analogy, a possible candidate for the molecular driveshaft (the physical coupling device) is the loop of the 19KD fragment predicted to lie within the cytoplasm (fig 4), adjacent to the central cytoplasmic loop and in a position to interact with the ATP-binding site.

REFERENCES

Glynn IM, Karlish SJD(1990) Occluded cations in active transport. Ann Rev Biochem (in press)

Jorgensen PL, Andersen JP(1988) Structural basis for E_1-E_2 conformational transitions in Na,K-pump and Ca-pump proteins. J Membran Biol 103:95-120

Ovchinnikov YA(1987) Probing the folding of membrane proteins. Trends Biochem Sci 12:434-438

Shani-Sekler M, Goldshleger R, Tal DM, Karlish SJD(1988) Inactivation of Rb^+ and Na^+ occlusion on (Na^+, K^+)-ATPase by modification of carboxyl groups. J Biol Chem 263:19331-19341

THE ACTIVITY OF THE BASOLATERAL MEMBRANE CALCIUM-PUMPING ATPase AND INTESTINAL CALCIUM TRANSPORT

Julian RF Walters
Gastroenterology Unit
Royal Postgraduate Medical School
London W12 0HS
United Kingdom

Introduction

The plasma membrane Ca^{2+}-pumping ATPase maintains the large, 10,000-fold difference in Ca^{2+} concentrations at the cell membrane. In Ca^{2+}-transporting tissues, extrusion of Ca^{2+} from the cell by the Ca^{2+}-pump is also the only active, energy-requiring step (Bronner, Pansu & Stein 1986). Although the first studies of Ca^{2+} transport by a calmodulin-sensitive Ca^{2+}-pump in intestinal basolateral membranes were published in 1981 (Nellans & Popovitch 1981), it is still not certain whether the activity of the Ca^{2+}-pump is regulated in the intestine. Regulation of activity would be expected in different states of Ca^{2+} absorption, such as on varying intakes of Ca^{2+}, or by vitamin D, and in different areas of the small intestine.

Several experimental findings describing properties of plasma membrane Ca^{2+}-pumps should be considered when assessing the evidence for regulation of Ca^{2+} extrusion by the Ca^{2+}-pump in Ca^{2+} absorption.

Experimental findings

1. The Ca^{2+}-pumping ATPase is only a small proportion of total Ca^{2+}-ATPase activity in intestinal basolateral membranes.

The assay of Ca^{2+}-stimulated ATP hydrolysis (ATPase) predominantly measures Ca^{2+}-stimulated nucleotide phosphatase activity (Moy, Walters & Weiser 1986). Unlike the Ca^{2+}-pumping ATPase responsible for vesicular Ca^{2+} transport it is not specific for ATP. Similar nucleotide phosphatase activities have been

NATO ASI Series, Vol. H 48
Calcium Transport and
Intracellular Calcium Homeostasis
Edited by D. Pansu and F. Bronner

shown in plasma membrane preparations from several tissues, and in the liver this has been shown to be due to an ecto-ATPase (Lin & Russell 1988).

	V max (nmol/min/mg)
Ca^{2+}-stimulated "ATPase" activity	233
Vesicular Ca^{2+} transport	11

Both activities were measured in similar preparations of rat duodenal membranes enriched for basolateral membrane markers. Ca^{2+}-stimulated ATPase activity was determined by phosphate release from ATP and vesicular Ca^{2+} transport by the uptake of $^{45}Ca^{2+}$. Values were calculated from a range of submicromolar Ca^{2+} concentrations obtained with EGTA as a Ca^{2+}-buffering agent (Moy, Walters & Weiser 1986).

2. The apparent affinity for Ca^{2+} of the Ca^{2+}-pumping ATPase is increased by EGTA (the "EGTA-effect").

Published estimates for the apparent affinity for Ca^{2+} of the Ca^{2+}-pumping ATPase in intestinal basolateral membranes have been between 0.1 and 0.3 μM (Nellans & Popovitch 1981, Ghijsen & van Os 1982, Walters & Weiser 1987). These estimates have been obtained in assay systems which used EGTA as a Ca^{2+}-buffering system.

However, when determined in the absence of EGTA, the affinity of the intestinal Ca^{2+}-pump was decreased by two orders of magnitude as shown in Figure 1 (Walters 1989). Similar values have been found with Ca^{2+}-pumps prepared from other tissues and have been interpreted as indicating that the Ca^{2+}/EGTA complex may activate the Ca^{2+}-pump. This EGTA-effect has not been previously considered when interpreting Ca^{2+}-transport kinetics in the intestine. It suggests that the true activity of the intestinal pump is low at physiological intracellular free Ca^{2+} concentrations (0.1 μM), and that the pump is very sensitive to changes in Ca^{2+} concentration and possibly to other stimulators.

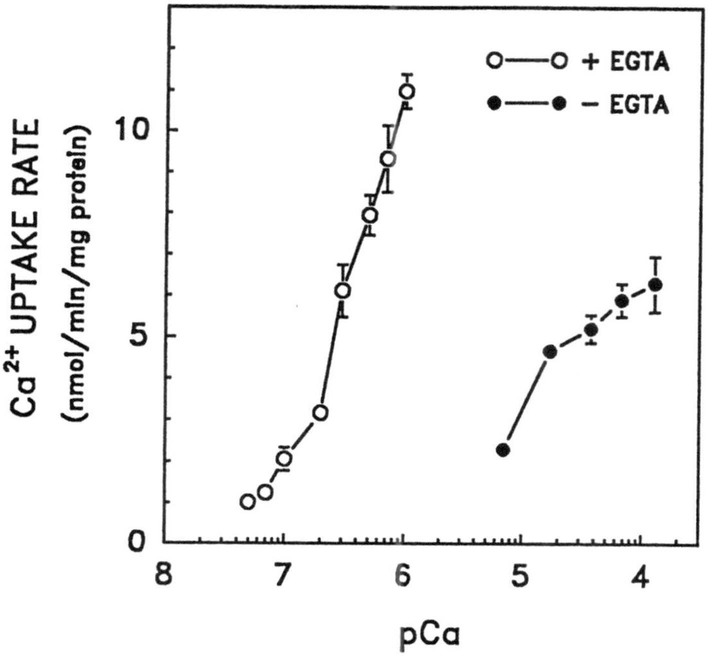

Figure 1. Rat duodenal basolateral membrane vesicular ATP-dependent Ca^{2+} uptake rates determined at various free-Ca^{2+} concentrations (pCa) in the presence and absence of 0.5 mM EGTA.

3. Correlation of Ca^{2+}-transport activity and the Ca^{2+}-pump protein.

Quantitation of the 130 kDa Ca^{2+}-pump phosphoprotein has been used to estimate the amount of Ca^{2+}-pump protein present in various intestinal cells (Wajsman, Walters & Weiser 1988). However, vesicular Ca^{2+} transport activity did not always correlate with the amount of phosphoprotein, which suggested that there may be other determinants of activity, in addition to the amount of pump protein.

	Ca^{2+}-transport activity	Phosphoprotein density
Aboral gradient	Duod. > Ileum	Duod. > Ileum
Differentiation	Villus > crypt	Villus = crypt
Vitamin D	Replete > defcnt.	Replete = defcnt.

4. Activation of the Ca^{2+}-pumping ATPase during preparation of intestinal cells.

Precise comparison of the activities of the Ca^{2+}-pump in membranes prepared from different areas of the intestine, or in various physiological states, is difficult as the pump is extremely sensitive to a range of activating enzymes (van Corwen, de Jong & van Os 1987). Enzymatic activities which have been described as affecting the various Ca^{2+}-pump isoforms in other tissues include proteases, protein kinases and phosphatases. Membrane phospholipids, which are altered by lipolytic activity, also affect the function of the Ca^{2+}-pump. These enzymatic activities are all present in intestinal cells and may affect the pump during membrane preparation. This following Table shows changes in Ca^{2+}-pump activation during three methods of cell preparation.

Cell preparation method	Basolateral membrane ATP-dependent Ca^{2+} uptake (nmol/min/mg)
Incubation in citrate/EGTA	6.7 ± 0.4
Perfusion + vibration (4°C)	2.4 ± 0.4
" (+ 10 min at 37°C)	9.0 ± 0.8

Rat duodenal intestinal cells were isolated by incubation of everted sacs in citrate/EGTA buffers at 37°C, or by intra-arterial perfusion with EGTA followed by vibration of everted sacs in buffer at 4°C. Further incubation of cells obtained by vibration for 10 min at 37°C significantly stimulated membrane vesicle Ca^{2+} transport activity ($p < 0.001$). Membrane purification was similar from all cells. ATP-dependent Ca^{2+} transport was measured at 0.5 μM free Ca^{2+} buffered by EGTA. Results are means ± SEM, n = 12.

5. Calbindin-D9k stimulates the intestinal Ca^{2+}-pump.

Intestinal basolateral membrane Ca^{2+}-transport activity was increased 3-fold by physiological concentrations (30 μM) of calbindin-D9k (Figure 2.) This stimulation occurred only in the absence of EGTA (Walters 1989).

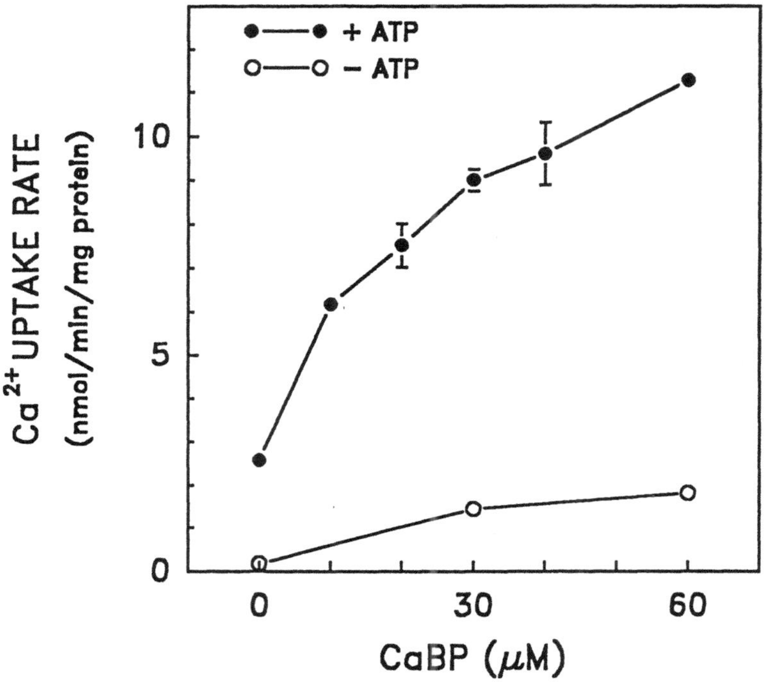

Figure 2. Stimulation by calbindin-D9k (CaBP) of Ca^{2+}-transport by rat duodenal basolateral membrane vesicles in the absence of EGTA.

In these experiments, extensive measures were taken to ensure that the addition of calbindin did not change the free Ca^{2+} concentration or the specific activity of $^{45}Ca^{2+}$ in the experimental system.

To investigate the mechanism for the stimulation by calbindin, experiments have recently been performed using various mutant recombinant calbindins in which specific domains had been modified (Linse et al 1988). Stimulation of transport was unaffected by changes in the affinity of the N-terminal binding site for Ca^{2+} or by neutralization of various surface charges (Walters et al 1989). Further experiments may indicate which properties of the molecule are essential for the stimulation and if calbindin-D9k presents Ca^{2+} to the pump so it binds with higher affinity, or whether it stimulates by interacting with a regulatory site in a manner similar to calmodulin.

Conclusions

1. The maximum rate of the Ca^{2+}-pumping ATPase is lower than that assumed from studies of Ca^{2+}-stimulated ATP hydrolysis activity. The affinity of the pump for Ca^{2+} is also decreased when measured in EGTA-free systems.

2. There appear to be differences in the amount of Ca^{2+}-pump protein in parts of the small intestine, but the effects of activators of the pump remain largely unexplored and may produce artefactual changes during cell preparation.

3. Stimulation of the Ca^{2+}-pumping ATPase by calbindin-D9k represents one way by which the rate of Ca^{2+} efflux from the enterocyte can be increased to respond to demands of Ca^{2+} absorption, though its mechanism is unclear at present.

4. As the concentration of calbindin correlates well with active, transcellular Ca^{2+} transport, this effect may be one of the major means of regulation of Ca^{2+} absorption.

Selected references

Bronner F, Pansu D, Stein WD (1986) An analysis of calcium transport across the rat intestine. Am J Physiol; **250**:G561-569

Ghijsen WEJM, van Os CH (1982) 1a,25-dihydroxy-vitamin D_3 regulates ATP-dependent calcium transport in basolateral plasma membranes of rat enterocytes. Biochim Biophys Acta; **689**:170-172

Lin SH, Russell WE (1988) Two Ca^{2+}-dependent ATPases in rat liver plasma membrane. The previously purified (Ca^{2+}-Mg^{2+})-ATPase is not a Ca^{2+}-pump but an ecto-ATPase. J Biol Chem; **263**:12253-12258

Linse S, Brodin P, Johannson C, Thulin E, Grundström T, Forsén S (1988) The role of protein surface charges in ion binding. Nature; **335**:651-652

Moy TC, Walters JRF, Weiser MM (1986) Intestinal basolateral membrane Ca-ATPase activity with properties distinct from those of the Ca-pump. Biochem Biophys Res Commun; **141**:979-985

Nellans HN, Popovitch JE (1981) Calmodulin-regulated, ATP-driven calcium transport by basolateral membranes of rat small intestine. J Biol Chem; **256**:9932-9936

Van Corven EJJM, de Jong MD, van Os CH (1987) The adenosine triphosphate-dependent calcium pump in rat small intestine: effects of vitamin D deficiency and cell isolation methods. Endocrinology; **120**:868-873

Wajsman R, Walters JRF, Weiser MM (1988) Identification and isolation of the phosphorylated intermediate of the calcium pump in rat intestinal baso-lateral membranes. Biochem J; **256**:593-598

Walters JRF, Weiser MM (1987) Calcium transport by rat duodenal villus and crypt basolateral membranes. Am J Physiol; **252**:G170-G177

Walters JRF (1989) Calbindin-D9k stimulates the calcium-pump in rat enterocyte basolateral membranes. Am J Physiol; **256**:G124-G128

Walters JRF, Charpin MV, Gniecko KC, Brodin P, Thulin E, Forsén S (1989) Stimulation of intestinal calcium-pump activity by recombinant synthetic calbindin-D9k and specific site-directed mutants. Gastroenterology; **96**:A535

SECTION III

CALCIUM EXTRUSION - THE Na/Ca EXCHANGER

SODIUM-CALCIUM EXCHANGE IN THE HEART

John P. Reeves, Diane C. Ahrens, Joo Cheon and John T. Durkin
Roche Institute of Molecular Biology
Roche Research Center
Nutley, N.J. 07110
U.S.A.

Most cells of higher eucaryotes maintain a low cytoplasmic concentration of Na^+ compared to the extracellular concentration by virtue of the activity of the Na^+, K^+ -ATPase. The transmembrane Na^+ gradient can be utilized for many specialized functions in these cells, including electrical activity, metabolite transport and Ca^{2+} homeostasis. The latter function is brought about by the action of the Na^+-Ca^{2+} exchange system, a carrier-mediated transport process in which the movement of Na^+ ions across the membrane is directly coupled to the movement of Ca^{2+} ions in the opposite direction. The Na^+-Ca^{2+} exchange system is found in many, but not all, cells and is particularly active in cells which use Ca^{2+} as a second messenger, i.e. muscle cells, neurons, secretory cells and transporting epithelial cells. In most cells, the stoichiometry of the exchange is 3 Na^+ per Ca^{2+}; this has been determined by several different approaches, including an assessment of thermodynamic driving forces in membrane vesicles (Reeves & Hale, 1984), measurements of exchange currents and their reversal potentials in heart cells (Kimura *et al.*, 1987; Lipp & Pott, 1988) and direct flux measurements in the barnacle muscle (Rasgado-Flores *et al.* , 1989). In retinal rods, Ca^{2+} fluxes mediated by the exchange system also involve the movement of K^+ ions, and in this system, the stoichiometry is 4 Na^+ per 1 Ca^{2+} + 1 K^+ (Cervetto *et al.* , 1989; Schnetkamp *et al.* , 1989).

The primary physiological role of the exchange system is to mediate Ca^{2+} efflux from cells, using the inwardly directed concentration gradient of Na^+ ions as an energy source. For a stoichiometry of 3 Na^+ per Ca^{2+}, the cytoplasmic Ca^{2+} concentration ($[Ca^{2+}]_i$) determined by the Na^+-Ca^{2+} exchange system is related to $[Na^+]_i$ by the following equation:

NATO ASI Series, Vol. H 48
Calcium Transport and
Intracellular Calcium Homeostasis
Edited by D. Pansu and F. Bronner

$$[Ca^{2+}]_i = [Ca^{2+}]_o\{[Na^+]_i/[Na^+]_o\}^3 \cdot \exp\,[E_mF/RT] \qquad (1)$$

where E_m is the membrane potential, F is the value of the Faraday, R is the gas constant and T is the absolute temperature. Note that for a constant $[Ca^{2+}]_o$ and $[Na^+]_o$, the cytoplasmic Ca^{2+} concentration is a *cubic* function of $[Na^+]_i$ and an *exponential* function of the membrane potential. Thus, in theory, large changes in $[Ca^{2+}]_i$ can be brought about by quite small changes in $[Na^+]_i$ or E_m. For a resting mammalian myocardial cell ($[Na^+]_i$* = 10 mM, $[Na^+]_o$ = 140 mM, $[Ca^{2+}]_o$ = 1.8 mM, E_m = - 80 mv), the value of $[Ca^{2+}]_i$ predicted by the above equation at 37° C is 33 nM. This value is similar to the values actually measured using fluorescent indicator dyes (≈100 nM).

These considerations indicate that the Na^+-Ca^{2+} exchange system is capable of reducing $[Ca^{2+}]_i$ to physiological levels and modulating $[Ca^{2+}]_i$ in a sensitive manner through small changes in the electrochemical Na^+ gradient. The cell types for which such a role has been most clearly demonstrated are the cardiac myocyte and the retinal rod. In the following paragraphs, we will consider the physiological role(s) of the Na^+-Ca^{2+} exchange system in the cardiac myocyte and describe the efforts of our laboratory to identify and purify the Na^+-Ca^{2+} exchange carrier from cardiac tissue.

CA2+ HOMEOSTASIS IN MYCARDIAL CELLS.

Contractile activity in the heart involves an exeedingly complex interplay of transmembrane Ca^{2+} fluxes. Ca^{2+} enters the heart cell through voltage-gated Ca^{2+} channels in the plasma membrane (or sarcolemma, SL) which open during each beat. This influx of Ca^{2+} is crucial for initiating the events that lead to the release of Ca^{2+} stored intracellularly in the sarcoplasmic reticulum (SR). The rise in $[Ca^{2+}]_i$ initiates cross-bridge formation between myosin and actin molecules in the myofibrillar apparatus, leading to a shortening of the muscle and tension development. During the relaxation phase of the cardiac cycle, cytoplasmic Ca^{2+} ions are either re-accumulated by the SR (through the action of a Ca^{2+}-activated ATPase) or

---

*Values of $[Na^+]_i$ are given here as concentrations; literature values for Na^+_i activity have been divided by an activity coefficient of 0.76.

pumped out of the cell by the Na^+-Ca^{2+} exchange system and the SL Ca^{2+}-ATPase (a different enzyme from the ATPase in the SR).

The strength of contraction of cardiac muscle is determined in large part by the amount of releasable Ca^{2+} stored in the SR. In the steady state, the amount of Ca^{2+} that enters the heart cell is equal to the amount that gets pumped out. Under conditions which call for a change in the force of cardiac contraction, e.g. during exercise, a temporary imbalance may develop between the amounts of Ca^{2+} entering and leaving the cell, leading to the development of a new steady state with an altered contractile force. The situation can be viewed as a "competition" between the SR Ca^{2+}-ATPase and the SL efflux mechanisms for available Ca^{2+}. Since the Na^+-Ca^{2+} exchange system is the predominant Ca^{2+} efflux mechanism in the cardiac myocyte (cf. below), the force of contraction is exquisitely sensitive to changes in the transmembrane Na^+ gradient. Indeed, recent measurements using Na^+-sensitive microelectrodes suggest that developed tension is directly related to the sixth or seventh power of $[Na^+]_i$ (Sonn & Lee, 1988). This provides an explanation for the well known effect of digitalis in increasing the force of cardiac contraction, since the active ingredient in digitalis is a cardiac glycoside which inhibits the Na^+,K^+-ATPase.

Ca^{2+} Efflux in Cardiac Myocytes. In amphibian hearts, which have less SR than mammalian hearts, the presence of extracellular Na^+ is necessary to bring about relaxation, indicating the crucial role of the Na^+-Ca^{2+} exchange system in mediating Ca^{2+} efflux (Vassort *et al.*, 1978). In mammalian hearts, the involvement of Na^+-Ca^{2+} exchange in bringing about relaxation of caffeine contractures (caffeine promotes Ca^{2+} release from the SR) is well established by numerous reports in the literature (e.g. Jundt *et al.*, 1975; Bridge *et al.*, 1988; Bers & Bridge, 1989; cf. Chapman, 1983). More recently, measurements of net trans-sarcolemmal Ca^{2+} fluxes during cardiac contraction using extracellular Ca^{2+}-indicating dyes provide clear-cut evidence for the importance of Na^+-Ca^{2+} exchange in removing Ca^{2+} from the cell during normal cardiac function (Hilgemann, 1986a,b). These findings indicate that the exchange system is the predominant Ca^{2+} efflux process in myocardial cells during periods of elevated $[Ca^{2+}]_i$.

Even during rest, when $[Ca^{2+}]_i$ is likely to be no greater than 100 nM, the exchange system appears to be the predominant Ca^{2+} efflux mechanism. This conclusion is based on studies conducted in rabbit ventricle in which the amount of Ca^{2+} remaining in the SR after a period of rest was assessed by measuring either the magnitude of the first post-rest contraction (Sutko *et al.*, 1986), or the magnitudes of contractures induced by either caffiene or rapid cooling (Bers, 1987). The results showed that the rate of decline in the SR Ca^{2+} content was greatly reduced when $[Na^+]_o$ was reduced, or when $[Na^+]_o$ and $[Ca^{2+}]_o$ were simultaneously reduced. These findings suggest that Na^+-Ca^{2+} exchange is an effective Ca^{2+} efflux mechanism at low values of $[Ca^{2+}]_i$, a conclusion that is consistent with measurements of the kinetics of Na^+-Ca^{2+} exchange currents in cardiac myocytes (cf. below). Thus, the earlier notion that the exchange system is merely a "back-up" system for the Ca^{2+}-ATPase (DiPolo & Beaugé, 1979) certainly does not apply to cardiac myocytes.

Ca^{2+} Influx in Cardiac Myocytes. Since Na^+-Ca^{2+} exchange is bidirectional and potential-dependent (Eq. 1), one would predict that at a sufficiently positive E_m, the exchanger would reverse direction and bring Ca^{2+} into the cell. The potential dependence of Na^+-Ca^{2+} exchange has been demonstrated very dramatically in the experiments of Eisner *et al.*, (1983), in which sheep cardiac Purkinje fibers were incubated in a K-free medium to inhibit the Na^+,K^+-ATPase. Under these conditions, $[Na^+]_i$ attained a value of nearly 40 mM and the resting (tonic) tension developed by the cells was highly potential-dependent, decreasing when the cells were hyperpolarized (using voltage-clamp pulses from a holding potential of -60 mv) and increasing markedly when the cells were depolarized. No potential-dependent changes in tonic tension were observed after $[Na^+]_i$ was reduced (to ≈12 mM) by restoration of Na^+,K^+-ATPase activity. These findings indicate that the Na^+-Ca^{2+} exchange system is capable of mediating large Ca^{2+} fluxes in either direction in a potential-dependent manner.

An interesting example of Ca^{2+} influx mediated by the exchanger comes from the rest-decay studies described in the previous section. As mentioned above, the Ca^{2+} content of the SR in rabbit ventricular cells declines during rest due to Ca^{2+} efflux mediated by the Na^+-Ca^{2+} exchanger. Rat ventricular cells behave differently, however, in that the SR Ca^{2+} content (measured as the magnitude of the contracture induced by rapid cooling) increases during rest (Bers, 1989). Subsequent studies with an extracellular Ca^{2+}-selective microelectrode showed that during rest there was a net loss of Ca^{2+} from the rabbit cells and a net gain of Ca^{2+} by the rat cells (Shattock & Bers, 1989). The difference in the behavior of the two species was correlated with a difference in the resting $[Na^+]_i$, which was 9.5 mM in the rabbit and

16.7 mM in the rat. Thus, the resting $[Na^+]_i$ in the rat was sufficiently high to bring about net Ca^{2+} influx by Na^+-Ca^{2+} exchange during rest, while in the rabbit, with the lower $[Na^+]_i$, the exchanger functioned in the direction of Ca^{2+} efflux. These results underscore the exquisite sensitivity of the exchange system to small changes in $[Na^+]_i$ (as indicated by Eq. 1) and the dramatic effects such a small change can have upon the physiology of the heart.

More recently, Leblanc & Hume (1990) have provided strong evidence that Na^+ entry into guinea-pig ventricular myocytes through voltage-dependent Na^+ channels can bring about the subsequent entry of Ca^{2+} via Na^+-Ca^{2+} exchange. They described the occurrence of a transient increase in $[Ca^{2+}]_i$ during depolarizing voltage-clamp pulses (in the presence of Ca^{2+}-channel blockers) that was dependent upon extracellular Ca^{2+} and blocked by the Na^+-channel antagonist tetrodotoxin. They suggested that a transient increase in $[Na^+]_i$ due to Na^+ channel activity activated a Ca^{2+} entry pathway that did not involve Ca^{2+} channels and was probably Na^+-Ca^{2+} exchange. Such a scenario has been discussed for years as a theoretical possibility, but the results of Leblanc and Hume (1990) provide the first direct experimental evidence supporting its occurrence. The extent to which Ca^{2+} entry via Na^+-Ca^{2+} exchange contributes to normal excitation-contraction coupling in the heart remains to be elucidated.

Na^+-Ca^{2+} Exchange Currents

By virtue of the 3-to-1 stoichiometry of the exchange process in cardiac cells, a current is generated during exhange activity. This can be demonstrated as a difference in current after an increase of either $[Na^+]$ or $[Ca^{2+}]$ on one side of the membrane (Kimura *et al.* , 1986, 1987; Mechmann & Pott, 1986), or as a gradual decay in a Na^+- and Ca^{2+}-dependent current after a sudden change in membrane potential ("creep currents")(Hume & Uehara, 1986; Barcenas-Ruiz *et al.*, 1987). Both inward (Ca^{2+} efflux) and outward (Ca^{2+} influx) currents have been measured. The characteristics of these currents and the criteria employed to identify them as Na^+-Ca^{2+} exchange currents have been described in a recent review (Reeves, 1990) and will not be discussed in detail here. In some cases, the current measurements have been combined with simultaneous measurements of $[Ca^{2+}]_i$ using intracellular indicator dyes (Barcenas-Ruiz *et al.*, 1987); these studies indicate that the current magnitudes are nearly linear with $[Ca^{2+}]_i$ up to 2-3 μM (higher concentrations could not be studied due to the contraction of the cell). Other studies, however, carried out in cells perfused internally with Ca^{2+} buffering agents, suggest that the current is saturable with a $K_m[Ca^{2+}]_i$ of 0.6 μM (Miura & Kimura, 1989). Although the differences between the two sets of studies have not been resolved, the data indicate that the exchange system is capable of pumping Ca^{2+} out of the cell at submicromolar concentrations of $[Ca^{2+}]_i$, in line with the conclusions of the functional studies discussed above.

Do Na^+-Ca^{2+} exchange currents contribute to electrical activity of the heart during normal cardiac function? This subject has been discussed extensively on theoretical grounds (see, for example, Mullins, 1979; DiFrancesco & Noble, 1985; Hilgemann & Noble, 1987) and recently, a number of papers have provided experimental evidence consistent with the presence of such a current. Fedida *et al.* (1987), for example, have described a delayed inward current that is correlated with contractile activity, and is probably activated by Ca^{2+} release from the SR. The authors suggested that it could be due to either Ca^{2+} efflux via Na^+-Ca^{2+} exchange activity or the activation of non-specific ion channel by $Ca^{2+}{}_i$. Kenyon & Sutko (1987) found that ryanodine (an agent that inhibits the function of the SR by holding the SR Ca^{2+}-release channels in an open configuration) abolishes a Na-dependent inward current during the plateau phase of the action potential in calf cardiac Purkinje fibers. Because the inward current was absent when extracellular Na^+ was replaced with Li^+ (an ion which does not activate Na^+-Ca^{2+} exchange but will carry current through the Ca^{2+}-activated non-specific cation channel) they suggested that this current was generated by the Na^+-Ca^{2+} exchanger mediating Ca^{2+} efflux during the $Ca^{2+}{}_i$ transient. Thus, the results obtained from functional studies, measurements of extracellular Ca^{2+} and from current measurements provide a consistent picture of the importance of Na^+-Ca^{2+} exchange in carrying out Ca^{2+} efflux during the normal cardiac cycle.

Measurements of Na^+-Ca^{2+} exchange currents also provide important insights into the regulation of Na^+-Ca^{2+} exchange activity in myocardial cells. The outward exchange current (Ca^{2+} influx) requires the presence of submicromolar concentrations of Ca^{2+} at the cytoplasmic surface (Kimura *et al.*, 1986). This requirement was first observed in studies of $^{45}Ca^{2+}$ fluxes in squid axons (Baker, 1970; Baker & McNaughton, 1976, DiPolo, 1979) and has also been observed in barnacle muscle (Rasgado-Flores *et al.*, 1989).

Although there is some controversy as to the proper interpretation of this observation (reviewed by Reeves, 1990), it is consistent with the idea that an activating site at a cytoplasmic domain of the exchange carrier must be occupied with Ca^{2+} for exchange activity to occur.

Recently, Hilgemann (1989) has described the measurement of Na^+-Ca^{2+} exchange currents in large membrane patches obtained from isolated guinea pig ventricular myocytes. With a Ca^{2+}-containing solution inside the patch pipette bathing the extracellular surface of the membrane, exchange currents were activated by Na^+ in the bath medium with a $K_{1/2}$ of 38 mM and a Hill coefficient of 2.6. The currents required the presence of Ca^{2+} (30 nM) in the bath medium and were blocked by Co^{2+} and dichlorobenzamil, agents known to inhibit Na^+-Ca^{2+} exchange activity. This important new development provides convenient experimental access to the cytoplasmic surface of the sarcolemma and will be invaluable for delineating the biochemical mechanisms regulating Na^+-Ca^{2+} exchange activity.

Na^+-Ca^{2+} Exchange in Cardiac Sarcolemmal Vesicles

Another biochemical approach to the Na^+-Ca^{2+} exchange system involves the use of plasma membrane vesicles. This was first described in 1979 using cardiac sarcolemmal vesicles (Reeves & Sutko, 1979). Briefly, the procedure for measuring exchange activity involves loading the vesicles with Na^+ by passive equilibration and then diluting the vesicles into an iso-osmotic Na^+-free medium containing $^{45}Ca^{2+}$. The outwardly-directed $[Na^+]$ gradient brings about the accumulation of $^{45}Ca^{2+}$ by the vesicles via Na^+-Ca^{2+} exchange, and this is easily measured by filtering the vesicles at various times after the dilution step. By appropriate loading and dilution steps, the procedure can be readily adapted to study Na^+-dependent Ca^{2+} efflux, Ca^{2+}-Ca^{2+} exchange and Na^+-Na^+ exchange (Reeves, 1988).

The basic kinetic features of the exchange system as studied in cardiac vesicles can be summarized as follows: The $K_m(Ca^{2+})$ averages about 25 µM, but varies over a broad range of values for different preparations. This variability probably reflects the large number of regulatory factors that can influence this parameter in the vesicles (cf. below). Na^+-Ca^{2+} exchange exhibits a sigmoidal dependence upon the Na^+ concentration with a $K_m(Na^+)$ of 20-30 mM and a Hill coefficient of 2-3. Na^+ in the external medium competitively inhibits Ca^{2+} uptake with a K_i of 16 mM and a Hill coefficient approaching 2 at high [Na^+]. The V_{max} for Na^+-Ca^{2+} exchange in cardiac vesicles is 20-40 nmol Ca^{2+} per mg protein per sec. This is a very high value, more than 50-fold higher than for vesicles from most other types of cells, and is consistent with the high levels of Na^+-Ca^{2+} exchange activity measured in intact cardiac tissue. The exchange system is electrogenic in cardiac vesicles and exhibits a stoichiometry of 3 Na^+ per Ca^{2+} (Reeves & Hale, 1984). These and other aspects of exchange activity in vesicles have been discussed in several recent reviews (Reeves, 1985, 1990; Philipson, 1985; Reeves & Philipson, 1989).

Site Density of Na^+-Ca^{2+} Exchange Carriers. Unfortunately, there are no toxins or inhibitors that are selective for Na^+-Ca^{2+} exchange or are effective at a sufficiently low concentration to serve as molecular probes for the Na^+-Ca^{2+} exchange carrier. Therfore, we must rely on functional activity for indentification and quantitation of the Na^+-Ca^{2+} exchange carrier. We have estimated the site density of the exchange carrier in reconstituted preparations of proteoliposomes by measuring the fraction of vesicles that exhibit exchange activity (Cheon & Reeves, 1988a). Reconstitution procedures involve solubilizing membrane proteins with a suitable detergent, adding exogenous phospholipids and then re-incorporating the solubilized proteins into membrane vesicles (proteoliposomes) that form spontaneously when the detergent is removed. We reasoned that proteoliposomes containing the Na^+-Ca^{2+} exchange carrier would equilibrate rapidly with Ca^{2+} under equilibrium conditions whereas vesicles without the carrier would equilibrate much more slowly. We measured the uptake of $^{45}Ca^{2+}$ in reconstituted proteoliposomes under conditions where both internal and external media had the following composition: 40 mM NaCl, 120 mM KCl, 0.1 mM $CaCl_2$. The rate of equilibration of $^{45}Ca^{2+}$ was corrected for the passive uptake of Ca^{2+} into proteoliposomes in which the Na^+-Ca^{2+} exchanger had been inactivated, and compared to the total uptake observed in the presence of the Ca^{2+} inophore A23187. The results indicated that under our reconstitution conditions, approximately 3% of the vesicles contained the exchange carrier. We assumed that the distribution of proteins among the proteoliposomes was random and that the proteoliposomes exhibiting exchange activity contained just one molecule of the exchange carrier. Based on this assumption, and the number and size of the reconstituted proteoliposomes, we calculated that the density of exchange carriers in the reconstituted preparations was 10-20 pmol/mg protein.

The V_{max} for Na^{+}-Ca^{2+} exchange in these reconstituted preparations was approximately 20 nmol/mg protein·sec; with an estimated site density of 10-20 pmol/mg protein, this yields a turnover number for the exchanger of 1,000 - 2,000 sec^{-1}. This value is similar to the turnover number that has been suggested for the Na^{+}-H^{+} exchanger (Vigne *et al.*, 1985; Dixon *et al.*, 1987).

Molecular Nature of the Na^{+}-Ca^{2+} Exchanger. Because there are no high affinity probes for the exchange carrier, purification procedures must rely on measuring functional activity in reconstituted preparations of proteoliposomes after separation of proteins by various procedures. This leads to considerable uncertainty in trying to correlate exchange activity with the presence or absence of a particular protein band in SDS gels. It is not surprising, therefore, that there have been a number of different reports which claim to have identified or purified the Na^{+}-Ca^{2+} exchange carrier, and that few of the reports agree. The molecular weights that have been suggested for the exchange carrier are 33, 70, 84, 120 and 220 kDa (Wakabayashi & Goshima, 1982; Barzilai *et al.*, 1984; Hale *et al.*, 1984; Soldati *et al.*, 1985; Philipson *et al.*, 1988; Cook & Kaupp, 1988; Nicoll & Applebury, 1989). In the following paragraphs, we will briefly summarize some selected recent reports on the identification and purification of exchange carrier(s) from neural and cardiac tissues, and from retinal rods, and then discuss the recent results from our own laboratory.

The Na^{+}-Ca^{2+} exchanger from rat brain synaptosomes was identified as a 70 kDa protein on the basis of "transport specificity fractionation" procedures (Barzilai *et al.*, 1984, 1987; Rahamimoff, 1989). This approach, developed by Goldin & Rhoden (1978), involves the use of a Ca^{2+}-precipitating agent, in this case phosphate, in the interior of reconstituted proteoliposomes. When Ca^{2+} is accumulated by the proteoliposomes during Na^{+}-Ca^{2+} exchange, it precipitates with the internal phosphate; this leads to an increase in density of the transporting proteoliposomes, allowing them to be separated from the other proteoliposomes by density gradient centrifugation. The transporting proteoliposomes isolated in this manner were greatly enriched in a 70 kDa protein when examined by SDS gel electrophoresis. Polyclonal antibodies prepared against this protein immuno-precipitated Na^{+}-Ca^{2+} exchange activity and reacted with a 70 kDa protein in rat heart sarcolemma on Western blots (Barzilai *et al.*, 1987). When the "transport specificity fractionation" procedure was applied to cardiac SL, the results were disappointing; in this case, the protein composition of the transporting subfraction of proteoliposomes was very similar to that of the entire population (Luciani 1984; Young & Reeves, unpublished observations). This is puzzling since the V_{max} for Na^+-Ca^{2+} exchange in cardiac SL is 50 times greater than that for brain synaptosomes and one would think that cardiac SL would therefore yield an even more dramatic enrichment in the exchange carrier than the synaptosomal membranes.

Two groups have purified a 220 kDa protein from retinal rod membranes that appears to be the Na^+-Ca^{2+} exchange carrier (Cook & Kaupp, 1988; Nicoll & Applebury, 1989). The exchanger is a glycoprotein which constitutes 1% or less of the total membrane protein. Na^+-Ca^{2+} exchange activity of the purified rod exchanger requires the presence of K^+ (Nicoll & Applebury, 1989), as does Na^+-Ca^{2+} exchange activity in the intact rod outer segments (see above). The turnover number of the purified rod exchanger has been estimated to be 30-60 sec^{-1}. This is considerably less than the 1,000 sec^{-1} estimated for the cardiac exchanger (see above). It seems likely that the requirement for the translocation of K^+ in addition to Na^+ and Ca^{2+} by the rod exchange could result in a lower turnover number compared to the cardiac exchanger. Other possible explanations for the low turnover number include losses in activity during purification and the use of suboptimal reconstitution conditions.

Philipson *et al.* (1988) have purified Na^+-Ca^{2+} exchange activity approximately 30-fold from from canine cardiac sarcolemma; their preparations consist prodominantly of two protein bands, at 120 and 70 kDa, although lesser amounts of a 160 kDa band are also seen. Under non-reducing conditions, only the 160 kDa band is seen in SDS gels. Polyclonal antibodies prepared against the partially purified preparation immunoprecipitate Na^+-Ca^{2+} exchange activity from solubilized extracts of SL membranes and react predominantly with the 120 and 70 kDa proteins on Western blots. Treatment of the purified preparation with chymotrypsin produces an increase in the intensity of the 70 kDa band and a decrease in the 120 kDa band. The authors suggest that the 70 kDa band may be a proteolytic fragment of the 120 kDa band, and have tentatively identified the 120 kDa band as the cardiac Na^+-Ca^{2+} exchanger. A turnover number of 150 sec^{-1} was estimated from the activity of the purified preparation, assuming a M_r of 120 kDa for the exchange protein. Recently, a partial cDNA clone of the 120 kDa protein was identified using the antibody and an expression library from dog ventricle (Philipson *et al.* 1989); probes from the clone hybridize in the 7 kb region in Northern blots of cardiac poly(A)+RNA. Sequencing of the cDNA and attempted isolation of a full length clone for the exchange carrier are now in progress.

In our laboratory, we have utilized various anion exchange procedures and wheat germ agglutinin (or lentil lectin) affinity chromatography to purify the Na^+-Ca^{2+} exchanger. Our most purified preparations are markedly enriched in a 150-160 kDa glycoprotein (Durkin *et al.* 1990) and yield an estimated turnover number of approximately 290 sec^{-1}, assuming that all the protein is 160 kDa. Most preparations also exhibit lesser amounts of a protein at 120 kDa. Our preparations react strongly at 160, 120 and 70 kDa on Western blots with an antibody prepared by Dr. Philipson against his partially purified 120/70 kDa exchanger (K. Philipson, personal communication). Thus it seems likely that the bands at 120 and 70 kDa are derived from the 160 kDa band by proteolytic clipping; the apparent proteolysis occurs despite the presence of 4 different protease inhibitors throughout the membrane preparation and purification procedures. Efforts are under way to obtain the sequence of peptides obtained by proteolysis of the 160 kDa band. These sequences will be compared with those of the partially cloned 120 kDa protein obtained by Philipson to determine if the two proteins are in fact related.

It is clear that there is no consensus as to the identity of a single Na^+-Ca^{2+} exchange protein that is common to all tissues. It seems likely, however, that some of the differences may be due to diversity among the exchange carriers themselves. Thus, as mentioned previously, the larger size and lower turnover rate of the retinal rod exchanger could be due to its requirement for K^+ translocation. The smaller size of the synaptosomal exchanger might reflect the absence of important regulatory domains that appear to be present in the cardiac exchanger (cf. Reeves, 1990). Other results, to be discussed below, are also consistent with a diversity in molecular size of exchange carriers from different sources. The resolution of such issues awaits the determination of the sequence of the exchange carrier, and the construction of molecular probes that would allow the examination of related sequences from different tissues.

Expression Cloning of the Na^+-Ca^{2+} Exchange Protein. Expression cloning has become a popular method for attempting to isolate cDNA clones for ion channels and transporters. The technique involves the injection of poly(A)$^+$RNA into oocytes from the frog *Xenopus laevis*. If expression of the desired activity is obtained, the RNA is fractionated according to size to localize the molecular weight range that yields the best expression of activity. A cDNA library containing a bacteriophage RNA polymerase promoter is then constructed from the size-selected RNA and used to transcribe RNA from the library. If the transcribed RNA expresses activity when injected into oocytes, the cDNA library is divided successively into smaller and smaller groups for transcription and analysis, until a single clone is isolated which yields RNA that expresses the desired activity. The technique was first utilized to clone interleukin 4 (Noma *et al.* 1986), and was subsequently used for the cloning of the substance-K receptor (Masu *et al.* 1987), the serotonin 1C receptor (Lübbert *et al.* 1987; Julius *et al.* 1987) and the intestinal Na^+-glucose co-transporter (Hediger *et al.* 1987). Several different groups are now applying this approach to the cloning of the Na^+-Ca^{2+} exchange carrier.

Sigel *et al.* (1988) and Longoni *et al.* (1988) have described the expression of Na^+-Ca^{2+} exchange activity in *Xenopus* oocytes injected with poly(A)$^+$RNA from cardiac tissue. In each case, the oocytes were loaded internally with Na^+ and $^{45}Ca^{2+}$ uptake was assayed in a Na^+-free medium compared to a high [Na^+] medium. Higher uptake was observed in the Na^+-free medium for oocytes injected with the poly(A)$^+$RNA, but not for uninjected or water-injected oocytes. Na^+-Ca^{2+} exchange activity was not observed with oocytes that had not first been loaded internally with high $[Na^+]_i$; the Na^+-loading procedure involved either treatment of the oocytes with nystatin, a monovalent ion ionophore (Longoni *et al.*, 1988), or incubating the oocytes in a hypertonic medium containing EGTA (Sigel *et al.*, 1988). Size-fractionation of the poly(A)$^+$RNA by sucrose gradient centrifugation revealed maximal activities in the 25 S region (≈3 kb; Sigel *et al.*, 1988) or the 2-5 kb region (Longoni *et al.*, 1988).

Expression studies in our laboratory have used a different tissue source. The brine shrimp *Artemia* lives in brine pools (e.g. in San Francisco Bay, the Great Salt Lake) and can tolerate salinities up to 8 times that of sea water. We have shown that crude membranes from *Artemia* exhibit surprisingly high levels of Na^+-Ca^{2+} exchange activity, nearly as much as is found in cardiac SL membranes (Cheon & Reeves, 1988b). Furthermore, we found that exchange activity increased markedly during development of hydrated *Artemia* cysts, exhibiting a maximal rate of increase at the time of larval hatching. We reasoned that hatching *Artemia* nauplii might be a particularly rich source of mRNA coding for the Na^+-Ca^{2+} exchange system and, indeed, when poly(A)+RNA from this source was injected into frog oocytes, a high level of expression was observed (Cheon & Reeves, 1990).

After size fractionation of the *Artemia* poly(A)+RNA, maximal levels of expression were found in the 0.5-2.0 kb region; *in vitro* translation of this RNA using the reticulocyte system revealed proteins less than 70 kDa in size. This size range is considerably smaller than observed for cardiac poly(A)+RNA (see above) and is not consistent with the size of the 160 kDa protein we have tentatively identified as the cardiac exchanger (see above). The results suggest that the *Artemia* exchanger may be considerably smaller than the cardiac exchanger. This conclusion is consistent with our previous observations that the *Artemia* exchange system does not respond to agents or conditions that stimulate the activity of the cardiac exchanger (e.g. chymotrypsin, intravesicular Ca^{2+}, etc). Moreover, we have found that the solubilized *Artemia* exchanger behaves very differently from the cardiac exchanger when subjected to the purification procedures described above. The results taken together suggest that the *Artemia* and cardiac exchangers may be quite different in their molecular properties and provide another example of the apparent molecular diversity in exchange carriers from different sources.

REFERENCES

Baker PF (1970) Sodium-calcium exchange across the nerve cell membrane. In: Cuthbert AW (ed) Calcium and Cellular Function. Macmillan, London.

Baker PF, McNaughton PA (1976) Kinetics and energetics of calcium efflux from intact squid giant axons. J Physiol (London) 259: 104-144.

Barcenas-Ruiz L, Beuckelmann DJ, Wier WG (1987) Sodium-calcium exchange in heart: membrane currents and changes in $[Ca^{2+}]_i$. Science 238: 1720-1722.

Barzilai A, Spanier R, Rahamimoff H (1984) Isolation, purification, and reconstitution of the Na^+ gradient-dependent Ca^{2+} transporter (Na^+-Ca^{2+} exchanger) from brain synaptic plasma membranes. Proc Nat Acad Sci, USA 81: 6521-6525.

Barzilai A, Spanier R, Rahamimoff H (1987) Immunological identification of the synaptic plasma membrane Na^+-Ca^{2+} exchanger. J Biol Chem 262: 10315-10320.

Bers, DM (1987) Ryanodine and the calcium content of cardiac SR assessed by caffeine and rapid cooling contractures. Am J Physiol 253: C408-C415.

Bers DM (1989) SR Ca loading in cardiac muscle preparations based on rapid-cooling contractures. Am J Physiol 256: C109-C120.

Bers DM, Bridge JHB (1989) Relaxation of rabbit ventricular muscle by Na-Ca exchange and sarcoplasmic reticulum calcium pump. Ryanodine and voltage sensitivity. Circ Res 65: 334-342.

Bridge JHB, Spitzer KW, Ershler PR (1988) Relaxation of isolated ventricular cardiomyocytes by a voltage-dependent process. Science 241: 823-825.

Cervetto L, Lagnado L, Perry RJ, Robinson DW, McNaughton PA (1989) Extrusion of calcium from rod outer segments is driven by both sodium and potassium gradients. Nature 337: 740-743.

Chapman RA (1983) Control of cardiac contractility at the cellular level. Am J Physiol 245: H535-H552.

Cheon J, Reeves JP (1988a) Site density of the sodium-calcium exchange carrier in reconstituted vesicles from bovine cardiac sarcolemma. J Biol Chem 263: 2309-2315.

Cheon J, Reeves JP (1988b) Sodium-calcium exchange in membrane vesicles from *Artemia*. Arch Biochem Biophys 267: 736-741.

Cheon J, Reeves JP (1990) Expression of *Artemia* Na/Ca exchange activity in *Xenopus laevis* oocytes. Biophys J 57: 184a.

Cook NJ, Kaupp UB (1988) Solubilization, purification and reconstitution of the sodium-calcium exchanger from bovine retinal rod outer segments. J Biol Chem 263: 11382-11388.

DiFrancesco D, Noble D (1985) A model of cardiac electrical activity incorporating ionic pumps and concentration changes. Phil Trans R Soc Lond B 307: 353-398.

DiPolo R (1979) Calcium influx in internally dialyzed squid giant axons. J Gen Physiol 73: 91-113.

DiPolo R, Beaugé L (1979) Physiological role of ATP-driven calcium pump in squid axon. Nature 278: 271-273.

Dixon SJ, Cohen S, Cragoe EJ Jr., Grinstein S (1987) Estimation of the number and turnover rate of Na^+/H^+ exchangers in lymphocytes. J Biol Chem 262: 3626-3632.

Durkin JT, Ahrens DC, Reeves JP (1990) Partial purification and identification of the sodium/calcium exchanger from bovine cardiac sarcolemma. Biophys J 57: 185a.

Eisner DA, Lederer WJ, Vaughan-Jones, RD (1983) The control of tonic tension by membrane potential and intracellular sodium activity in the sheep cardiac Purkinje figer. J Physiol (London) 335: 723-73.

Fedida D, Noble D, Shimoni Y, Spindler AJ (1987) Inward current related to contraction in guinea-pig ventricular myocytes. J Physiol 385: 565-589.

Goldin SM, Rhoden V (1978) Reconstitution and "transport specificity fractionation" of the human erythrocyte glucose transport system. J Biol Chem 253: 2575-2583.

Hale CC, Slaughter RS, Ahrens DC, Reeves JP (1984) Identification and partial purification of the cardiac sodium-calcium exchange protein. Proc Natl Acad Sci, USA 81: 6569-6573.

Hediger MA, Coady MJ, Ikeda TS, Wright EM (1987) Expression cloning and cDNA sequencing of the Na^+/glucose co-transporter. Nature 330: 379-381.

Hilgemann DW (1986a) Extracellular calcium transients and action potential configuration changes related to post-stimulatory potentiation in rabbit atrium. J Gen Physiol 87: 675-706.

Hilgemann DW (1986b) Extracellular calcium transients at single excitations in rabbit atrium measured with tetramethylmurexide. J Gen Physiol 87: 707-735.

Hilgemann DW (1989) Giant excised cardiac sarcolemmal membrane patches: sodium and sodium-calcium exchange currents. Pflügers Arch 415: 247-249.

Hilgemann DW, Noble D (1987) Excitation-contraction coupling and extracellular calcium transients in rabbit atrium: reconstruction of basic cellular mechanisms. Proc Roy Soc Lond B 230: 163-205.

Hume JR, Uehara A (1986) "Creep currents" in single frog atrial cells may be generated by electrogenic Na/Ca exchange. J Gen Physiol 87: 857-884.

Julius D, MacDermott AB, Axel R, Jessel TM (1988) Molecular characterization of a functional cDNA encoding the serotonin 1C receptor. Science 241: 558-564.

Jundt H, Porzig H, Reuter H, Stucki JW (1975): The effect of substances releasing intracellular calcium ions on sodium-dependent calcium efflux from guinea-pig auricles. J Physiol (London) 246: 229-253.

Kenyon JL, Sutko JL (1987) Calcium- and voltage-activated plateau currents of cardiac Purkinje fibers. J Gen Physiol 89: 921-958.

Kimura J, Miyamae S, Noma A (1987) Identification of sodium-calcium exchange current in single ventricular cells of guinea-pig. J Physiol 384: 199-222.

Kimura J, Noma A, Irisawa H (1986) Na-Ca exchange current in mammalian heart cells. Nature 319: 596-597.

Leblanc N, Hume JR (1990) Sodium current-induced release of calcium from cardiac sarcoplasmic reticulum. Biophys J 57: 136a.

Lipp P, Pott L (1988) Voltage dependence of sodium-calcium exchange current in guinea-pig atrial myocytes determined by means of an inhibitor. J Physiol 403: 355-366.

Longoni S, Coady MJ, Ikeda T, Philipson KD (1989) Expression of cardiac sarcolemmal Na^+-Ca^{2+} exchange activity in *Xenopus laevis* oocytes. Am J Physiol 255: C870-C873.

Lübbert H, Hoffman BJ, Snutch TP, van Dyke T, Levine AJ, Hartig PR, Lester HA, Davidson N (1987) cDNA cloning of a serotonin 5-HT_{1C} receptor by electrophysiological assays of mRNA-injected *Xenopus* oocytes. Proc Nat Acad Sci, USA 84: 4332-4336.

Luciani S (1984) Reconstitution of the sodium-calcium exchanger from cardiac sarcolemmal vesicles. Biochim Biophys Acta 772: 127-134.

Masu Y, Nakayama K, Tamaki H, Harada Y, Kuno M, Nakanishi S (1987) cDNA cloning of bovine substance-K receptor through oocyte expression system. Nature 329: 836-838.

Mechmann S, Pott L (1986) Identification of Na-Ca exchange current in single cardiac myocytes. Nature 319: 597-599.

Miura Y, Kimura J (1989) Sodium-calcium exchange current: dependence on internal Ca and Na and competitive binding of external Na and Ca. J Gen Physiol 93: 1129-1145.

Mullins LJ (1979) The generation of electric currents in cardiac fibers by Na/Ca exchange. Am J Physiol 236: C103-C119.

Nicoll DA, Applebury ML (1989) Purification of the bovine rod outer segment Na^{+}/Ca^{2+} exchanger. J Biol Chem 264: 16207-16213.

Noma Y, Sideras P, Naito T, Bergstedt-Lingquist S, Azuma C, Severinson E, Tanabe T, Kinashi T, Matsuda F, Yaoita Y, Honjo T (1986) Cloning of cDNA encoding the murine IgG1 induction factor by a novel strategy using SP6 promoter. Nature 319: 640-646.

Philipson KD (1985) Sodium-calcium exchange in plama membrane vesicles. Ann Rev Physiol 47: 561-571.

Philipson KD, Longoni S, Ward R (1988) Purification of the cardiac Na^{+}-Ca^{2+} exchange protein. Biochim Biophys Acta 945: 298-306.

Philipson KD, Longoni S, Ward R, Scott B (1989) Possible identification of a cDNA clone for the cardiac sarcolemmal Na/Ca exchange protein. Biophys J 55: 165a.

Rahamimoff H (1989) The molecular biochemistry of the sodium-calcium exchanger. In: Allen TJA, Noble D, Reuter H (eds) Sodium-Calcium Exchange. Oxford University Press, Oxford, pp 153-177.

Rasgado-Flores H, Santiago EM, Blaustein MP (1989) Kinetics and stoichiometry of coupled Na efflux and Ca influx (Na/Ca exchange) in barnacle muscle cells. J Gen Physiol 93: 1219-1241.

Reeves JP (1985) The sarcolemmal sodium-calcium exchange system. Curr Top Membr Transp 25: 77-127.

Reeves JP (1988) Measurement of sodium-calcium exchange activity in plasma membrane vesicles. Methods Enzymol. 157: 505-510.

Reeves JP (1990) Sodium-calcium exchange. In: Bronner F (ed) Intracellular Calcium Regulation. Alan R Liss, Inc., New York, in press.

Reeves JP, Hale CC (1984) The stoichiometry of the cardiac sodium-calcium exchange system. J Biol Chem 259: 7733- 7739.

Reeves JP, Philipson KD (1989) Sodium-calcium exchange activity in plasma membrane vesicles. In: Allen TJA, Noble D, Reuter H (eds) Sodium-Calcium Exchange. Oxford University Press, Oxford pp 27-53.

Reeves JP, Sutko JL (1979) Sodium-calcium ion exchange in cardiac membrane vesicles. Proc Natl Acad Sci, USA 76: 590-594.

Schnetkamp PPM, Basu DK, Szevenscei RT (1989) Na^{+}-Ca^{2+} exchange in bovine rod outer segments requires and transport K^{+}. Am J Physiol 257: C153-C157.

Shattock MJ, Bers DM (1989) Rat vs. rabbit ventricle: Ca flux and intracellular Na assesed by ion-selective microelectrodes. Am J Physiol 256: C813-C822.

Sigel E, Baur R, Porzig H, Reuter H (1988) mRNA-induced expression of the cardiac Na^{+}-Ca^{2+} exchanger in *Xenopus* oocytes. J Biol Chem 263: 14614-14616.

Soldati L, Longoni S, Carafoli E (1985) Solubilization and reconstitution of the Na^{+}/Ca^{2+} exchanger of cardiac sarcolemma. J Biol Chem 260: 13321-13327.

Sonn JK, Lee CO (1988) Na^{+}-Ca^{2+} exchange in regulation of contractility in canine cardiac Purkinje fibers. Am J Physiol 255: C278-C290.

Sutko JL, Bers DM, Reeves JP (1986) Postrest inotropy in rabbit ventricle: Na^{+}-Ca^{2+} exchange determined sarcoplasmic reticulum Ca^{2+} content. Am J Physiol 250: H654-H661.

Vassort GM, Roulet M-J, Mongo KG, Ventura-Clapier RF (1978) Control of the frog heart relaxation by Na-Ca exchange. Eur J Cardiol 7 (Suppl): 17-25.

Vigne P, Jean R, Barbry P, Frelin C, Fine LG, Lazdunski M (1985) [^{3}H]ethyl-propylamiloride, a ligand to analyze the properties of the Na^{+}/H^{+} exchange system in the membranes of normal and hypertrophied kidneys. J Biol Chem 260: 14120-14125.

Wakabayashi S, Goshima K (1981b) Partial purification of Na-Ca antiporter from plasma membrane of chick heart. Biochim Biophys Acta 693: 125-133.

THE ROLE OF SODIUM/CALCIUM EXCHANGE IN THE REGULATION OF VASCULAR CONTRACTILITY

M.P. Blaustein, S. Bova[1], X.-J. Yuan and W.F. Goldman
Departments of Physiology and Medicine
and the Hypertension Center
UNIVERSITY OF MARYLAND
School of Medicine
655 West Baltimore Street
Baltimore, Maryland 21201
USA

A rise in the cytosolic free calcium concentration, $[Ca^{2+}]_c$, is the immediate trigger for an increase in tension in vascular smooth muscle (VSM). Ca^{2+} can enter the cytosol from the sarcoplasmic reticulum (SR) when it is released by inositol 1,4,5-trisphosphate (IP_3) and, perhaps, by Ca^{2+}-induced Ca^{2+} release as a result of agonist activation. Ca^{2+} can also enter the cytosol from the extracellular fluid via voltage-gated and/or agonist receptor-operated Ca^{2+}-sensitive channels, as well as via Na/Ca exchange (see below). These features have led to several interesting and controversial, unresolved questions about Ca^{2+} metabolism in VSM: Is the SR store of Ca^{2+} modulated? And if so, how? How is Ca^{2+} extruded and Ca^{2+} balance maintained? And, how is supra-threshold $[Ca^{2+}]_c$ and tonic tension maintained in VSM that exhibits tonic contractions? The answers to these questions are essential for understanding how tension is regulated in VSM.

To address some of the aforementioned issues, we measured tension in rings of rat aorta and a small branch of the mesenteric artery, and Ca^{2+} transients in cultured A7r5 cells (derived from fetal rat aorta). The methods employed for the contraction experiments are published (Ashida & Blaustein,

[1]Present Address: Pharmacology Institute, University of Padua, Padua, Italy

NATO ASI Series, Vol. H 48
Calcium Transport and
Intracellular Calcium Homeostasis
Edited by D. Pansu and F. Bronner

1987). Details of the digital imaging methods used for the measurement of the Ca^{2+} transients are also published (Goldman et al., 1990).

Effects of Reducing the Na^+ Electrochemical Gradient on Agonist-Evoked Contractions

Fig. 1A shows data from a representative experiment in which we determined the effects of low external Na^+, $[Na^+]_o$, on the tension in a ring of rat aorta evoked by various concentrations of serotonin (5-HT). The 5-HT (or, in other experiments, arginine vasopressin, AVP) was applied as a bolus injection in the fluid flowing into the tissue chamber, so that the agonist was present for only 15-20 sec. In this experiment, NaCl in the HEPES-buffered physiological salt solution (PSS) was replaced by N-methyglucamine [=PSS(NMG)] so that $[Na^+]_o$ was reduced to 1.2 mM. At all submaximal 5-HT concentrations tested, the 5-HT evoked tension was augmented, reversibly, in PSS(NMG). As these data indicate, low $[Na^+]_o$ shifted the 5-HT dose-response (tension) curve to the left (toward lower 5-HT concentrations). Comparable results were obtained with AVP as the agonist, with Li^+ as the Na^+ replacement (although Li^+ is known to augment IP_3 transients by inhibiting IP_3 degradation), and with bicarbonate-buffered solutions (in which $[Na^+]_o$ was only reduced to 26 mM; see below). The dose-response curve for K^+ depolarization-evoked contractions was also shifted to the left in low $[Na^+]_o$.

The augmentation of the 5-HT response was inversely related to $[Na^+]_o$. A 25% reduction in $[Na^+]_o$ (from 141 to 105 mM) often induced a slight amplification of the response, but this effect did not become statistically significant until $[Na^+]_o$ was reduced to 70 mM or less. These augmented responses to reduced $[Na^+]_o$ are not mediated by endothelial factors - either reduced endothelial relaxing factor(s) or increased endothelin secretion. Comparable results were obtained in rings that were denuded of endothelium.

The data in Fig. 1B indicate that the dose-response curve for the 5-HT activated contraction is also shifted to the left when intracellular Na^+, $[Na^+]_i$, is increased by inhibiting the Na^+ pump with ouabain. Augmented responses were observed even in rings treated with 10 μM verapamil, to block voltage-gated Ca^{2+} channels that may be opened if the VSM cells are depolarized by ouabain. Furthermore, phentolamine, which should block the action of endogenous norepinephrine, also did not inhibit the augmentation of the AVP-evoked response.

These effects of reducing the Na^+ gradient were totally dependent upon extracellular Ca^{2+}: Agonist-evoked responses were attenuated in Ca-free media, but reduction of $[Na^+]_o$ then no

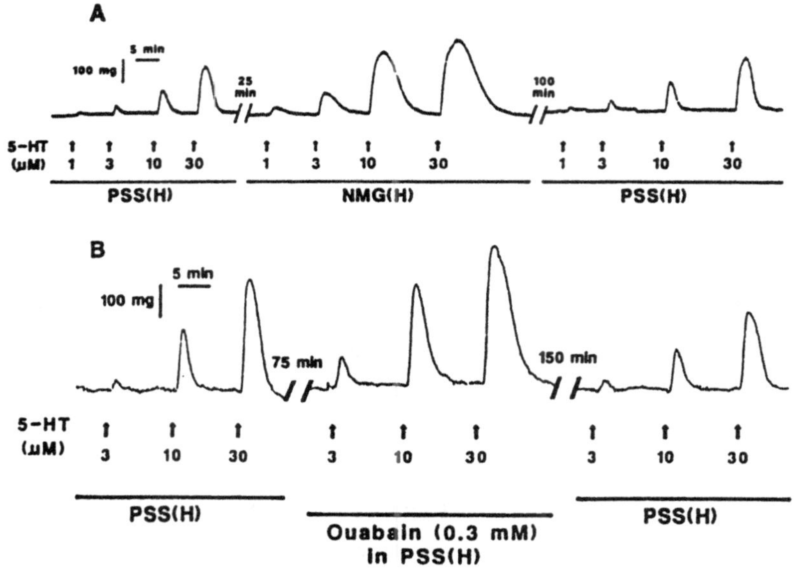

FIG. 1. A. Effect of reduced $[Na^+]_o$ on the responses of a ring of rat aorta to 5-HT. Records show responses to 1, 3, 10 and 30 μM 5-HT, before, during and after reduction of $[Na^+]_o$ from 141 to 1.2 mM; Na^+ was replaced by NMG. Original resting tension = 560 mg; tissue wet weight = 1.44 mg; temp. = 37°C. Reprinted from Bova et al., 1990, with permission.

B. Effect of ouabain on the contractile responses of a ring of rat aorta to 5-HT. Original records show the response to 3, 10 and 30 μM 5-HT before ouabain, after 15 min in ouabain, and following a 115 min washout in ouabain-free PSS. Tissue wet weight = 2.65 mg; original resting tension = 580 mg; temp. = 37°C. (Reprinted from Bova et al., 1990, with permission.)

longer augmented the responses. This focuses on the key role of Ca^{2+} in the generation of the vasoconstrictor-induced contractile augmentation under reduced Na^+ gradient conditions. Two possible explanations for this augmentation of contraction are: Either the contractile apparatus develops an increased sensitivity to the available Ca^{2+}, or an increased amount of Ca^{2+} is delivered to the contractile apparatus when the Na^+ gradient is reduced. Subsequent studies were carried out in an effort to distinguish between these possibilities.

Effects of Caffeine on VSM: Influence of a Reduced Na^+ Gradient

Caffeine evokes contraction by releasing Ca^{2+} from the SR. Therefore, caffeine may be used to obtain information about the size of the SR Ca^{2+} store because caffeine-evoked contractions should be graded with the amount of stored Ca^{2+} (Hashimoto et al., 1986). Furthermore, inhibition of Ca^{2+} sequestration by the SR may be useful for isolating, functionally, the plasma membrane Ca^{2+} transport mechanisms (Ashida & Blaustein, 1987). Na^+ pump inhibition, either by removal of external K^+, or by addition of strophanthidin, even in the presence of a Ca^{2+} channel blocker, is associated with augmentation of caffeine-evoked contractions (Ashida & Blaustein, 1987 and 1988). Likewise, reduction of $[Na^+]_o$ also causes a reversible increasein caffeine-activated tension (Fig. 2 Top).

Caffeine also induced a progressive increase in steady (resting, unstimulated) tension in rings exposed to $\leq$30 mM $[Na^+]_o$ (Fig. 2 Top), and it increased the rate of rise of unstimulated tension in K^+-free media (Ashida & Blaustein, 1987) - presumably because of enhanced (net) Ca^{2+} entry (since these effects were external Ca^{2+} dependence). In the presence of caffeine, entering Ca^{2+} can't be sequestered in the SR, and thus remains in the cytosol and activates contraction.

When tonic contractions were induced by the combination of caffeine and low $[Na^+]_o$, tension could be decreased either by removing caffeine, or by adding back external Na^+ (Fig. 2 Bottom). Removal of caffeine permits the entering Ca^{2+} to be sequestered in the SR, whereas raising $[Na^+]_o$ reduces Ca^{2+} influx and stimulates Ca^{2+} efflux via the Na/Ca exchanger. The data indicate that these two transport systems, the SR Ca^{2+} pump and the sarcolemmal Na/Ca exchanger, play the dominant roles in the removal of Ca^{2+} from the cytosol following VSM activation.

Effects of $[Na^+]_o$ on VSM Relaxation

The aforementioned observations indicate that an important mechanism of Ca^{2+} extrusion from the VSM cells depends upon

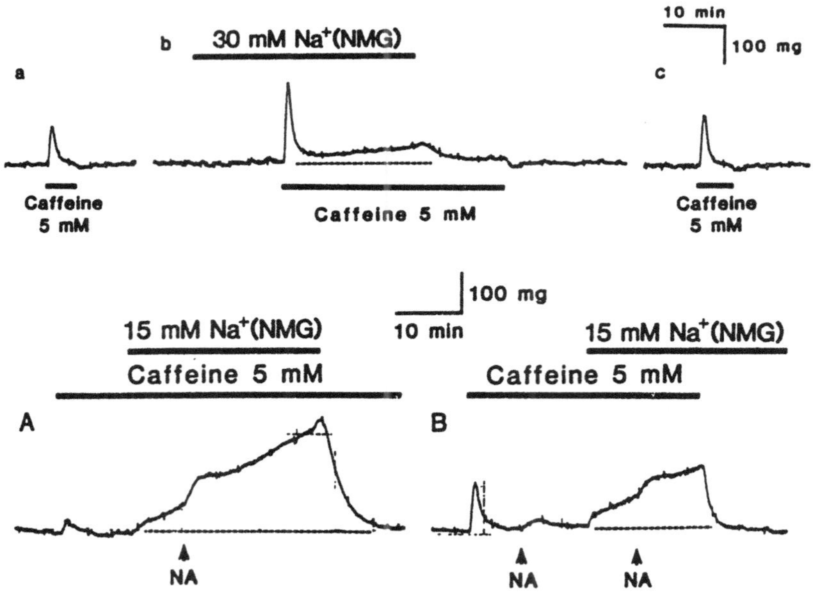

FIG. 2. Top. Contractile responses of a ring of rat aorta to 5 mM caffeine during incubation in media containing the normal concentration of Na^+ (140 mM; "a" and "c"), or only 30 mM Na^+ (replaced by NMG; "b"). Reprinted from Ashida & Blaustein, 1987, with permission.

Bottom. Effect of Na^+ replacement and removal of caffeine on tension of a ring of rat aorta contracted in media containing 15 mM Na^+ and 5 mM caffeine. In "A" the ring was relaxed by restoring the normal $[Na^+]_o$ (140 mM) to permit more rapid extrusion of Ca^{2+} via Na/Ca exchange. In "B" the ring was relaxed by removing the caffeine to permit sequestration of the Ca^{2+} in the SR. "NA" = stimulation with noradrenaline. Reprinted from Blaustein, 1989a, with permission.

external Na^+. Fig. 3 shows the relationship between $[Na^+]_o$ and relaxation rate (as a measure of the rate of removal of Ca^{2+} from the SR). In these experiments, tension was induced by reduction of $[Na^+]_o$, and external Ca^{2+} was then removed to promote relaxation; this also eliminated complications due to simultaneous Ca^{2+} entry. In low $[Na^+]_o$, relaxation was relatively slow, and increasing $[Na^+]_o$ increased the rate of relaxation: 25-30 mM Na^+ induced half-maximal stimulation of relaxation, as compared to the rate in media with $[Na^+]_o$ = 140 mM. These data demonstrate that the dominant mechanism of Ca^{2+} removal from the VSM cell cytosol in rat aorta (with caffeine present to inhibit Ca^{2+} sequestration in the SR), and in bovine tail artery (without

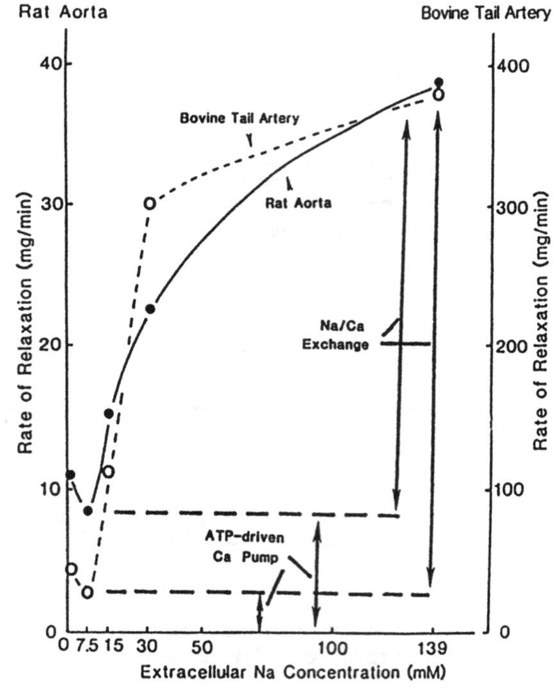

FIG. 3. Relationship between $[Na^+]_o$ and the rate of relaxation of rings of rat aorta (solid circles, left-hand ordinate scale) and bovine tail artery (open circles, right-hand ordinate scale). Na^+ was replaced isosmotically by NMG in the low $[Na^+]$ solutions (see abscissa scale). All solutions contained 10 μM phentolamine and 10 μM verapamil; solutions for the rat aorta also contained 5 mM caffeine. The Na^+-dependent (Na/Ca exchange) and Na^+-independent (ATP-driven Ca^{2+} pump) mediated components of the relaxation are indicated. Reprinted from Blaustein, 1989a, with permission.

caffeine) is dependent upon external Na^+, and may involve Na/Ca exchange (Ashida & Blaustein, 1987).

These experiments are all consistent with the idea that an increased amount of Ca^{2+} is delivered to the contractile apparatus under reduced Na^+ gradient conditions. However, they do not rule out the alternative possibility, namely, that a reduced Na^+ gradient somehow increases the sensitivity of the contractile apparatus to the available Ca^{2+}. This uncertainty can be resolved by direct measurement of intracellular free Ca^{2+} levels.

A Reduced Na^+ Gradient Augments Ca^{2+} Transients in VSM Cells

In order to study the effects of a reduced Na^+ gradient on vasoconstrictor-evoked transient increases in $[Ca^{2+}]_c$ ("Ca^{2+} transients"), we employed digital imaging methods to measure the apparent intracellular free Ca^{2+} concentration, $[Ca^{2+}]_{App}$, in cultured A7r5 cells (Goldman et al., 1990). The cells were loaded with the membrane-permeable acetoxymethyl ester form of the fluorochrome, fura-2. This dye enters intracellular organelles as well as the cytosol, and is then cleaved to the Ca^{2+}-sensitive free acid. We report the observed intracellular free Ca^{2+} values as $[Ca^{2+}]_{App}$ because they include contributions from both the cytosol and the organelles (Goldman et al., 1990).

A7r5 cells have receptors for various vasoconstrictors, including serotonin (5-HT) and (arginine) vasopressin (AVP), which are known to open Ca^{2+} channels in these cells (Van Renterghem et al., 1988). These agonists evoke transient increases in A7r5 cells (Goldman et al., 1990; Bova et al., 1990). The time-course of the mean non-nuclear change in $[Ca^{2+}]_{App}$ in response to 1 μM 5-HT, over a 5 min period, is shown by the open circles in Fig. 4.

When $[Na^+]_o$ was reduced to 6.4 mM, by replacement with either NMG or Li^+, $[Ca^{2+}]_{App}$ increased significantly: in eight cells studied in detail (see Fig. 4), non-nuclear $[Ca^{2+}]_{App}$ rose, over 5 min, from 206 nM to 318 nM ($p<0.02$) when the external Na^+ was replaced by NMG. When these cells were stimulated with 1 μM 5-HT, the Ca^{2+} transients were significantly augmented (Fig. 4, filled circles): $[Ca^{2+}]_{App}$ rose faster, the peak was much higher, and substantially elevated levels were maintained for 5 min (the period during which the 5-HT was applied). These effects were completely reversible. Moreover, similar effects were obtained when AVP was used as an agonist, and when Na^+ was replaced by Li^+ (although, as mentioned above, Li^+ augments IP_3 transients).

When the Na^+ pump was inhibited by K^+-free media or by ouabain, non-nuclear $[Ca^{2+}]_{App}$ also rose significantly; e.g. from 137 nM to 219 nM ($p<0.02$) in 4 cells in media containing 1-3 mM ouabain. Under these circumstances, the Ca^{2+} transients evoked by AVP and 5-HT were also augmented. Fig. 5 shows data from an A7r5 cell treated with 1 nM AVP. In this cell the Ca^{2+} transient rose more rapidly, and to a higher peak (open circles), after ouabain treatment for 15 min (solid circles) and 30 min (solid triangles). Moreover, in the presence of ouabain, during

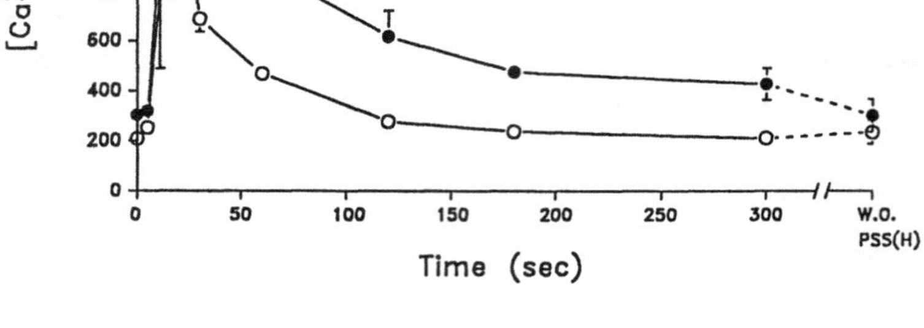

FIG. 4. Effect of reduced $[Na^+]_o$ (NaCl replaced by NMG) on the $[Ca^{2+}]_{App}$ transient in response to 5-HT in A_7r_5 cells. Graph shows the mean data for the time course of the changes in average non-nuclear $[Ca^{2+}]_{App}$ for four cells (each from a different experiment) that were activated with 1 μM 5-HT just after the "0"-time images were obtained. Reprinted from Bova et al., 1990, with permission.

continued exposure to AVP, $[Ca^{2+}]_{App}$ then rose again, before finally declining to a low level. This behavior is reminiscent of the secondary rises in $[Ca^{2+}]_c$ that occur in cardiac muscle cells following treatment with high concentrations of cardiac glycosides (Stern et al., 1988).

These data from the A7r5 cells demonstrate that a reduced Na^+ gradient raises the level of Ca^{2+} in the cytoplasm. Presumably, both cytosolic free Ca^{2+} and the Ca^{2+} stored in intracellular organelles including the SR rise, since caffeine reduces the $[Ca^{2+}]_{App}$ (Bova et al., 1990). Furthermore, the Ca^{2+} transients are augmented under reduced Na^+ gradient conditions. Therefore, we conclude that the augmented contractile responses observed under these conditions are due, at least in part, to an increase in the amount of Ca^{2+} delivered to the contractile proteins when the cells are activated by vasoconstrictors. A large fraction of this Ca^{2+} is derived from intracellular stores (i.e., the SR) when arterial cells are activated by agonists (Kowarski et al., 1985; Ashida et al., 1988). Therefore, these results suggest

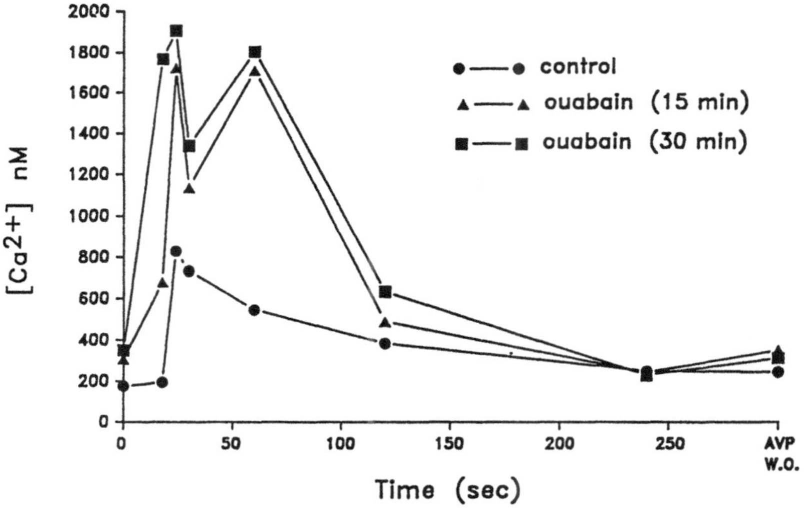

FIG. 5. Effect of ouabain on the $[Ca^{2+}]_{App}$ in an unstimulated A_7r_5 cell and during exposure to 1 nM AVP. Graph shows the time course of the changes in average non-nuclear $[Ca^{2+}]_{App}$ for a single cell in standard PSS (circles) and after 15 min (triangles) and 30 min (squares) exposure to 3 mM ouabain. The cell was activated with 1 nM AVP just after each of the "0"-time images was obtained. Data from Fig. 16 of Bova et al., 1990.

that the Na^+ gradient, via an Na/Ca exchange mechanism (Fig. 1, and see Blaustein, 1989a and b), plays an important role in the control of, not only cytosolic free Ca^{2+}, but (indirectly) SR Ca^{2+} as well. In this way, the exchanger may modulate tonic tension as well as contractility in response to agonists. Furthermore, the Na/Ca exchanger provides a unique link between Na^+ metabolism and vascular contractility. Thus, it may play a critical role in the pathogenesis of salt-dependent hypertension (Blaustein, 1977 and 1989c).

References

Ashida T, Blaustein MP (1987) Regulation of cell calcium and contractility in mammalian arterial smooth muscle: the role of sodium-calcium exchange. J Physiol 392:617-635.

Ashida T, Schaeffer J, Goldman WF, Wade JB, Blaustein MP (1988) Role of sarcoplasmic reticulum in arterial contraction: comparison of ryanodine's effect in a conduit artery and a muscular artery. Circ Res 62:854-863.

Blaustein MP (1977) Sodium ions, calcium ions, blood pressure regulation and hypertension: a reassessment and a hypothesis. Am J Physiol 232:C165-C173.

Blaustein, MP (1989a) Sodium/calcium exchange in cardiac, smooth and skeletal muscles: key to the control of contractility. Curr Topics Membranes Transp 34:289-330.

Blaustein MP (1989b) Sodium-calcium exchange in mammalian smooth muscles. In:Allen TJA, Noble D, Reuter H (eds) Sodium-Calcium Exchange. Oxford University Press, Oxford pp 208-232.

Blaustein MP (1989c) The pathogenesis of essential hypertension: the Na pump inhibitor (natriuretic hormone) - Na/Ca exchange - hypertension hypothesis. Japan J Hypertension 11:107-117.

Bova S, Goldman WF, Yuan X-J, Blaustein MP (1990) Influence of the Na^+ gradient on Ca^{2+} transients and contraction in vascular smooth muscle. Am. J. Physiol., in press.

Goldman WF, Bova S, Blaustein MP (1990) Measurement of intra cellular Ca^{2+} in cultured arterial smooth muscle cells using fura-2 and digital imaging microscopy. Cell Calcium 11:221-231.

Hashimoto, T., Hirata, M., Itoh, T., Kanamura, Y, Kuriyama, H. (1986) Inositol 1,4,5-trisphosphate activates pharmacomechanical coupling in smooth muscle of the rabbit mesenteric artery. J. Physiol. 370:605-618.

Kowarski D, Shuman H, Somlyo AP, Somlyo AV (1985) Calcium release by noradrenaline from central sarcoplasmic reticulum in rabbit main pulmonary artery smooth muscle. J Physiol (Lond) 366:153-175.

Van Renterghem C, Romey G, Lazdunski M (1988) Vasopressin modulates the spontaneous electrical activity in aortic cells (line A7r5) by acting on three different types of ionic channels. Proc Natl Acad Sci USA 85:9365-9369.

SECTION IV

REGULATION AND REGULATORY ROLE OF INTRACELLULAR CALCIUM

FUNCTION AND REGULATION OF INTRACELLULAR Ca IN RENAL CELLS

E.E. Windhager
Department of Physiology
Cornell University Medical College
1300 York Avenue
New York, N.Y. 10021

As in all other cells of the mammalian body, $[Ca^{2+}]$ within renal epithelial cells is maintained at a level far below that in extracellular fluid. Whereas extracellular fluid $[Ca^{2+}]$ is less than 1 mM, $[Ca^{2+}]$ in intracellular fluid is approximately four orders of magnitude lower. Measurements of cytosolic Ca ion concentration have been obtained in vivo with ion-selective microelectrodes (Lee et al., 1980, Lorenzen et al., 1984, Yang et al., 1988) in proximal tubules of Necturus, and with a variety of Ca sensitive indicators in mammalian renal epithelia in culture or in tubule suspensions or isolated nephron segments (Bonventre et al., 1986, Borle and Snowdowne 1982, Bourdeau and Lau 1989, Burnatowska-Hledin and Spielman 1987, Chase and Wong 1988, Dolson et al., 1985, Dominguez et at, 1989, Hruska et al., 1987, Mandel and Murphy 1984, Murphy et al., 1986, Murphy and Mandel 1982, Sakhraui et al., 1985, Shayman et al., 1986, Tang and Weinberg 1986, Taniguchi et al., 1989a and 1989b, Teitelbaum and Berl 1986). Values obtained, range between 70 and 450 nM, with the majority of studies reporting average concentrations of 100 to 200 nM. In addition to the low cytosolic Ca ion concentration, the interior of the renal cell is electrically negative by

NATO ASI Series, Vol. H 48
Calcium Transport and
Intracellular Calcium Homeostasis
Edited by D. Pansu and F. Bronner

some 70 to 80 mV with respect to extracellular fluid. Renal cells are known to be permeable to Ca ions (Borle 1981, Borle and Uchikawa 1978) but the low intracellular Ca ion activity is maintained by the activity of primary ATP-driven, and secondary active transport of Ca^{2+} (Na/Ca exchange) across the basolateral cell membrane of renal epithelial cells (Gmaj and Murer 1988).

Comparison of the total cell [Ca] of 10^{-3} M with the cytosolic $[Ca^{2+}]$ of 10^{-7} M indicates that the major fraction of cell Ca is buffered or contained within intracellular organelles. Cytosolic buffering is accomplished mainly by Ca binding proteins such as calmodulin or the vitamin D-dependent Ca-binding protein (Wasserman et al., 1982) or to polyvalent anions such as phosphates.

The major portion of non-ionized Ca is located within the endoplasmic reticulum and in mitochondria. Thevenod and his collaborators (1986) have used permeabilized cells from rat kidney cortex to evaluate the relative role of cell organelles in the regulation of a steady state Ca ion concentration. They found that mitochondrial inhibitors of oxydative phosphorylation blocked Ca uptake rate, but did not affect the steady state $[Ca^{2+}]$. In contrast, vanadate, an inhibitor of endoplasmic reticulum ATPase (Ortiz et al., 1984, Simons 1979), led to an increase in $[Ca^{2+}]$. It was therefore concluded that, in the absence of an intact functioning plasma membrane, steady state levels of intracellular Ca are maintained by

ATP-driven Ca uptake into a non-mitochondrial pool, probably the endoplasmic reticulum. Only when cells are loaded with unphysiologically large amounts of Ca, exceeding the transport capacity of the endoplasmic reticulum, will mitochondria assume a major role in the buffering of intracellular Ca (Cheung et al., 1986, Thevenod et al., 1986). Thevenod et al. (1986) and others (Cheung et al., 1986, Hruska et al., 1987) have reported that inositol-1,4,5-triphosphate (IP_3) may cause a release of Ca from the endoplasmic reticulum and thus serve as an intracellular messenger for various hormonal stimuli.

Steady state changes in intracellular [Ca^{2+}] can influence the rate of tubular reabsorption of salt and water (for review see ref. 1). Our own studies in rabbits have shown that reabsorption of Na and water in proximal tubules (Friedman et al., 1981), and the reabsorption of Na (Frindt and Windhager 1990) and the vasopressin-induced increase in osmotic water permeability (Frindt et al., 1982, Lorenzen et al., 1987, Jones et al., 1988) in cortical collecting tubules are inhibited by experimental maneuvers thought to produce steady state increases in intracellular [Ca^{2+}].

The inhibitory effect of high [Ca^{2+}] on epithelial Na transport has led several investigators (for review see Taylor and Windhager 1985 and Palmer et al., 1989) to propose that Ca^{2+} mediates a negative feedback process that links apical Na entry to basolateral Na extrusion in transporting epithelia, including the nephron. At first, it was believed that intracellular [Na^+] was directly responsible for the downregulation of Na transport, observed in a number of transporting epithelia but studies on rabbit and toad urinary bladder (Eaton 1981, Palmer 1985) and data on Na uptake in apical membrane vesicles of toad bladder epithelium (Chase and AlAwqati 1983, Garty et al., 1987) have shown that changes in cell [Na] per se do not alter the apical Na conductance. Instead, if Na/Ca exchange operates in the basolateral cell membrane, changes in cell [Ca^{2+}] might play the mediatory role previously assigned to alterations in cell [Na^+].

It is obvious that Ca ions can act as mediators only if Na/Ca exchange is highly effective in adjusting cell [Ca^{2+}] in parallel to changes in cell [Na^+]. We have therefore examined this question experimentally under a variety of experimental conditions (low peritubular [Na], high peritubular [K], zero peritubular [K], peritubular ouabain, luminal gramicidin, low luminal [Na], and zero luminal organic solutes) in isolated perfused Necturus proximal tubules by measuring intracellular and extracellular ion activities of sodium and calcium by means of Na^+- and Ca^{2+}-selective microelectrodes whereas membrane voltage was measured with conventional microelectrodes (Yang et al., 1988). When the electrochemical potential gradient for Na^+ (y in KJ/Mol) was reduced over a 5 fold range from control conditions by the above listed maneuvers, the intracellular calcium ion activity (x in nM) rose in nearly linear fashion ($y = 797 - 75.7\ x$; $r=0.9$) from approximately 80 nM in control tubules to more than 800 nM in tubules exposed 10^{-4} M ouabain. The only exception to this pattern was when tubules were superfused with 42.5 mM K. In this condition a^{i}_{Ca} failed to rise despite a marked reduction in the electrochemical driving force for Na ions. It is possible that the strong depolarization of the basolateral cell membrane observed in this situation may have reduced the activity of Ca channels or increased the activity of the ATP-driven Ca pump. In all other conditions tested, intracellular $[Ca^{2+}]$ was clearly a function of the electrochemical potential gradient for Na ions across the basolateral cell membrane, in accord with the operation of a Na/Ca exchange process at this site.

Na/Ca exchange may play a crucial role in a number of critical function of renal epithelium. In order to obtain information on the molecular nature of this process, we have recently attempted to obtain the expression of rabbit kidney mRNA coding for this function in Xenopus laevis oocytes (Milovanovic et al., 1990). Defolliculated oocytes were injected with 50 ng total poly(A)$^{+}$RNA or an equivalent volume of water. The activity of Na^{+}/Ca^{2+} exchange was estimated by measuring the Ca uptake, using ^{45}Ca. This uptake was measured in Na-loaded (nystatin) and non-Na-loaded oocytes, during exposure to a Na- or K-Ringer solution. There was no Na-dependent Ca-uptake in water-injected Na-loaded oocytes. Injection of poly(A)$^{+}$RNA induced a larger Na-gradient-dependent Ca uptake in Na-loaded oocytes (2.85± 0.2 [SEM], n=17 in K-Ringer; 1.29± 0.15, n=18 in Na Ringer; thus, the Na-dependent Ca uptake was 1.56) than in oocytes not loaded with Na (1.47± 0.12, n=15 in K-Ringer; 1.17± 0.06, n=16 in Na-Ringer; thus, the Na-gradient-dependent Ca uptake was 0.30). The Na-gradient-induced enhancement of Ca uptake in mRNA injected oocytes is consistent with the expression of Na^+/Ca^{2+} exchange. The magnitude of of the Na-gradient-dependent Ca uptake in oocytes injected with kidney-derived mRNA is nearly the same as that obtained by others (Longoni et al., 1988, Sigel et al., 1988) after injection of mRNA derived from cardiac tissue. In recent fractionation of rat kidney mRNA by sucrose density gradient centrifugation (Milovanovic, Frindt, Windhager and Tate, unpublished) the approximate size of the mRNA coding for Na/Ca exchange was estimated. Peak activity, comparable in magnitude to that of total mRNA, was obtained with injection of fraction K_3, corresponding to 18S to 23S.

In addition to a Na/Ca exchange process in the basolateral cell membrane, a cell-[Ca^{2+}]-mediated downregulation of the Na permeability of the apical cell membrane is a crucial prerequisite for the operation of a Ca^{2+}-mediated negative feedback mechanism. Palmer and Frindt (1987) have therefore patch-clamped apical amiloride-sensitive Na channels in rat cortical collecting tubules (CCTs) that were exposed to the Ca^{2+} ionophore ionomycin. CCTs, freshly dissected from rat kidneys were split longitudinally to expose the apical cell surface to the tip of a patch-clamp microelectrode. A cell-attached patch was then formed and the pipette voltage maintained at +60 mV with respect to the bath. A representative result is shown in Fig. 1.

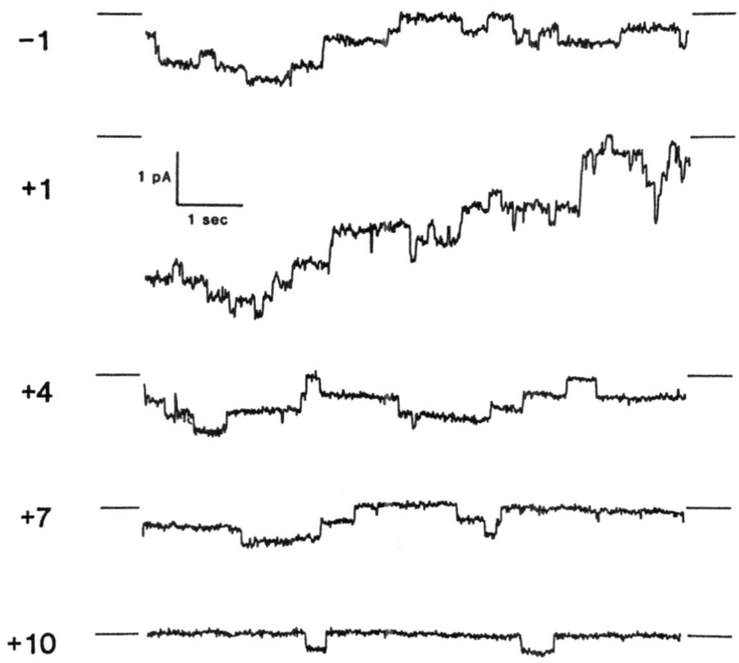

Figure 1. Effect of ionomycin on Na channel activity in cell-attached patches (Palmer and Frindt, 1987). Pipette-voltage was at +60 mV with respect to bath.

Channel activity was recorded under control conditions for 3 minutes before superfusing the tubule with fluid containing 10 uM ionomycin. Horizontal lines indicate the current level at which all channels were closed. A channel opening is shown by a downward deflection. Time in minutes before and after addition of ionomycin is indicated by the numbers on the left side of Fig. 1. Channel activity was increased during the first minute after addition of ionomycin. Afterwards, channel activity was diminished and practically abolished after 10 minutes. The height of the current transients was not significantly affected by the Ca ionophore, suggesting that the single channel conductance was not altered by ionomycin. Nearly identical results were obtained in six such experiments whereas time controls showed no change in Na channel activity. The initial transient increase in Na channel activity is probably due to a transient alkalinization of the cytoplasm that occurs in response to this dose of ionomycin (Yang et al., 1988). When only 1 uM ionomycin was added, the initial activation of Na channels was absent. However, the degree and time course of Na channel inactivation was similar as with the high dose of ionomycin. Removal of Ca in the bathing solution (omission of Ca and addition of 1 mM EGTA) abolished the ionomycin-induced inhibition of Na channels.

In contrast to these observations on cell-attached patches, changes in [Ca^{2+}] failed to influence directly the activity of Na channels in isolated, inside-out patches. This finding is in apparent disagreement with the results of Chase and

Al-Awqati (1983) and Garty and Asher (1985) who found that Ca^{2+} reduces the amiloride-sensitive Na flux in vesicles from toad bladder membranes. It is possible that in inside-out patches of the apical membrane of rat CCTs, Ca ions were ineffective because the studies were carried out at room temperature. In fact, Garty and Asher (1985) have suggested that a Ca^{2+}-dependent enzymatic reaction may mediate the Ca^{2+}-induced downregulation of Na channel activity. Such a mechanism is entirely consistent with the ionomycin-induced inhibition of Na channel activity observed by Palmer and Frindt (1987).

Although most available observations are consistent with the view that Ca ions play a role in the negative feedback regulation of Na transport, a similar role for intracellular H ions cannot be excluded. Palmer and Frindt (1987) found that acidification diminished, whereas alkalinization increased the activity of amiloride- sensitive Na channels in isolated, inside-out patches from apical membranes of rat CCTs. Harvey et al., (1988) and Harvey and Ehrenfeld (1988) have pointed out that the presence of Na/H exchange in the basolateral cell membrane of tight epithelia provides a mechanism for intracellular acidification to occur in response to an increase in Na transport. The degree to which either or both of the two ion species mediate the feedback response under physiological conditions remains to be elucidated.

Intracellular Ca ions also regulate the activity of K channels in the apical membrane of the proximal tubule (Kawahara et al., 1987, Sackin 1989), the diluting segment (Taniguchi and Guggino 1989), and cortical collecting tubules (Hunter et al., 1984, Koeppen et al., 1984, Frindt and Palmer 1987). In proximal tubules, Ca activated K channels may play a role in cell volume regulation (Sackin 1989) but also in the regulation of apical membrane voltage. Furthermore, the increase in cell $[Ca^{2+}]$ that follows a sudden increase in the rate of Na entry across the apical membrane (Yang et al., 1988) may cause a hyperpolarization of the apical membrane which in turn would diminish the rate of apical Na entry via cotransport mechanisms (Hunter et al. 1988).

In the limited space available, some studies have been presented which illustrate the potential importance of intracellular calcium ions in the regulation of salt and water transport by the kidneys. Steady state elevations of cell $[Ca^{2+}]$ inhibit renal sodium and water transport. The molecular mechanisms responsible for this effect and the importance of physiological variations in the level of cell $[Ca^{2+}]$ for the control of transport remain to be clarified by future studies.

REFERENCES

Bonventre,J.V., and Cheung, J.Y. (1986). Cytosolic free calcium concentration in cultured renal epithelial cells. Am. J. Physiol.250:F329-F338.

Borle, A.B. and Snowdowne, K.W. (1982). Measurement of intracellular free calcium in monkey kidney cells with aequorin. Science 217:252-254.

Borle, A.B., and Uchikawa,T. (1978). Effects of parathyroid hormone on the distribution and transport of calcium in cultured kidney cells. Endocrinology 102:1725-1732.

Bourdeau,J.E., and Lau, K. (1989). Effects of parathyroid hormone on cytosolic free calcium concentration in individual rabbit connecting tubules. J. Clin. Invest. 83:373-379.

Burnatowska-Hledin, M.A., and Spielman, W.S. (1987). Vasopressin increases cytosolic free calcium in LLC-PK1 cells through a V_1 receptor. Am. J. Physiol. 253:F328-F332.

Chase,H.S.,Jr., and Al-Awqati, Q. (1983). Calcium reduces the sodium permeability of luminal membrane vesicles from toad bladder. J. Gen. Physiol. 81:643-665.

Chase, H.S.,Jr., and Wong, S.M.E. (1988). Isoproterenol and cyclic AMP increase intracellular free [Ca] in MDCK cells. Am. J. Physiol. 254:F374-F384.

Cheung,J.Y., Constantine, J.M., and Bonventre, J.V. (1986). Regulation of cytosolic free calcium concentration in cultured renal epithelial cells. Am. J. Physiol. 251:F690-F701.

Dolson, G.M., Hise, M.K., and Weinman. (19855). Relationship among parathyroid hormone, cAMP and calcium on proximal tubule sodium transport. Am. J. Physiol. 249: F409-F416.

Dominuez, J.H., Rothrock, J.K., Macias, W.L. and Price, J. (1989). Na^+ electrochemical gradient and Na^+-Ca^{2+} exchanger in rat proximal tubule. Am. J. Physiol. 257:F531-F538.

Eaton, D.C. (1981). Intracellulat sodium ion activity and sodium transport in rabbit urinary bladder. J. Physiol. 316:527-544.

Friedman, P.A, Figueiredo, J.F., Maack, T., and Windhager, E.E. (1981). Sodium-calcium interactions in the renal proximal convoluted tubule of the rabbit. Am. J. Physiol. 240:F558-F568.

Frindt, G., Windhager, E.E., and A. Taylor (1982). Hydroosmotic response of collecting tubules to ADH or cAMP at reduced peritubular sodium. Am. J. Physiol. 243:F503-F513.

Frindt, G. and L.G. Palmer. (1987). Ca-activated K channels in apical membrane of mammalian CCT, and their role in K secretion. Am. J. Physiol. 252:F458-F467.

Frindt, G., and Windhager, E.E. (1990). Ca^{2+}-dependent inhibition of sodium transport in rabbit cortical collecting tubules. Am. J. Physiol. 258:F568-F582.
Garty, H., and Asher, C. (1985). Ca^{2+}-dependent, temperature-sensitive regulation of Na^+ channels in tight epithelia. A study using membrane vesicles. J. Biol. Chem. 260:8330-8335.
Garty, G., Asher, C., and Yeger, O. (1987). Direct inhibition of epithelial Na^+ channels by Ca^{2+} and other divalent cations. J. Membrane Biol. 95:151-162.
Gmay, P. and Murer, H. (1988). Calcium transport mechanisms in epithelial cell membraneas. Miner. Electrolyte Metab. 14:22-30.
Guggino, S.E., Suarez-Isla, B.A., Guggino, W., and Sacktor, B. (1985). Forskolin and antidiuretic hormone stimulate Ca^{2+}-activated K^+ channels in cultured rabbit kidney cells. Am. J. Physiol. 249:F448-F455.
Harvey, B.J., Thomas, S.R., and Ehrenfeld, J. (1988). Intracellular pH controls cell membrane Na^+ and K^+ conductances and transport in frog skin epithelium. J. Gen. Physiol. 92:767-792.
Harvey, B.J. and Ehrenfeld, J. (1988). Role of Na^+/H^+ exchange in the control of intracellular pH and cell membrane conductances in frog skin epithelium. (1988). J. Gen Physiol. 92:793-810.
Hruska, K.A., Moskowitz, D., Esbrit, P., Civittelli, R., Westbrook, S. and Huskey, M. (1987). Stimulation of inositol triphosphate and diacylglycerol production in renal tubular cells by parathyroid hormone. J. Clin. Invest. 79:230-239.
Hunter, M., Lopes, A.G., Boulpaep, E. and Giebisch, G. (1983). Single channel recordings of calcium-activated potassium channels in the apical membrane of rabbit-cortical collecting tubules. Proc. Natl. Acad. Sci. USA 81:4237-4239.
Jones, S.M., Frindt, G., and Windhager, E.E. (1988). Effect of peritubular [Ca] or ionomycin on hydroosmotic response of CCTs to ADH or cAMP. Am. J. Physiol. 254:F240-F253.
Kawahara, K., Hunter, M., and Giebisch, G. (1987). Potassium mechannels in Necturus proximal tubule. Am. J. Physiol. 253: F488-F494.
Koeppen, B.M., Beyenbach, K.W., and Helman, S.I. (1984). Single channel currents in renal tubules. Am. J. Physiol. 247:F380-F384.
Lee, C.O., Taylor, A., and Windhager, E.E. Cytosolic calcium ion activity in epithelial cells of Necturus kidney. (1980). Nature 287:859-861.
Longoni, S., Coady, M.J., Ikeda, T., and Philipson, K.D. (1988). Expression of cardiac sarcolemmal Na^+-Ca^{2+} exchange activity in Xenopus laevis oocytes. Am. J. Physiol. 255:C870-C873.
Lorenzen, M., Lee, C.O., and Windhager, E.E. (1984). Cytosolic Ca^{2+} and Na^+ activities in perfused proximal tubules of Necturus kidney. Am. J. Physiol. 247:F93-F102.
Mandel, L.J., and Murphy, E. (1984). Regulation of

cytosolic free calcium in rabbit proximal renal tubules. J. Biol. Chem. 259:11188-11196.
Milovanovich, S., Frindt, G., and Windhager, E.E. Expression of renal cortical Na^{+}/Ca^{2+} exchange activity in Xenopus laevis oocytes (1990). FASEB (abstr.) in press.
Murphy, e., Chamberlin, M.E., and L.J. Mandel. (1986). Effects of calcitonin on cytosolic Ca in a suspension of rabbit medullary thick ascending limb tubules. Am. J. Physiol. 251:C491-C495.
Murphy, E., and Mandel, L.J. (1982). Cytosolic free calcium levels rabbit proximal kidney tubules. Am. J. Physiol. 242:C124-C128.
Ortiz, A., Garzia-Carmona, F., Garcia-Canovos, F., and Gomez-Fernandez, J.C. (1984). A kinetic study of the interaction of vanadate with the Ca^{2+}- and Mg^{2+}-dependent ATPase from sarcoplasmic reticulum. Biochem. J. 221:213-222.
Palmer, L.G. (1985). Modulation of apical Na permeability of the toad urinary bladder by intracellular Na, Ca and H. J. Membrane Biol. 83:57-69.
Palmer, L.G. and Frindt, G. (1987). Effects of cell Ca and pH on Na channels from rat cortical collecting tubule. Am. J. Physiol. 253:F333-F339.
Palmer, L.G., Frindt, G., Silver, R., and Strieter, J. (1989). Feedback regulation of epithelial Na channels. In: Current Topics in Membrane and Transport. Edt.: S. Schultz. Academic Press, New York, 34:45-60.
Sackin, H. (1989). A stretch-activated K^{+} channel sensitive to cell volume. Proc. Natl. Acad. Sci. USA 86:1731-1735.
Sakhraui, L.M., Tessitore, N., and Massry, S.G. (1985). Effect of calcium on transport characteristics of cultured proximal renal cells. Am. J. Physiol.249:F346-F355.
Shayman, J.A., Hruska, K.A., and Morrison, A.R. Bradykinin stimulates increased intracellular calcium in papillary collecting tubules of the rabbit. (1986). Biochem. Biophys. res. Commun. 134:299-304.
Sigel, E.R., Baur, R., Porzig, H., and Reuter, H. (1988). mRNA-induced expresion of the cardiac Na^{+}/Ca^{2+} exchanger in Xenopus laevis. J. Biol. Chem. 263:13614-14616.
Simons, T.J.B. (1979). Vanadate - a new tool for biologists. Nature 281:337-338.
Tang, M.J., and Weinberg, J.M. (1986). Vasopressin-induced increases of cytosolic calcium in $LLC-PK_1$ cells. Am. J. Physiol. 251:F1090-F1095.
Taniguchi,J., and W.B.Guggino. (1989). Membrane stretch: a physiological stimulator of Ca-activated K channels in thick ascending limb. Am. J. Physiol. 257:F347-F352.
Taniguchi, S., Marchetti, J., and F. Morel. (1989a). Cytosolic free calcium in single microdissected rat cortical collecting tubules. Pfluger's Arch. 414:125-133.
Taniguchi, S. Marchetti, J., and Morel, F. (1989b). Na/Ca

exchangers in collecting cells of rat kidney. A single tubule fura-2 study. Pfluger's Arch. 415:191-197.

Taylor, A., and Windhager, E.E. (1985). Cytosolic calcium and its role in the regulation of transepithelial ion and water transport. In: Physiology and Pathophysiology of Electrolyte Metabolism, edt. by D. Seldin and G. Giebisch. New York: Raven, pp. 1297-1379.

Teitelbaum, I., and Berl, T. (1986). Effects of calcium on vasopressin-mediated cyclic adenosine monophosphate formation in cultured rat inner medullary collecting tubule cells. Evidence for the role of intracellular calcium. J. Clin. Invest. 77:1574-1583.

Thevenod, F., Streb,H., Ullric, K.J., and Schulz, I.(1986). Inositol-1,4,5-triphosphate releases Ca^{2+} from nonmitochondrial store site in permeabilized rat cortical kidney cells. Kidney Int. 29:695-702.

Wasserman, R.H., and Fullmer, C.S. (1982). Vitamin D-induced calcium binding proteins. In: Calcium and Cell Function, vol. II. Edt. by W.Y.Cheung. Academic Press, New York, pp.175-216.

Yang,J.M., Lee, C.O., andWindhager, E.E. (1988). Regulation of cytosolic free calcium in isolated perfused proximal tubules of Necturus. Am. J. Physiol. 255:F787-F799.

Na^+-Ca^{2+} AND Na^+-H^+ ANTIPORTER INTERACTIONS. RELATIONS BETWEEN CYTOSOLIC FREE Ca^{2+}, Na^+ AND INTRACELLULAR pH

André B. Borle
Department of Physiology
University of Pittsburgh School of Medicine
Pittsburgh, PA 15261, USA

There is increasing evidence that Na^+-Ca^{2+} and Na^+-H^+ antiporters are present in practically all cells. However, many questions regarding their importance, their role in the control of cellular ion homeostasis and their mode of operation remain to be answered. For instance, a) what is the prevailing mode, forward or reverse, of the antiporter in a particular cell? b) can Na^+-Ca^{2+} exchange be activated in both modes depending on the ionic balance and the thermodynamic conditions of the cell? and c) is there an interaction between Na^+-Ca^{2+} and Na^+-H^+ exchange since both depend on the Na^+ electrochemical potential?

A few years ago, we demonstrated that in renal proximal tubular cells (LLC-MK_2), lowering the Na^+ potential by decreasing Na_o^+ causes a rise in the cytosolic free Ca^{2+} (Ca_i^{2+}) by increasing Ca^{2+} influx and Na^+ efflux, i.e by stimulating the reverse mode of Na^+-Ca^{2+} exchange (Snowdowne and Borle, 1985). We recently repeated these experiments in cells derived from the distal tubule (MDCK) and we obtained identical results. We then tested whether, in MDCK cells, the forward mode of Na^+-Ca^{2+} exchange could be activated, and whether we could detect any interaction between the Na^+-Ca^{2+} and Na^+-H^+ antiporters. We measured Ca_i^{2+} with aequorin, Na_i^+ and pH_i with the fluorescent dyes SBFI and BCECF respectively, in perfused cells imbedded in agar gel threads. Ca^{2+} influx and efflux were measured with ^{45}Ca, Na^+ influx and efflux with ^{22}Na and H^+ secretion by the pH-stat method (Borle et al, 1990).

In resting control conditions, MDCK cells were found to have a cytosolic free Ca^{2+} of 120 ± 29 nM, an intracellular Na^+ (Na_i^+) of 16.7 ± 5.4 mM and an intracellular pH (pH_i) of 7.32 ± 0.04 (Table 1). In the same cells and in similar conditions, Paulmichl et al (1986) found a membrane potential E_m of -50 mV. Thus, the reversal potential E_R ($E_R = 3E_{Na^+} - 2E_{Ca^{2+}}$) for the Na^+-Ca^{2+}

NATO ASI Series, Vol. H 48
Calcium Transport and
Intracellular Calcium Homeostasis
Edited by D. Pansu and F. Bronner

Table 1. Intra- and extracellular ion concentrations and thermodynamic conditions in MDCK cells.

	ECF	ICF	E	$\Delta\mu$
	mM	nM or mM	mV	kJ/mol
Ca^{2+}	1.3	120 ± 5.4	+121	-32.8
Na^{+}	140	16.7 ± 5.4	+55.4	-10.1
pH	7.40	7.32 ± .04	-5.8	-4.23

E_m = -50 mV (From Paulmichl et al, 1986)
E_R = $3E_{Na^+} - 2E_{Ca^{2+}}$ = -76 mV
$E_m - E_R$ = +26 mV (reverse mode)
$E_{Na^+} - E_{H^+}$ = +61.2 mV

antiporter can be calculated to be -76 mV. The reversal potential is the membrane potential at which the antiporter shifts from one mode to the other. Since in control conditions E_R is more negative than E_m, the Na^+-Ca^{2+} antiporter must operate in the reverse mode as a Ca^{2+} influx pathway with a driving force $E_m - E_R$ of +26 mV (Table 1). Furthermore, H^+ secretion on the Na^+-H^+ antiporter is energized by the Na^+ potential with a driving force $E_{Na^+} - E_{H^+}$ of +61 mV.

Theoretically, decreasing the Ca^{2+} potential $\Delta\mu Ca^{2+}$ should activate the Na^+-Ca^{2+} antiporter in the forward mode i.e. Ca^{2+} efflux vs Na^+ influx. We tested this hypothesis by perfusing MDCK cells with Ca^{2+}-free media containing 0.1 mM EGTA (free Ca_o^{2+} = 10^{-8} M). Fig. 1 shows the results of these experiments. First, Ca^{2+} efflux increased 2-3 folds and Na^+ influx rose 400%. As a result Ca_i^{2+} fell from 120 to 35 nM and Na_i^+ rose from 16.7 to 26.1 mM. That strongly suggests that the forward mode of Na^+-Ca^{2+} exchange was activated by the collapse of the Ca^{2+} potential. Since the Na^+-Ca^{2+} antiporter is electrogenic, one would predict a fall in the cell membrane potential, and indeed, Paulmichl et al (1986) showed that the membrane potential of MDCK cells drops by 36.7 mV within 40 seconds when the cells are incubated in 0 Ca_o^{2+}, supporting the idea of a stimulation of the Na^+-Ca^{2+} antiporter in the forward mode. Obviously, the rise in Na_i^+ observed in Ca^{2+}-

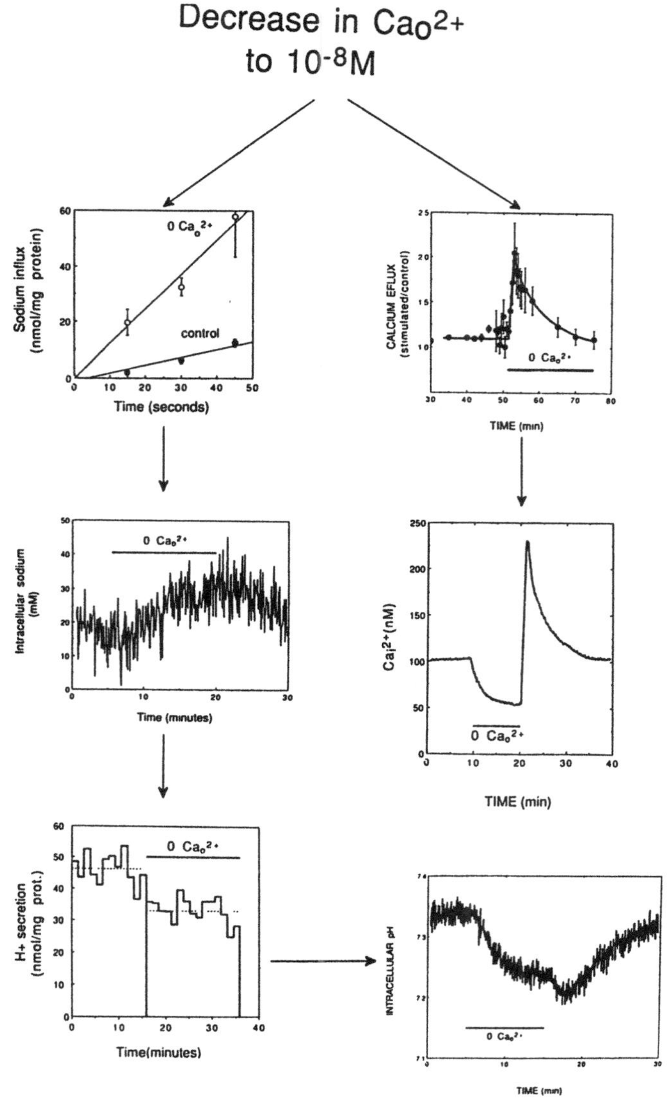

Figure 1. Effects of decreasing extracellular Ca^{2+} to 0 (Ca^{2+}-free medium + 0.1 mM EGTA) on Na^+, Ca^{2+} fluxes, H^+ secretion and cytosolic Na^+, Ca^{2+} and pH in MDCK cells. Recordings of ion concentration are representative of 5-6 separate experiments.

free media must decrease the Na^+ potential and E_{Na^+}-E_{H^+}, the driving force for H^+ secretion on the Na^+-H^+ antiporter. Fig. 1 shows that, indeed, H^+ secretion decreased 38% from 17.7 to 11.0 nmol min^{-1} mg^{-1} protein in Ca-free media, and, as a result, pH_i fell from 7.32 to 7.22.

When the perfusate calcium concentration was restored to its normal value, from 10^{-8} M to 1.3 mM, Ca_i^{2+} rose dramatically and rapidly from 35 to 234 nM in less than 1 min. This was caused by a reversal of the Na^+-Ca^{2+} antiporter from forward to reverse mode because 1) Ca^{2+} influx increased more than two folds, 2) Na^+ efflux increased 20%, and 3) according to Paulmichl et al (1986), the cell membrane potential repolarizes to its control level in less than 60 s. Indeed, when Ca_o^{2+} was restored to normal, the Ca^{2+} potential was larger than control, because of the low Ca_i^{2+}, and the Na^+ potential was less than control because of the high Na_i^+. Knowing E_m and E_R before, during and after the exposure of the cells to Ca^{2+}-free media, we were able to calculate E_m-E_R, giving us the operating mode and the driving forces controlling Na^+-Ca^{2+} exchange at the various time points of the experiment, as shown in Fig. 2. During the control period, the antiporter operated in the reverse mode with a driving force for Ca^{2+} influx of +26 mV. When the extracellular Ca^{2+} was removed, the antiporter suddenly shifted to the forward mode with a driving force for Ca^{2+} efflux of -190 mV. During the Ca^{2+}-free period, the driving force for Ca^{2+} efflux progressively decreased to -129 mV because Ca_i^{2+} fell, Na_i^+ rose and the cell membrane depolarized.

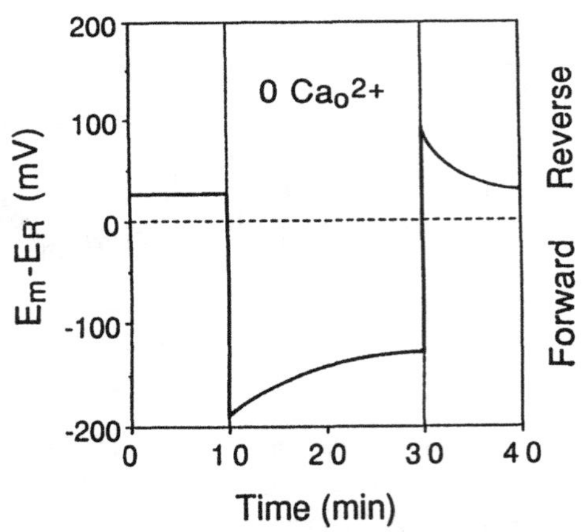

Figure 2. Calculated change in the mode and driving force (E_m-E_R) of the Na^+-Ca^{2+} antiporter during exposure of MDCK cells to Ca^{2+}-free media + 0.1 mM EGTA.

When Ca_o^{2+} was restored to normal, the antiporter flipped again to the reverse mode with a driving force for Ca^{2+} influx of +92 mV. Within 10 min, all parameters were back to control levels with the cells operating in the reverse mode with a driving force for Ca^{2+} influx of +26 mV, as in the resting conditions. These results demonstrate that the Na^+-Ca^{2+} antiporter can rapidly shift from one mode to the other, that it interacts with the Na^+-H^+ antiporter, and that its activation alters not only Ca_i^{2+} but Na_i^+ and pH_i as well.

Since the activation of the Na^+-Ca^{2+} antiporter influences the activity of the Na^+-H^+ antiporter by altering the Na^+ potential, the question arises whether the opposite might possibly occur i.e. whether the activation of the Na^+-H^+ antiporter similarly influences Na^+-Ca^{2+} exchange. Table 1 shows that $\Delta\mu Na^+$ is more than twice $\Delta\mu H^+$ indicating that H^+ secretion on the Na^+-H^+ antiporter can occur against its thermodynamic gradient with a driving force E_{Na^+}-E_{H^+} of +61.2 mV. Theoretically, decreasing E_{H^+} by raising the extracellular pH (pH_o) should increase the driving force E_{Na^+}-E_{H^+} and stimulate H^+ secretion. And indeed, Fig. 3 shows that when pH_o was increased to 7.7 or 8.0 by raising the medium bicarbonate concentration, H^+ secretion and Na^+ influx increased 58% and 53% respectively. As a result, pH_i rose to 7.76 and Na_i^+ increased from 15.6 to 20 mM. At the same time, Ca_i^{2+} more than doubled from 120 to 270 nM. This rise in Ca^{2+} was not caused by mobilization of intracellular Ca^{2+} because it did not occur in the absence of extracellular Ca^{2+}. Rather it was caused by the activation of the reverse mode of Na^+-Ca^{2+} exchange evoked by the rise in Na_i^+. Indeed, the increase in Na_i^+ from 15 to 19.7 mM depressed E_{Na^+} from +57.2 to +51.1 mV, the reversal potential became more negative so that the driving force for Ca^{2+} influx practically doubled from +20.4 to +39 mV. The stimulation of the reverse mode of Na^+-Ca^{2+} exchange should hypopolarize the cell. And in fact, Paulmichl <u>et al</u> (1985) reported that increasing the extracellular pH does increases the membrane potential of MDCK cells.

The reverse occured when the extracellular pH was decreased to 6.8. E_{H^+} increased from -1.5 to +40 mV which depressed E_{Na^+}-E_{H^+}, the driving force for H^+ secretion on the Na^+-H^+ antiporter, from -55.7 to -30 mV. Fig. 4 shows that both Na^+ influx and H^+ secretion decreased 42% and 63% respectively. As a result, pH_i fell to 7.11 and Na_i^+ decreased from 15.6 to 9.5 mM. At the same time, Ca^{2+} efflux increased 36% while Ca_i^{2+} fell from 120 to

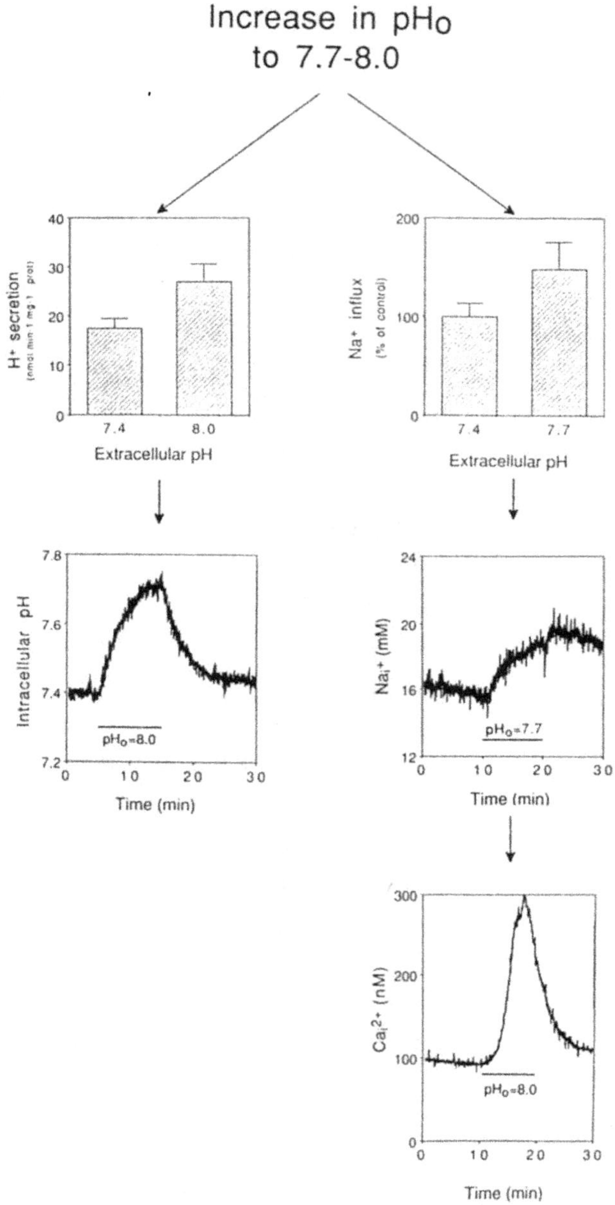

Figure 3. Effects of increasing extracellular pH to 7.7-8.0 on H^+ secretion, Na^+ influx and cytosolic Na^+, pH and Ca^{2+} in MDCK cells. Recordings of ion concentration are representative of 5-6 separate experiments.

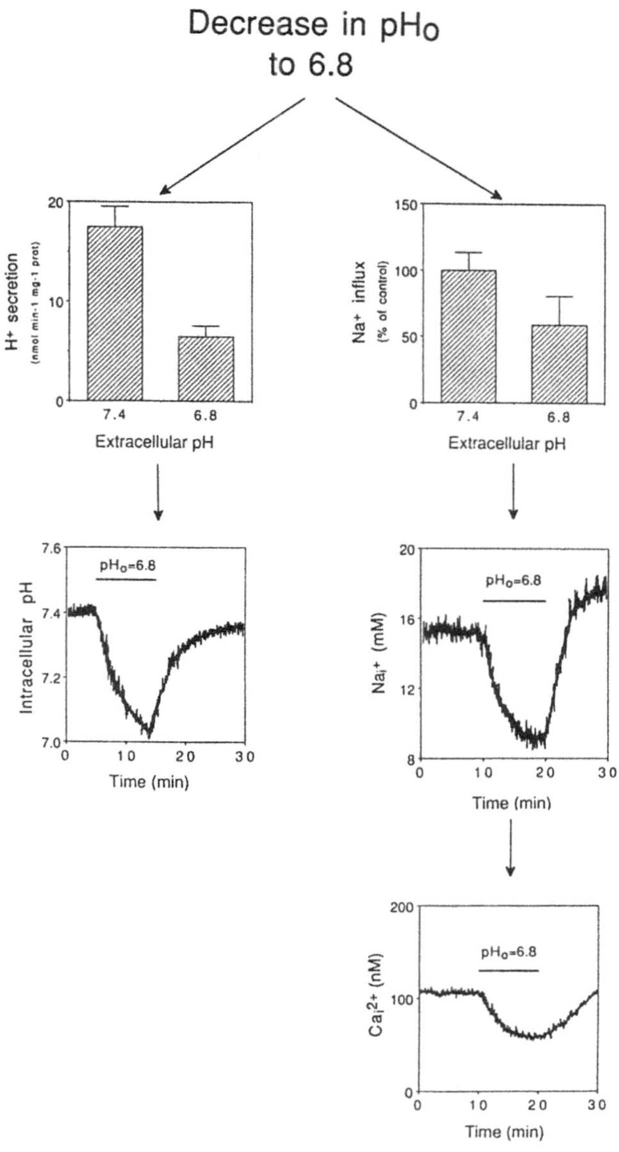

Figure 4. Effects of decreasing extracellular pH to 6.8 on H^+ secretion, Na^+ influx, and cytosolic Na^+, pH and Ca^{2+} in MDCK cells. Recordings of ion concentration are representative of 5-6 separate experiments.

14 nM, suggesting that the forward mode of Na^+-Ca^{2+} exchange was activated. This was caused by the fall in Na_i^+. The decrease in Na_i^+ increased E_{Na^+} from +57.2 to +70 mV, and the reversal potential fell from -70.4 to -32 mV, a value less negative than the membrane potential E_m of -50 mV. As a result E_m-E_R shifted from a positive value, +20.4 mV, to a negative value of -18 mV. That triggered the reversal of the Na^+-Ca^{2+} antiporter mode from reverse to forward with a driving force for Ca^{2+} efflux of -18 mV. Activation of the forward mode Na^+-Ca^{2+} antiporter should depolarize the cell. And indeed, Paulmichl et al (1985) showed that lowering extracellular pH decreases the membrane potential of MDCK cells. Fig. 5 shows the calculated changes in the Na^+-Ca^{2+} antiporter mode and the driving force for Ca^{2+} influx or efflux before, during and after a change in extracellular pH. These experiments demonstrate that stimulation or inhibition of the Na^+-H^+ antiporter indirectly influences Na^+-Ca^{2+} exchange and the cytosolic free Ca^{2+}. They further show that the concentration of intracellular Ca^{2+}, Na^+ and H^+ are interrelated through the interaction of the Na^+-Ca^{2+} and Na^+-H^+ antiporters.

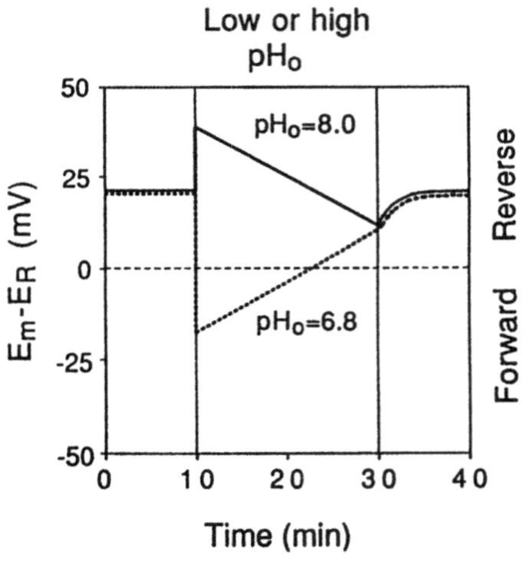

Figure 5. Calculated change in the mode and driving force (E_m-E_R) of the Na^+-Ca^{2+} antiporter during exposure of MDCK to high or low extracellular pH.

Several conclusions can be drawn from these experiments: 1) in resting control conditions, the Na^+-Ca^{2+} antiporter of MDCK cells operates in the reverse mode as a Ca^{2+} influx pathway; 2) lowering $Ca_o{}^{2+}$ shifts the mode of Na^+-Ca^{2+} exchange from reverse to forward, i.e. from Ca^{2+} influx to Ca^{2+} efflux. We have calculated that, in MDCK cells, the concentration of extracellular Ca^{2+} at which the antiporter reverses its mode, i.e. when $E_m = E_R$, is 0.4 mM; 3) incubation in Ca^{2+}-free medium not only prevents Ca influx into the cells and lowers cytosolic free Ca^{2+}, but it also increases intracellular Na^+, decreases H^+ secretion, and lowers intracellular pH. These changes in the intracellular ionic composition may have important secondary effects on many cellular functions; 4) when the cells are placed in a normal 1.3 mM $Ca_o{}^{2+}$ after a period in Ca^{2+}-free media, there is a sudden, large but transient increase in $Ca_i{}^{2+}$. That large peak in $Ca_i{}^{2+}$ is analogous to the Ca^{2+} paradox seen in myocardial cells; 5) a change in extracellular pH or the stimulation or inhibition of the Na^+-H^+ antiporter not only affects H^+ secretion, intracellular Na^+ and pH, but cytosolic free Ca^{2+} and Ca^{2+} transport as well.

These results have one more interesting but speculative implication for Ca^{2+} transport in the renal distal tubule: because of the difference in membrane potential between the apical and basal side of the distal tubular cells, transepithelial Ca^{2+} transport from the lumen to the interstitial fluid compartment, could use Na^+-Ca^{2+} exchange across both cell membrane barriers, theoretically without direct expenditure of energy. At the luminal side, the antiporter would operate in the reverse mode, while, at the basal side, it would operate in the forward mode. The energy cost of this indirect active transport would be born by the Na^+-K^+ ATPase-dependent Na^+ pump. Fig. 6 shows a purely speculative model of this Ca^{2+} transport.

This model rests on 3 assumptions: 1) there is a Na^+-Ca^{2+} antiporter in the brush border of the distal tubule; 2) because of the low osmolarity of the distal tubule fluid (200 mOsm), the concentration of Na^+ is 30% lower than in the plasma, and 3) the Na^+-K^+ pump maintains a low intracellular Na^+ concentration of 10 mM. In these conditions, and mainly because of the large difference in membrane potential between luminal and apical membranes (-30 vs -60 mV respectively), Ca^{2+} can enter the cell on the Na^+-Ca^{2+} antiporter operating the reverse mode with a driving force E_m-E_R of +30 mV. At the basal membrane, Ca^{2+} can be transported out of the cell on Na^+-Ca^{2+} antiporters operating in the forward mode with a driving force of -30 mV.

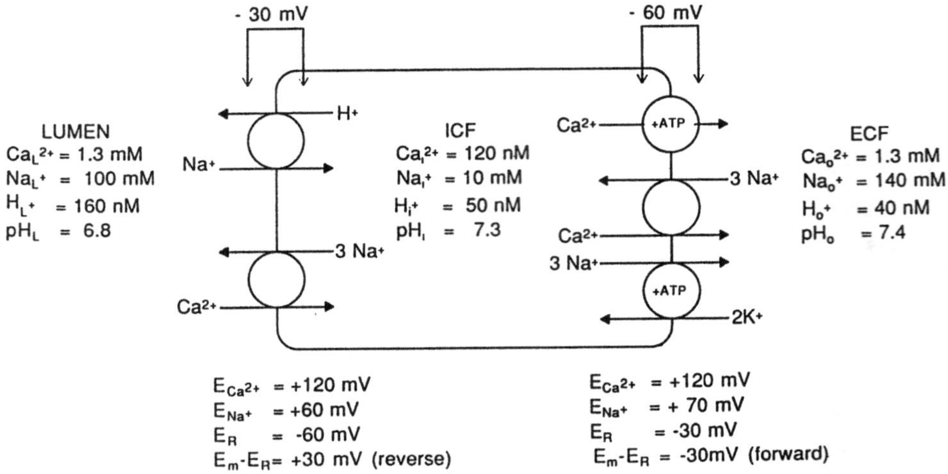

Figure 6. Theoretical model of Ca^{2+} transport across an epithelial cell of the renal distal tubule using the Na^+-Ca^{2+} antiporter in the reverse mode for Ca^{2+} influx at the luminal membrane, and the same antiporter in the forward mode for Ca^{2+} efflux at the basal membrane.

There is, of course, no compelling reason for the two antiporters on both sides of the cell to transport Ca^{2+} at the same rate, since the (Ca^{2+}-Mg^{2+})-ATPase-dependent Ca^{2+} pump of the basal membrane carries a significant fraction of Ca^{2+} efflux out of the cell. Obviously, these considerations are purely theoretical. But they raise the interesting possibility that, in the distal tubule, the Na^+-Ca^{2+} antiporter may operate in different modes at each side of the cell. They further suggest that Ca^{2+} efflux out of the distal tubular cells can be increased not only by a rise in cytosolic Ca^{2+}, but also by a decrease in intracellular Na^+.

In conclusion, we have shown that in renal cells derived from the distal tubule, the Na^+-Ca^{2+} and Na^+-H^+ antiporters are interdependent because they are both a function and a determinant of the cell Na^+ electrochemical potential. As a result, cytosolic free Ca^{2+}, intracellular Na^+ and intracellular pH are closely interrelated. These interrelations may be relevant to the reabsorption of Na^+ and Ca^{2+}, to H^+ secretion and to their endocrine regulation in the renal distal tubule.

Acknowledgement: This work was funded by the National Dairy Board and administered in cooperation with the National Dairy Council.

References

Borle AB, Borle CJ, Dobransky P, Gorecka-Tisera AM, Bender C, Swain K (1990) Effects of low extracellular Ca^{2+} on cytosolic Ca^{2+}, Na^+ and pH of MDCK cells, Am J Physiol, in press.

Paulmichl M, Friedrich F, Long F (1986) Electrical properties of Madin-Darby-canine-kidney cells. Effects of extracellular sodium and calcium. Pflugers Arch 407, 258-263.

Paulmichl M, Gstraunthaler G, Lang F (1985) Electrical properties of Madin-Darby-canine-kidney cells. Effects of extracellular potassium and bicarbonate. Pflugers Arch 405, 102-107.

Snowdowne KW, Borle AB (1985) Effects of low extracellular sodium on cytosolic ionized calcium. Na^+-Ca^{2+} exchange as a major calcium influx pathway in kidney cells, J Biol Chem 260, 14998-15007.

REGULATION OF INTRACELLULAR CALCIUM IN CULTURED RENAL EPITHELIOID (MDCK-) CELLS

F. Lang, Paulmichl M., Friedrich F., Pfeilschifter J.*, Woell E., Weiss H.
Department for Physiology
University of Innsbruck
Fritz-Pregl-Str. 3
A-6010 Innsbruck
Austria

Stimulation of electrolyte transport by hormones involves calcium dependent activation of ion channels in a variety of epithelia. To elucidate the cellular mechanisms and ion channels involved, we have studied the activation of ion channels by epinephrine, bradykinin and ATP in Madin Darby canine kidney (MDCK-) cells [13], a permanent cell line from dog kidney. Chloride secretion in MDCK-cell monolayers has been shown to be stimulated by epinephrine [2,19] and ATP [18].

In subconfluent MDCK-cells epinephrine [15], bradykinin [16] and ATP [8,11] hyperpolarize the cell membrane. The halfmaximal effects are observed at 5 nmol/l epinephrine [15], 1 nmol/l bradykinin [16], and 300 nmol/l ATP [8,11] respectively. The hyperpolarizing effect of epinephrine is blocked by phentolamine, pointing to involvement of α-receptors. Isoproterenol slightly depolarizes the cell membrane, due to activation of chloride channels [9]. The effect of bradykinin requires the full length of the peptide and is thus likely to be mediated by B2-receptors [16]. The effect of ATP is mimicked by UTP, ITP, ADP (half maximal concentration $\approx$ 300 nmol/l), and to a lesser extent by UDP, GTP and GDP (half maximal concentration $\approx$ 10 μmol/l). Up to 1 mmol/l AMP, UMP, GMP, TTP or CTP do not significantly alter cell membrane potential. The effect of the nucleotides is probably mediated by P2 purinergic receptors. The hyperpolarization following application of either hormone is paralleled by a decrease of cell membrane resistance and an increase of the potassium selectivity of the cell membrane, reflecting an increase of the potassium conductance of the cell membrane. At extracellular calcium activities reduced to less than 1 μmol/l, the hyperpolarizing effects of either, epinephrine [15], ATP [8,11], acetylcholine [10] and bradykinin [16] are transient and can be elicited only once.

We report here about our attempts to (1) define the role of intracellular calcium activity for the hormone induced hyperpolarization and (2) to elucidate the mechanisms accounting for hormone induced alterations of intracellular calcium activity.

*Ciba Geigy AG, RI 1056 P23, Post Box, CH-4002 Basel, Switzerland

NATO ASI Series, Vol. H 48
Calcium Transport and
Intracellular Calcium Homeostasis
Edited by D. Pansu and F. Bronner

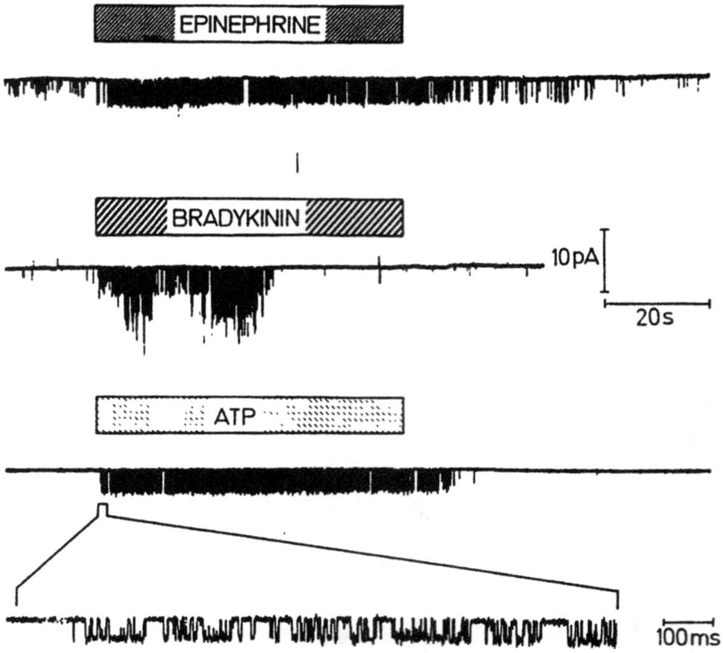

Fig. 1: Original recordings from cell attached patches demonstrating ion channel activation by epinephrine (1 μmol/l), bradykinin (100 nmol/l) and ATP (10 μmol/l).
The patch clamp experiments were performed according to Sackmann and Neher [17]. Patch pipettes were approached to the cell and allowed to seal with the cell membrane by gentle suction. Single channel current events were measured by means of a L/M-EPC-7 amplifier (LIST-Electronics, Darmstadt, FRG), stored on a VHS- video-tape recorder (ELIN-6101, Vienna, Austria) via pulse code modulation (SONY PCM-501ES). The pipette solution was composed of 145 mmol/l KCl, 10 mmol/l Hepes-KOH (pH 7.4), 45 μmol/l Phenol Red and 1 mmol/l $CaCl_2$. The holding potential was 0 mV between pipette and bath.
Prior to the experiments MDCK-cells [7,13] were maintained in a 1:1 mixture of Dulbecco's modified Eagle's medium (DMEM) with 10 % fetal calf serum, 100 U/ml Penicillin and 100 μg/ml Streptomycin [6,20] at 5 % CO_2, 95 % air and 37°C. Cells were dispersed by incubation in a calcium and magnesium free, trypsin-EDTA containing balanced salt-solution (pH 7.4), plated on cover glasses and incubated again in the same medium as above for at least 48 hours. During the experiments temperature was again 37°C. Control extracellular fluid was composed of (in mmol/l) 114 NaCl, 5.4 KCl, 0.8 $MgCl_2$, 1.2 $CaCl_2$, 0.8 Na_2HPO_4, 0.2 NaH_2PO_4, 16 $NaHCO_3$, 5.5 glucose and continuously bubbled with 95 % air and 5 % CO_2 (pH 7.4).

HORMONE AND CALCIUM ACTIVATED K^+-CHANNELS

Patch clamp studies reveal that epinephrine, bradykinin and ATP activate ion channels in the cell membrane (Fig.1). Further analysis (Fig.2) discloses the properties of these ion channels: They are inwardly rectifying potassium channels with single channel conductance of 34 pS at zero potential difference and 60 pS at -50 mV (cell negative) across the patch (150 mmol/l K^+ on either side). The channels are activated following exposure of the cells to A 23187 or by increasing the calcium activity at the intracellular side of excised cell membrane patches from 0.1 to 1 μmol/l (Fig.2).

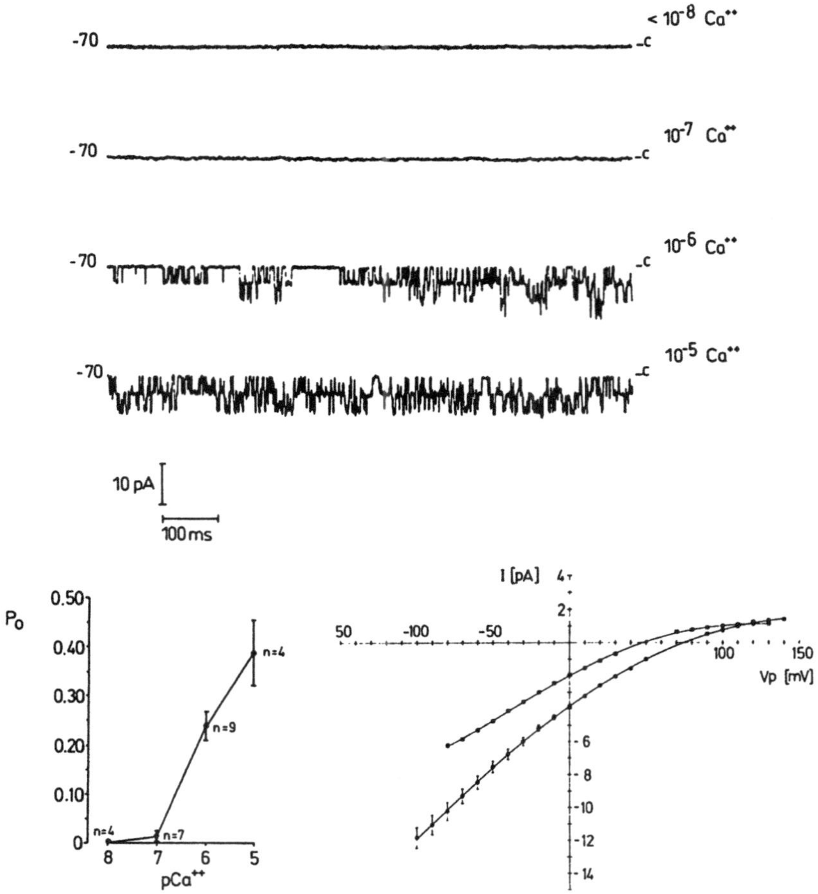

Fig. 2: Calcium activated K^+ channels in subconfluent MDCK cells (Data from [3,4,12]). Above: Original tracing from an excised inside out patch exposed to different calcium activities on the intracellular side of the cell membrane (holding potential -70 mV). Below left: Mean values ± SEM of open probabilities at different "intracellular" calcium activities. Below right: Relation of single channel current (I) and voltage (V) across an excised inside out patch at 150 KCl (closed symbols) and 50 mmol/l KCl + 100 mmol/l NaCl in the bath. Further experimental details see Fig.1.

INTRACELLULAR CALCIUM ACTIVITY

Fura2 and Fluo3 fluorescence reveals that epinephrine (Fig.3), bradykinin (Fig.4) and ATP (Fig.5) lead to an increase of intracellular calcium activity (to 555 ± 43, 462 ± 51 and 1055 ± 103 nmol/l, respectively). The effect of bradykinin is transient, the effects of epinephrine and ATP are sustained. In the nominal absence of extracellular calcium the effect of either hormone is transient (348 ± 38, 213 ± 32, and 557 ± 162 nmol/l, resp.).

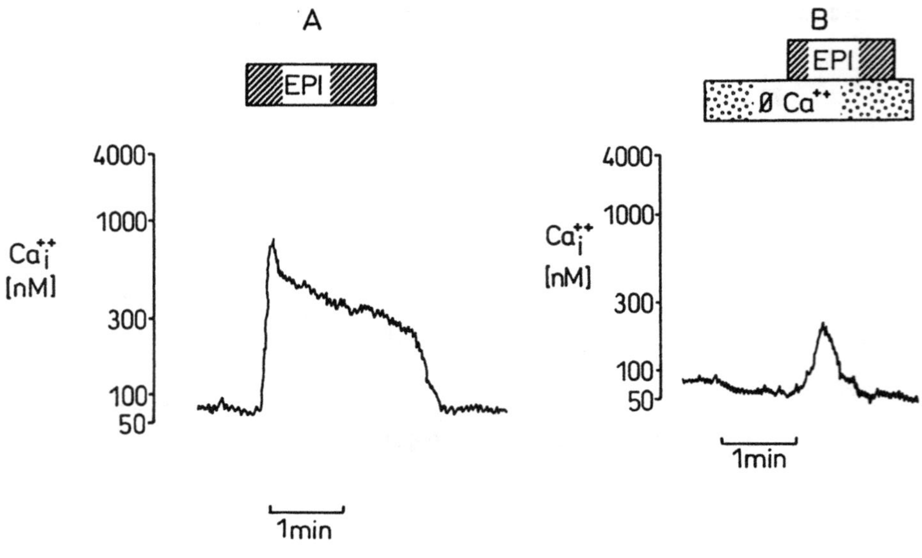

Fig. 3: Effect of epinephrine (1μmol/l) on intracellular calcium activity of MDCK cells in both, the presence (A) and absence (B) of extracellular calcium.
The cells were loaded with fura2 by 45 min exposure to 2.5 μmol/l fura2-AM (Molecular Probes, Junction City, OR, and Calbiochem, Genf, CH). Measurements were made under an inverted phase- contrast microscope (IM-35, Zeiss, FRG) equipped for epifluorescence and photometry (Hamamatsu, Herrsching, FRG) [1]. Light from a xenon arc lamp (XBO75, Osram) was direct through a grey-filter (nominal transmission 3.16 %, Oriel, Darmstadt, FRG), a 340 nm interference filter (Halfwidth 10nm, Oriel, Darmstadt, FRG) and a diaphragm and was deflected by a dichroic mirror (FT425, Zeiss, FRG) into the objective (Plan-Neofluar 63 x oil immersion, Zeiss, FRG). Emitted fluorescence was directed through a 420 nm cutoff filter to a photomultiplier tube (R4829, Hamamatsu, Herrsching, FRG). The diameter of the observed area was 60 μm. Cai in nmol/l was calculated from the observed fluorescence intensity of fura2 under experimental (F) and calcium saturated conditions (F_{max}, determined by addition of 20 μmol/l digitonin from Sigma, Munich, FRG) [14,21]:

$$[Ca^{2+}]_i = 225 \cdot [(F - 0.33 \cdot F_{max})/(F_{max} - F)]$$

Fluorescence values were corrected for cellular autofluorescence, 225 nmol/l is the apparent K_d for calcium-fura2 at cytoplasmatic ionic conditions [5], and $0.33 \cdot F_{max}$ is the fluorescence intensity of the calcium free fura2 [5].
For further experimental details see Fig.1.

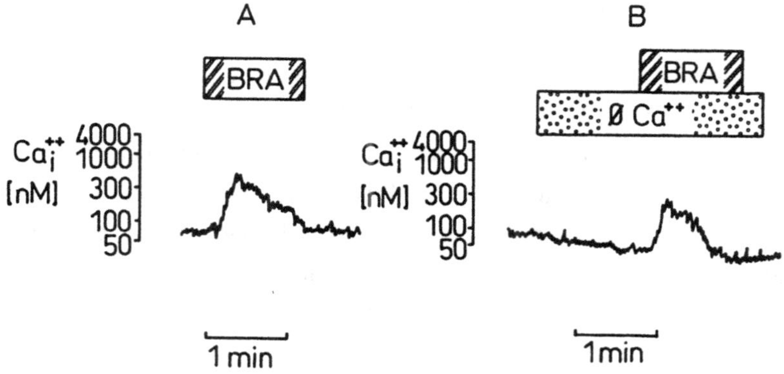

Fig. 4: Effect of bradykinin (100 nmol/l) on intracellular calcium activity of MDCK cells in both, the presence (A) and absence (B) of extracellular calcium. For further experimental details see Fig. 1.

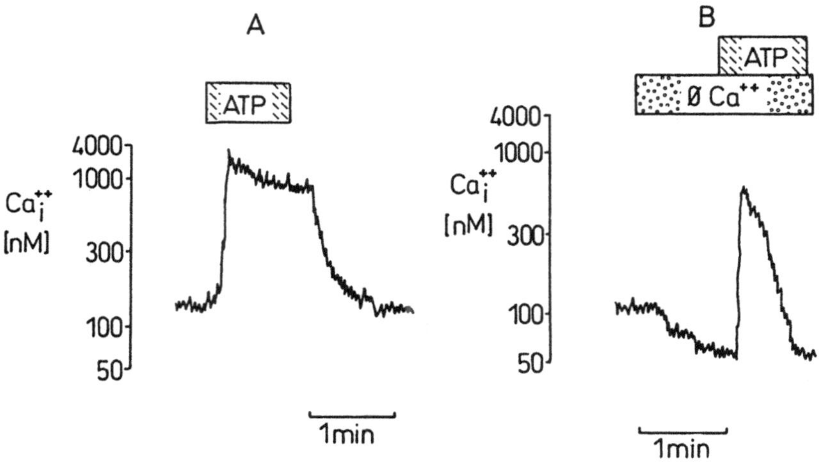

Fig. 5: Effect of ATP (10 μmol/l) on intracellular calcium activity of MDCK cells in both, the presence (A) and absence (B) of extracellular calcium. For further experimental details see Fig. 1.

INOSITOLPHOSPHATE FORMATION

Epinephrine, bradykinin and ATP stimulate formation of 1,4,5 inositoltrisphosphate (IP3) (by 125 ± 15 %, 175 ± 22 %, and 425 ± 68 %, resp.) and of 1,3,4,5 IP4 (by 133 ± 11 %, 188 ± 11 %, and 477 ± 110 %, resp.), which may at least partially account for the increase of intracellular calcium activity. Pretreatment of the cells with pertussis toxin almost abolishes the ATP- and bradykinin-induced, but not the epinephrine-induced stimulation of inositoltrisphosphate (IP3) formation, indicating that both, pertussis toxin sensitive and pertussis toxin insensitive activation of phospholipase C occurs in MDCK-cells. Pretreatment of the cells with phorbolester TPA blunts the stimulation of IP3 formation by either hormone.

However, pretreatment of the cells with pertussis toxin does not prevent the increase of intracellular calcium activity by either hormone in the presence or absence of extracellular calcium. This observation points to a pertussis toxin insensitive mechanism parallel to IP3 formation allowing for increase of intracellular calcium activity or for a mechanism

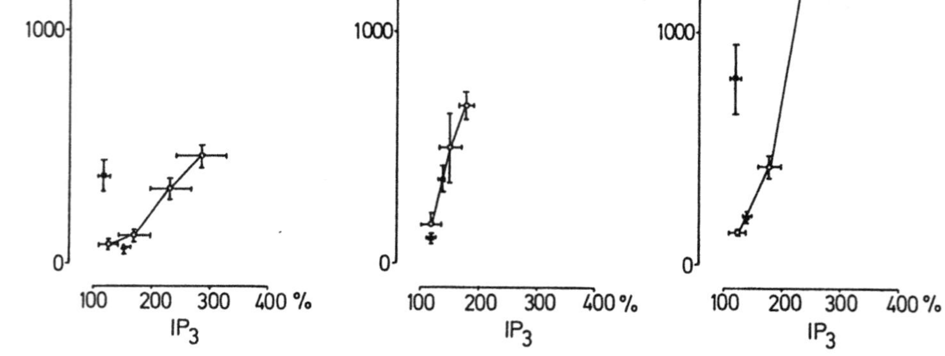

Fig. 6: Correlation between inositoltrisphosphate (IP3) formation and intracellular calcium activity following application of epinephrine, bradykinin and ATP. Open circles: Respective values at different hormone concentrations (epinephrine from 1 - 100 nmol/l, bradykinin from 1 - 100 nmol/l and ATP from 10 - 1000 nmol/l). Closed symbols: effect of the hormones (epinephrine 1 μmol/l, bradykinin 100 nmol/l, ATP 10 μmol/l) in cells pretreated with either, pertussis toxin (closed triangles) or phorbolester TPA (closed squares).

modulating the IP3 sensitivity of calcium recruitment. The mechanism is apparently inhibited by pretreatment with phorbolester. Pretreatment of the cells with phorbolesters abolishes the effect of EPI and BK and blunts the effect of ATP on intracellular calcium activity.

Accordingly, in cells pretreated with pertussis toxin, the increase of Cai following application of bradykinin or ATP clearly exceeds the values expected from the minimal changes of IP3 formation (Fig.6).

CONCLUSIONS

All three hormones (epinephrine, bradykinin and ATP) activate K^+ channels in subconfluent MDCK cells by increasing intracellular calcium activity. All three hormones are capable to mobilise calcium from intracellular stores and to stimulate entry of calcium from the extracellular space. The release of calcium is in part secondary to formation of 1,4,5 inositoltrisphosphate. Bradykinin and ATP activate the phospholipase C via pertussis toxin sensitive G-proteins, whereas epinephrine stimulated increase of IP3 formation is not sensitive to pertussis toxin. Bradykinin and ATP are seemingly capable to increase intracellular calcium activity without appropriate stimulation of IP3 formation.

REFERENCES

1. Almers W, Neher E (1985) The Ca signal from fura-2 loaded mast cells depends strongly on the method of dye loading. FEBS Lett 192:13-18
2. Brown CDA, Simmons NL (1981) Catecholamine stimulation of Cl- secretion in MDCK cell epithelium. Biochim Biophys Acta 649:427-435
3. Friedrich F, Paulmichl M, Kolb H-A, Lang F (1988) Inward rectifier K channels in renal epithelioid cells (MDCK) activated by serotonin. J Membr Biol 106:149-155
4. Friedrich F, Weiss H, Paulmichl M, Lang F (1989) Activation of potassium channels in renal epithelioid cells (MDCK) by extracellular ATP. Am J Physiol 256:C1016-1021
5. Grynkiewicz G, Poenie M, Tsien RY (1985) A new generation of Ca2+ indicators with greatly improved fluorescence properties. J Biol Chem 260:3440-3450
6. Gstraunthaler G, Pfaller W, Kotanko P (1985) Biochemical characterization of renal epithelial cell cultures (LLC-PK1 and MDCK). Am J Physiol 248:F536-F544
7. Gstraunthaler GJA (1988) Epithelial cells in tissue culture. Renal Physiol Biochem 11:1-42
8. Jungwirth A, Lang F, Paulmichl M (1989) Effect of extracellular adenosine triphosphate on electrical properties of subconfluent Madin-Darby canine kidney cells. J Physiol 408:333-343
9. Lang F, Defregger M, Paulmichl M (1986) Apparent chloride conductance of subconfluent Madin-Darby canine kidney cells. Pflügers Arch 407:158-162
10. Lang F, Klotz L, Paulmichl M (1988) Effect of acetylcholine on electrical properties of subconfluent Madin-Darby canine kidney cells. Biochim Biophys Acta 941:217-224

11. Lang F, Plöckinger B, Häussinger D, Paulmichl M (1988) Effects of extracellular nucleotides on electrical properties of subconfluent Madin-Darby canine kidney cells. Biochim Biophys Acta 943:471-476
12. Lang F, Paulmichl M, Friedrich F, Pfeilschifter J, Jungwirth A, Steidl M, Schobersberger W, Weiss H, Woell E, Tschernko E, Hallbrucker C, Deetjen P (1990) Regulation of potassium channels in cultured renal epitheloid (MDCK) cells. In: Diuretics III, Chemistry, Pharmacology, and Clinical Applications (Püschett JB, Greenberg A, eds). Elsevier New York, pp. 685-694
13. Madin SH, Darby NB (1958) As catalogued in: American Type Culture Collection Catalogue of Strains 2:574-576
14. Moolenaar WH, Tertoolen LGJ, deLaat SW (1984) Growth factors immediately raise cytoplasmic free Ca2+ in human fibroblasts. J Biol Chem 259:8066-8069
15. Paulmichl M, Defregger M, Lang F (1986) Effects of epinephrine on electrical properties of Madin-Darby canine kidney cells. Pflügers Arch 406:367-371
16. Paulmichl M, Friedrich F, Lang F (1987) Effects of bradykinin on electrical properties of Madin-Darby canine kidney epithelioid cells. Pflügers Arch 408:408-413
17. Sakmann B, Neher E (1983) Single channel recording. Plenum Press New York London
18. Simmons NL (1981) Stimulation of Cl- secretion by exogenous ATP in cultured MDCK epithelial monolayers. Biochim Biophys Acta 646:231-242
19. Simmons NL, Brown CDA, Rugg EL (1984) The action of epinephrine on Madin-Darby canine kidney cells. Fed Proc 43:2225-2229
20. Taub M, Chuman L, Saier MHJr, Sato G (1979) Growth of Madin- Darby canine kidney epithelial cell (MDCK) line in hormone- supplemented, serum-free medium. Proc Natl Acad Sci USA 76:3338-3342
21. Tsien RY, Pozzan T, Rink TJ (1982) Calcium homeostasis in intact lymphocytes: cytoplasmic free calcium monitored with a new, intracellularly trapped fluorescent indicator. J Cell Biol 94:325-334

ASSESSMENT OF THE ROLE OF CALCIUM IN NEUTROPHIL ACTIVATION USING ELECTROPERMEABILIZED CELLS

G. P. Downey*, S. Trudel¶, W. Furuya¶ and S. Grinstein¶
¶Division of Cell Biology
Hospital for Sick Children
555 University Ave.
Toronto, Canada, M5G 1X8

Neutrophils are one of the first lines of defense against invading microorganisms. When exposed to bacteria or to chemoattractant peptides derived thereof, neutrophils undergo a complex series of coordinated responses intended to eliminate the microorganisms, thereby preventing infection. The neutrophils initially approach the site of infection by a migratory process called chemotaxis. The bacteria are then phagocytosed and killing results from the generation of superoxide and the secretion of lytic enzymes into the phagosomal space. Chemotaxis as well as phagocytosis are thought to involve extensive cytoskeletal reorganization, requiring conversion of actin from its monomeric, globular (G) form, to a polymeric, filamentous (F) form. Superoxide radicals are generated by reduction of molecular oxygen, with NADPH as the electron donor. This reaction, known as the respiratory burst, is catalyzed by the NADPH-oxidase, a multimeric plasma membrane membrane enzyme that is quiescent before stimulation, but is greatly activated by chemotactic peptides and other stimuli.

Because the generation of superoxide and actin polymerization are stimulated in parallel by chemoattractants, similar signalling pathways are thought to mediate the activation of both processes. It is generally believed that formation of a complex between the stimulating ligand and an exofacial receptor leads to the dissociation of the α-subunit of a G protein from its heterotrimeric complex. The activated, GTP-bound G_α proceeds to stimulate a phosphoinositide-specific phospholipase C, which hydrolyzes primarily phosphatidylinositol -4,5-bisphosphate, releasing diacylglycerol and inositol-1,4,5-trisphosphate. The latter product is known to increase the free cytosolic Ca^{2+} concentration ($[Ca^{2+}]_i$) by releasing Ca^{2+} from intracellular stores. This rapid $[Ca^{2+}]_i$ increase is followed by a more sustained component, due largely to entry of Ca^{2+} across the plasma membrane (see Sha'afi and Molski, 1988 for review).

The mechanism(s) underlying actin polymerization and activation of the NADPH oxidase are not well understood, but increased $[Ca^{2+}]_i$ has been proposed to play a central role. This suggestion is based on the following evidence: a) chemoattractants and other stimuli elicit a

*Respiratory Division , Toronto General Hospital and Clinical Sciences Division, Department of Medicine, University of Toronto.

NATO ASI Series, Vol. H 48
Calcium Transport and
Intracellular Calcium Homeostasis
Edited by D. Pansu and F. Bronner

rapid rise in $[Ca^{2+}]_i$, which precedes or at least coincides with the development of the actin and oxidative responses (Pozzan et al., 1983; Lew, 1989); b) the induction of actin assembly and of the respiratory burst by chemoattractants can be mimicked by addition of the divalent cation ionophores A23187 or ionomycin, which similarly increase $[Ca^{2+}]_i$ (Hallett and Campbell, 1984); c) superoxide generation elicited by physiological stimuli can be inhibited by "intracellular calcium antagonists" such as trifluoperazine and 8-(diethylamino)octyl-3,4,5-trimethoxybenzoate (TMB8; Hallett et al., 1981; Smolen et al, 1981); d) the respiratory burst can be partially inhibited by omission of external calcium, or by chelation of internal calcium using trapped ethylenebis(oxyethylenenitrilo)tetraacetic acid (EGTA) or bis-(2-amino-5-methylphenoxy)-ethane-N,N,N',N' tetraacetic acid (MAPTAM) (Campbell and Hallett, 1983; Korchack et al., 1988).

Suspension of cells in calcium-free solutions or incorporation of chelators into the cytoplasm not only prevent the stimulus-induced increases in $[Ca^{2+}]_i$, but generally also result in decreased basal $[Ca^{2+}]_i$ levels. Therefore, these criteria cannot define whether inhibition of the functional responses is due to elimination of the $[Ca^{2+}]_i$ increase promoted by the chemoattractant, or to reduction of $[Ca^{2+}]_i$ below a critical "permissive" level. These alternative hypotheses can be differentiated by stimulating the cells under conditions where $[Ca^{2+}]_i$ remains constant at the basal (unstimulated) level throughout the course of the experiment. This can be accomplished, in principle, by suspending a broken cell preparation in a calcium-buffered medium. Unfortunately, cell-free preparations do not retain responsiveness to physiological stimuli, nor do they preserve the cytoarchitecture required to study actin assembly. To circumvent these problems, we have used a mild permeabilization technique to study signal transduction in human neutrophils. Pores allowing the passage of molecules of $\leq$700 Da were generated reproducibly by exposing the cells to intense electric fields (5 kV/cm). Unlike the cell-free preparation, such electroporated cells maintain a relatively normal cytoarchitecture and remain sensitive to stimulation by a variety of physiological stimuli (Grinstein and Furuya, 1988).

Several criteria were used to confirm that successful permeabilization was accomplished. First, measurements of cell volume demonstrated that the electroporated cells had swollen, as expected from the unrestricted, colloidosmotically driven influx of ionorganic ions and accompanying water. Secondly, light microscopy and flow cytometry indicated that, while less than 4% of the intact cells are permeable to trypan blue or propidium iodide, more than 90% of the electroporated cells were stained by these dyes. Thirdly, leakage of Rb^+ or of fluorescein derivatives trapped inside intact cells was greatly accelerated upon electroporation. Lastly, unlike intact cells, which utilize cytoplasmic NADPH, electroporated cells required the addition of exogenous NADPH to mount a respiratory burst, indicating that the endogenous nucleotide had leaked out of the cell.

Electropermeabilized cells suspended in EGTA-containing media were used to study the role of $[Ca^{2+}]_i$ in neutrophil activation (Grinstein and Furuya, 1988). The respiratory burst was measured by polarographically monitoring the disappearance of oxygen from the cell suspension, using a Clark-type electrode. We found that exposure of the permeabilized cells to media containing micromolar levels of free calcium elicited a substantial increase in the rate of oxygen consumption. This finding resembles the stimulatory effect of calcium ionophores and provides further evidence of the success of the permeabilization protocol. As in the case of ionophore-treated cells, however, these findings only indicate that elevated $[Ca^{2+}]_i$ suffices to stimulate the oxidase, but do not prove that this is the physiological mechanism of activation. In fact, the concentrations of internal calcium required to generate a significant respiratory burst in the absence of chemoattractants are considerably higher than those attained during physiological stimulation.

To test whether increased $[Ca^{2+}]_i$ is required to induce the respiratory burst, permeabilized cells were suspended in media containing 100 nM free $[Ca^{2+}]$, which approximates the levels in resting (unstimulated) cells. Under these conditions $[Ca^{2+}]_i$ is expected to remain virtually unchanged since: a) there is no driving force for transmembrane calcium influx and b) calcium released from internal stores will be rapidly chelated by EGTA in the cytoplasm and/or exit the cell through the pores. To our surprise, the response of such permeabilized cells to formyl-methionyl-leucyl-phenylalanine (fMLP) was at least as large and generally more sustained than that recorded in intact cells. Though similar results were obtained using up to 10 mM EGTA, it can still be argued that release of calcium from stores in the vicinity of the plasma membrane can exceed the buffering capacity of the chelator, producing a transient, localized change in $[Ca^{2+}]_i$. To address this possibility, cells suspended in low $[Ca^{2+}]$ medium were treated with ionomycin to deplete the intracellular calcium stores prior to stimulation with the chemoattractant. As shown in figure 1A, the ionophore effectively permeabilized the internal stores, as evidenced by the large calcium transient recorded in otherwise intact cells using the fluorescent indicator indo-1. Subsequent stimulation of these cells with fMLP failed to increase $[Ca^{2+}]_i$. When electroporated cells were subjected to a similar protocol while measuring oxygen consumption (Fig. 1B), ionomycin itself did not activate the oxidase, despite the mobilization of internal calcium. In contrast, the subsequent addition of fMLP elicited a pronounced respiratory burst.

The data presented above indicate that an increase in $[Ca^{2+}]_i$ is not essential for the fMLP-induced activation of the NADPH oxidase. However, further experiments revealed that, though an increase is not necessary, maintenance of $[Ca^{2+}]_i$ at or near the resting physiological level is required for successful signal transduction. This became apparent when cells were permeabilized and suspended in media containing less than 10 nM free calcium (nominally calcium-free medium plus 1 mM EGTA). Under these conditions the response to fMLP was greatly inhibited. By comparison, the effects of phorbol esters or exogenous diacylglycerol were unaffected by omission of calcium, indicating that the oxidase remained functional and responsive in calcium-free conditions. Since phorbol esters and diacylglycerol are thought to stimulate by binding to protein kinase C, the calcium-sensitive component must lie upstream of this step.

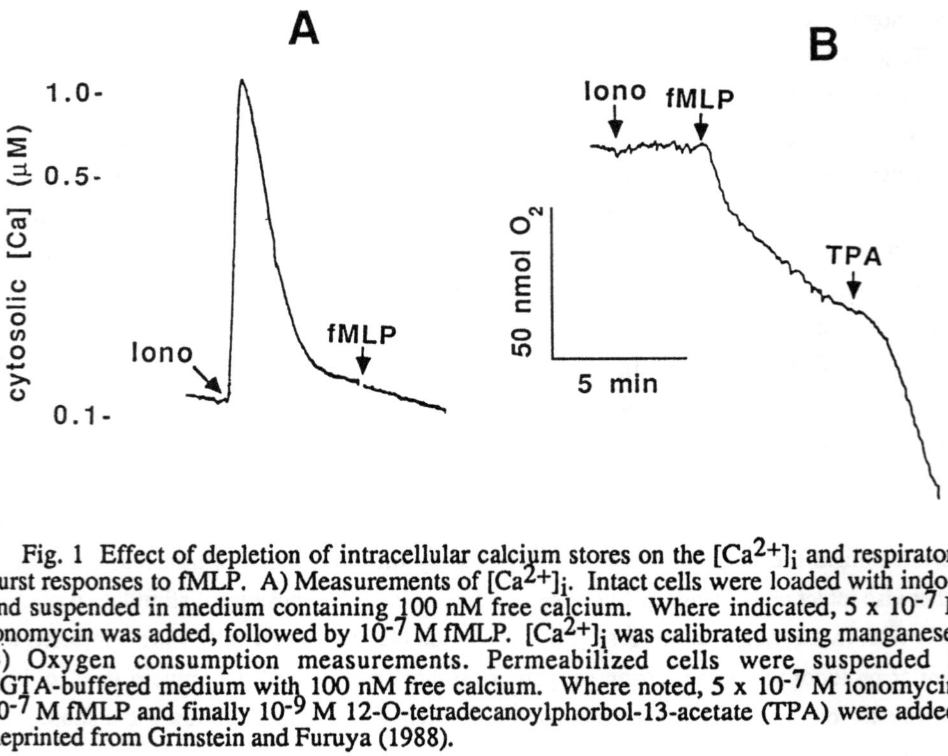

Fig. 1 Effect of depletion of intracellular calcium stores on the $[Ca^{2+}]_i$ and respiratory burst responses to fMLP. A) Measurements of $[Ca^{2+}]_i$. Intact cells were loaded with indo-1 and suspended in medium containing 100 nM free calcium. Where indicated, 5 x 10^{-7} M ionomycin was added, followed by 10^{-7} M fMLP. $[Ca^{2+}]_i$ was calibrated using manganese. B) Oxygen consumption measurements. Permeabilized cells were suspended in EGTA-buffered medium with 100 nM free calcium. Where noted, 5 x 10^{-7} M ionomycin, 10^{-7} M fMLP and finally 10^{-9} M 12-O-tetradecanoylphorbol-13-acetate (TPA) were added. Reprinted from Grinstein and Furuya (1988).

Permeabilized neutrophils can also be used to study the signalling pathways leading to actin polymerization. The transition of G to F actin can be monitored indirectly, by measuring the right-angle light scattering decrease that results from the associated shape changes, or directly, by quantitating the F actin content by means of 7-nitrobenz-2-oxa-1,3-diazole phallacidin (NBD-phallacidin). The fluorescence of this compound, which binds selectively to F actin, can be conveniently monitored in large numbers of single cells by flow cytometry. As was the case for the respiratory burst, we found that fMLP-induced actin assembly persisted in electropermeabilized cells equilibrated with 100 nM free $[Ca^{2+}]_i$ (Fig. 2; Downey and Grinstein, 1989) and in cells pretreated with ionophore in low $[Ca^{2+}]$ media to deplete intracellular stores (Downey, G.P., unpublished observations). Thus, a rise in $[Ca^{2+}]_i$ is not required to promote actin polymerization, consistent with earlier observations using intracellular calcium chelators (Sha'afi et al., 1986). The permeabilized cells also enabled us to study directly the effects of controlled elevations of $[Ca^{2+}]_i$ on the state of actin assembly. Unexpectedly, increasing $[Ca^{2+}]$ was found to promote actin disassembly. Conversely, reducing $[Ca^{2+}]_i$ below the physiological range induced spontaneous actin polymerization. The polymerization observed at low $[Ca^{2+}]_i$ superficially resembles that induced by chemotactic factor, but closer inspection revealed some differences. Unlike the fMLP-stimulated polymerization, which is mediated by a pertussis toxin-sensitive G protein,

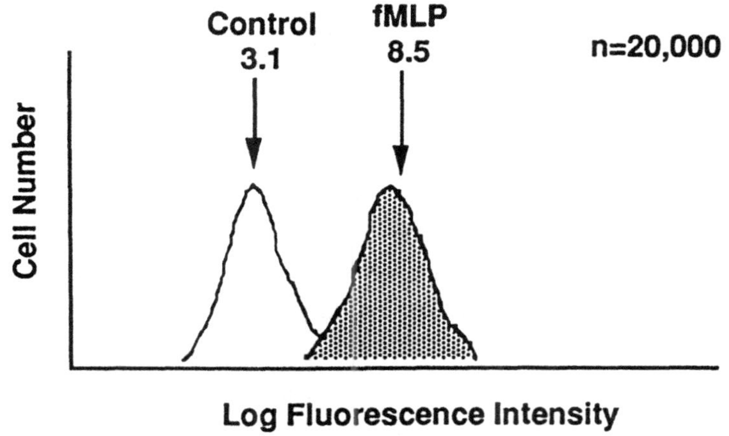

Fig. 2 Increase in F-actin content of permeabilized neutrophils as determined by staining with NBD-phallacidin and analysis of fluorescence by flow cytometry. Each histogram represents 50,000 cells. The ordinate represents cell number. The abscissa indicates the fluorescence intensity and has a log scale. The peak channel numbers for populations of control and fMLP (10^{-8} M) treated cells are indicated.

the polymerization observed at low $[Ca^{2+}]_i$ was not affected by pertussis toxin, nor was it inhibited by guanosine 5'-O-(2-thiodiphosphate) (GDP[S]). More importantly, the effects of fMLP and of reduced $[Ca^{2+}]_i$ were additive, suggesting that they take place through at least partially independent pathways. The precise mechanism underlying the assembly recorded at low $[Ca^{2+}]_i$ remains to be elucidated.

The actin depolymerization noted when $[Ca^{2+}]$ is increased in permeabilized cells is diametrically opposed to the results obtained when $[Ca^{2+}]_i$ is elevated in otherwise intact cells using ionophores. As discussed above, both ionomycin and A23187 promote a somewhat delayed actin assembly. Several experiments were undertaken in an attempt to reconcile these seemingly discrepant observations. Pertussis toxin blocks coupling between the receptor and G proteins, which occurs prior to hydrolysis of phosphoinositides and calcium release. Therefore, the effects of increasing $[Ca^{2+}]_i$ by means of ionophores are not anticipated to be affected by the toxin. Contrary to this expectation, and in accordance with earlier data by Shefcyk et al.(1985), we found that pretreating the cells with pertussis toxin blocked the actin polymerizing effects of both A23187 and ionomycin.

The observed inhibition by pertussis toxin is suggestive of an effect of the calcium ionophores at a stage preceding the activation of GTP-binding proteins, most likely involving surface receptors. It is conceivable that, because of the comparatively high $[Ca^{2+}]$ in the extracellular and/or intravesicular spaces, supraphysiological $[Ca^{2+}]_i$ is attained in the vicinity of the plasma membrane upon treatment with the ionophores. This abnormally high $[Ca^{2+}]_i$ could stimulate phospholipase A_2, a calcium-sensitive enzyme (Matsumoto et al., 1988). The products of phospholipid hydrolysis by phospholipase A_2, e.g. arachidonic acid, or one of their metabolites, such as leukotriene B_4 (LTB_4) or platelet activating factor (PAF), could in principle be responsible for actin assembly through a receptor-mediated (and therefore pertussis toxin-sensitive) process (Naccache, 1987). To analyze the possibility that PAF was signalling actin assembly following addition of ionophores, the cells were treated with two unrelated PAF-antagonists: WEB-2086 and L659,989. These compounds effectively antagonized stimulation of the cells by PAF, but did not attenuate the responses induced by A23187 or ionomycin. Thus, PAF is unlikely to mediate the effects of the ionophores on actin polymerization.

Using a similar approach, we also investigated whether formation of LTB_4 and subsequent interaction with its receptor could account for the polymerization of actin triggered by the ionophores. A specific antagonist of LTB_4, compound LY-223982 (Shappell et al., 1989), was utilized for this purpose. Treatment of the cells with either A23187 or ionomycin in the presence of LY-223982 failed to increase the F actin content of the cells. The inhibitory effect of the LTB_4 antagonist was specific since, under comparable conditions, stimulation of actin polymerization by fMLP was unaffected. Moreover, the antagonist did not interfere with the $[Ca^{2+}]_i$ rise produced by ionomycin. Taken together, these data suggest that the polymerization of actin induced by either A23187 or ionomycin is secondary to an ionophore-induced stimulation of phospholipase A_2 in response to the elevated $[Ca^{2+}]_i$, resulting in arachidonic acid release and its oxidation via the lipoxygenase pathway to LTB_4. According to this model, the leukotriene would then interact with plasma membrane receptors of the same or vicinal cells, stimulating actin assembly via a pertussis toxin-sensitive GTP-binding protein.

In summary, two main lines of evidence suggest that physiological, chemoattractant-induced activation of the respiratory burst and of actin assembly does not require an increase of $[Ca^{2+}]_i$: a) significant oxidative and polymerizing responses to fMLP were recorded in cells pretreated with ionophores in low $[Ca^{2+}]$ medium, under conditions where the chemotactic peptide had little effect on $[Ca^{2+}]_i$ and b) the responses were also preserved in permeabilized cells equilibrated with EGTA-buffered solutions adjusted to free $[Ca^{2+}]$ similar to that found in the cytoplasm of resting (unstimulated) neutrophils. Though the occurrence of small, localized changes in $[Ca^{2+}]_i$ cannot be rigorously ruled out, further evidence argues against this possibility, particularly in the case of actin polymerization. In permeabilized cells, increasing $[Ca^{2+}]_i$ by raising external $[Ca^{2+}]$ in fact resulted in actin depolymerization. The latter effect may be mediated through calcium-sensitive, actin-severing proteins such as gelsolin, which is known to be activated by calcium in the range of concentrations attained in the cytosol of activated neutrophils (Yin et al., 1980). It is possible that, following actin assembly induced by chemoattractants by a mechanism heretofore not understood, the accompanying $[Ca^{2+}]_i$ increase would result in a subsequent disassembly. The ensuing sequential assembly and disassembly of actin filaments could be essential for chemotaxis or phagocytosis. In conclusion, increases of $[Ca^{2+}]_i$ are apparently not essential to initiate neutrophil activation. However, calcium plays a permissive role in the activation of the NADPH oxidase and may be important in reversing stimulus-induced actin polymerization.

ACKNOWLEDGMENTS

This work was supported by the Medical Research Council of Canada, the Ontario Thoracic Society and the National Sanatorium Association. S. Trudel holds a graduate Studentship from the Canadian Cystic Fibrosis Foundation. G.P. Downey is the recipient of a Career Scientist award from the Ontario Ministry of Health and S. Grinstein a Scientist award from the Medical Research Council.

REFERENCES

Campbell, A.K. and M.B. Hallett (1983) Measurement of intracellular calcium ions and oxygen radicals in polymorphonuclear leucocyte-erythrocyte"ghost" hybrids. J. Physiol. 338: 537-550.

Downey, G.P. and S. Grinstein (1989) Receptor-mediated actin assembly in electropermeabilized neutrophils: role of intracellular pH. Biochem. Biophys. Res. Commun. 160: 18-24.

Grinstein, S. and W. Furuya (1988) Receptor-mediated activation of electropermeabilized neutrophils. J. Biol. Chem. 263: 1779-1783.

Hallett, M.B. and A.K. Campbell (1984) Is intracellular calcium the trigger for oxygen radical production by polymorphonuclear leucocytes? Cell Calcium 5: 1-19.

Hallett, M.B., J.P. Luzio and A.K. Campbell (1981) Stimulation of calcium-dependent chemiluminescence in rat leukocytes by polystyrene beads and the non-lytic action of complement. Immunology 44: 509-576.

Korchack, H.M., L.B. Vosshall, K.A. Haines, K.F. Lundquist and G. Weissmann. (1988) Activation of the human neutrophil by calcium-mobilizing ligands. J. Biol. Chem. 263: 11098-11105.

Lew, D. (1989) Receptor signalling and intracellular calcium in neutrophil activation. Eur. J. Clin. Invest. 19: 338-346.

Matsumoto, T., T. Weng and R.I. Sha'afi. (1988) Demonstration of calcium-dependent phospholipase A_2 activity in a membrane preparation of rabbit neutrophils. Biochem. J. 250: 343-348.

Naccache, P.H. (1987) Signals for actin polymerization in neutrophils. Biomed. Pharmacother. 41: 297-304.

Pozzan, T., D.P. Lew, C. B. Wollheim and R.Y. Tsien (1983) Is cytosolic ionized calcium regulating neutrophil activation? Nature 221: 1413-1415.

Sha'afi, R.I. and T.F.P. Molski (1988) Activation of the neutrophil. Prog. Allergy 42:1-64.

Sha'afi, R.I., J. Shefcyk, R. Yassin, T.F.P. Molski, M. Volpi, P.H. Naccache, J.R. White, M.B. Feinstein and E.L. Becker (1986) Is a rise in intracellular free calcium necessary or sufficient for actin polymerization? J. Cell Biol. 102: 1459-1463.

Shappell, S. B., C.W. Smith, A.C. Gasic, A.A. Taylor and J.R. Mitchell (1989) Effects of a specific LTB_4 antagonist on human neutrophil function. FASEB J. 3: A2145.

Shefcyk, J., R. Yassin, M. Volpi, T.F.P. Molski, P.M. Naccache, J.J. Munoz, E.L. Becker, M.B. Feinstein and R.I. Sha'afi (1985) Pertussis toxin but not cholera toxin inhibits the stimulated increase in actin association with the cytoskeleton in rabbit neutrophils. Biochem. Biophys. Res. Commun. 126: 1174-1181.

Smolen, J.E., H.M. Korchack and G. Weissmann (1981) The roles of extracellular and intracellular calcium in lysosomal enzyme release and superoxide anion generation in neutrophils. Biochem. Biophys. Acta 677: 512-520.

Yin, H.L., K.S. Zaner and T. Stossel (1980) Calcium control of actin gelation. Interaction of gelsolin with actin filaments and regulation of actin gelation. J. Biol. Chem. 255: 9494-9500.

THE ROLE OF CALCIUM IN REGULATING THE AGONIST-EVOKED INTRACELLULAR ACIDOSIS IN RABBIT SALIVARY GLAND ACINI

K.R. Lau & R.M. Case
Dept. of Physiological Sciences & Epithelial Membrane Research Centre
University of Manchester
M13 9PT
United Kingdom

Introduction

In isolated acini from rabbit mandibular salivary glands stimulation with acetylcholine (ACh) results in a transient intracellular acidosis of about 0.1 pH units. This intracellular acidosis appears to depend upon the transport of HCO_3^- out of the cell as judged by the following observations: (i) the acidosis is absent if the acini are suspended in HCO_3^--free solution; (ii) it is inhibited if the acini are pre-incubated with the carbonic anhydrase inhibitor acetazolamide; (iii) it is also prevented by Cl^- channel inhibitors such as diphenylamine-2-carboxylic acid (DPC) or 5-nitro-2-(3-phenylpropylamino)-benzoic acid (NPPB) (Lau et al., 1989; Brown et al., 1989). We have also shown that it is possible to evoke an acidosis if HCO_3^- is substituted with a range of weak acid anions such as acetate (Brown et al., 1989). These and other results indicate that stimulation of these acini by ACh causes the activation of a relatively non-selective anion channel through which HCO_3^- ions pass (Lau et al., 1989).

We have also begun to study the intracellular regulation of this activation process and describe here some of our more recent findings.

Methods

Acini are isolated from whole mandibular glands by digestion of the minced glands in a modified Krebs' buffer containing collagenase (60 units/ml) and hyaluronidase (1 mg/ml) for 60 min. at 37°C. After the digestion, acini are released by passing the gland pieces through pipettes of decreasing tip diameter. The released acini are then separated

NATO ASI Series, Vol. H 48
Calcium Transport and
Intracellular Calcium Homeostasis
Edited by D. Pansu and F. Bronner

from tissue debris by centrifuging the mixture through solutions containing bovine serum albumin at concentrations ranging from 4% to 1% and collecting the pellets. The pellets are then resuspended in a HCO_3^- containing buffer and loaded with fluorescent dye: either with BCECF or fura 2. Loading is achieved by incubating the acinar suspension (20 min. and 30 min. respectively at 37°C) with the esters of the dyes (0.5 uM and 5 uM final concentrations respectively). After the loading period the acini are washed free of remaining extracellular dye using fresh buffer.

BCECF fluorescence is used to monitor intracellular pH (pH_i) as an indicator of HCO_3^- activity. Fura 2 fluorescence is used to monitor intracellular calcium activity ($Ca^{2+}{}_i$). The experiments are carried out in cuvettes using a Perkin-Elmer LS-3 fluorimeter. Full details of the methods used are published elsewhere (Lau et al., 1989; Brown et al., 1989).

Results

The role of Ca^{2+}

Fig. 1 illustrates typical changes in pH_i and $Ca^{2+}{}_i$ in response to ACh stimulation. A rapid, transient fall in pH_i is followed by a slow recovery. As reported previously (Lau et al., 1989), the acidosis is about 0.1 pH units and the time taken to recover by half of the initial acidosis is about 2.5 min. At the same time as the acidosis (within the resolution of these experiments), there is a rapid, transient rise in $Ca^{2+}{}_i$ followed by a partial recovery towards the resting value. The partial recovery appears to be a true plateau that is maintained for as long as stimulation is continued. The subsequent blockade of receptors with atropine causes both pH_i and $Ca^{2+}{}_i$ to return to their resting values. Thus, in summary, activation of muscarinic cholinoceptors in these acini leads to mobilisation of $Ca^{2+}{}_i$ and anion channel activation.

Is the mobilisation of $Ca^{2+}{}_i$ a necessary event in anion channel activation? This question is clearly answered in the experiment illustrated in Fig. 2. When acini are suspended in a nominally Ca^{2+}-free solution, addition of ACh does not evoke

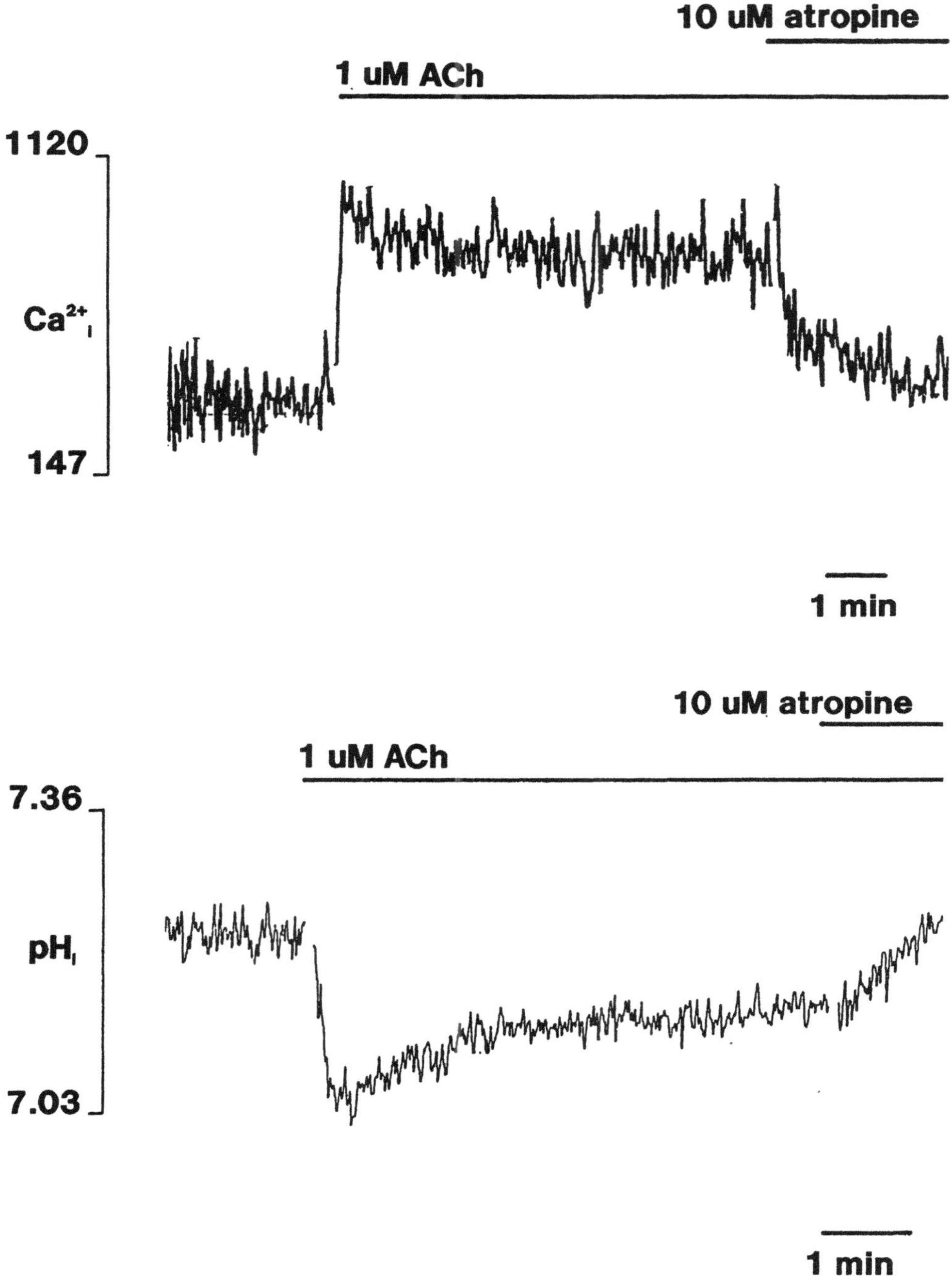

Fig. 1 The effect of ACh followed by atropine on intracellular Ca^{2+} (upper trace) and intracellular pH (lower trace).

the usual acidosis until Ca^{2+} is added to the cuvette. The acidosis is clearly dependent upon the availability of $Ca^{2+}{}_i$.

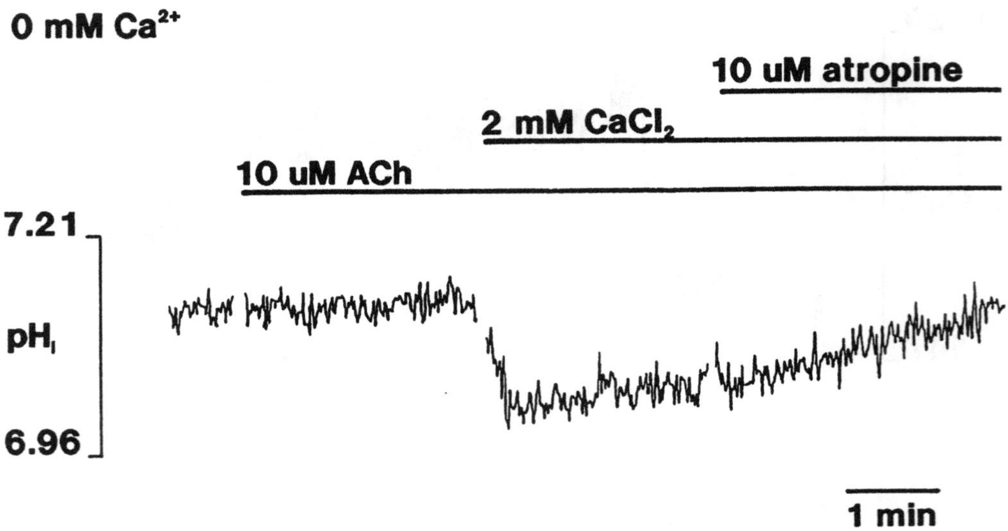

Fig. 2 The effect of ACh followed by $CaCl_2$ on intracellular pH in acini suspended in Ca^{2+}-free solution.

In the experiment illustrated in Fig. 3, receptor activation is by-passed by using the ionophore ionomycin. At the concentrations shown, ionomycin causes a rise in $Ca^{2+}{}_i$ similar to that evoked by ACh. However, ionomycin does not cause any change in pH_i. If ACh is subsequently added, then the acidosis is evoked. These data indicate that although Ca^{2+} is necessary for the acidosis it is not sufficient. Other factors must also be present.

The role of diacylglycerols

Cholinoceptor activation leads not only to mobilisation of $Ca^{2+}{}_i$ through the action of inositol polyphosphates but also to activation of protein kinase C through diacyl glycerol release. However, simultaneous addition of ionomycin with one of a number of synthetic diacyl glycerols (1,2-dioctanoyl-sn-glycerol; 1-stearoyl-2-arachidonoyl-sn-glycerol; and 1-oleoyl-2-acetyl-sn-glycerol; all at 100 uM) or with a phorbol ester

(1 uM) does not reproduce the acidosis observed with ACh. Furthermore, pre-incubation of acini with the protein kinase C inhibitor H-7 (50 uM) for up to an hour does not affect the ACh-induced acidosis.

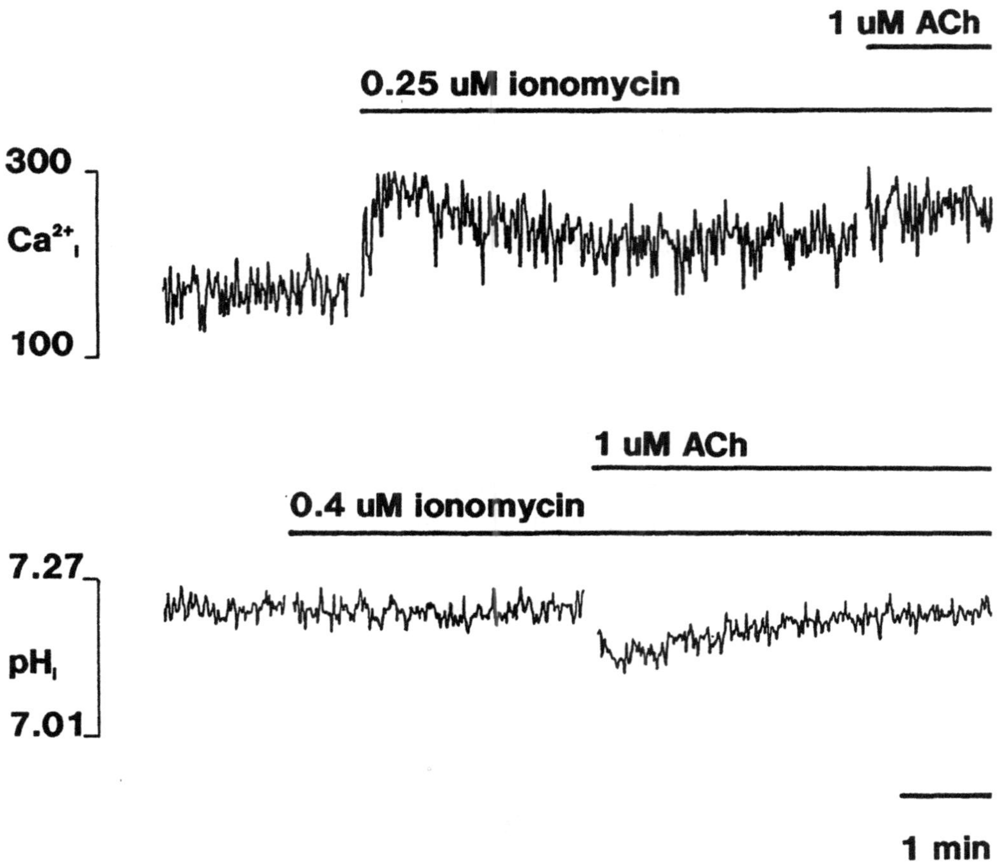

Fig. 3 The effect of ionomycin followed by ACh on intracellular Ca^{2+} (upper trace) and on intracellular pH (lower trace).

Thus, it seems that protein kinase C activation is not an major factor in eliciting the acidosis.

The role of cyclic nucleotides

In many cell types it has been reported that there is

cross-talk between the cyclic nucleotide signalling pathway and the Ca^{2+} signalling pathway (e.g. Warhurst et al., 1988). We have examined such a possible interaction by pre-incubating acini with the dibutyryl derivatives of cyclic AMP or cyclic GMP (0.1 mM) together with theophylline (0.1 mM). In such acini, ionomycin is still unable to evoke an acidosis (i.e. a rise in cyclic nucleotide concentration has no stimulatory effect), while the acidosis induced by ACh is unaffected (i.e. a fall in cyclic nucleotide concentration is not necessary for channel activation). Thus, cyclic nucleotides appear to play no role in mediating the receptor response. This confirms previous observations we have made in the isolated, perfused mandibular gland preparation which show that forskolin has no effect on fluid secretion (Case et al., 1988).

The role of archidonic acid

In our most recent experiments we have studied the potential role of arachidonic acid as an intracellular messenger in these acini. It is known that arachidonic acid metabolism is important as an intracellular signalling pathway in secretory cells such as mast cells, and it is known that there is stimulation of this pathway by some cholinoceptor sub-types (Bonner, 1989).

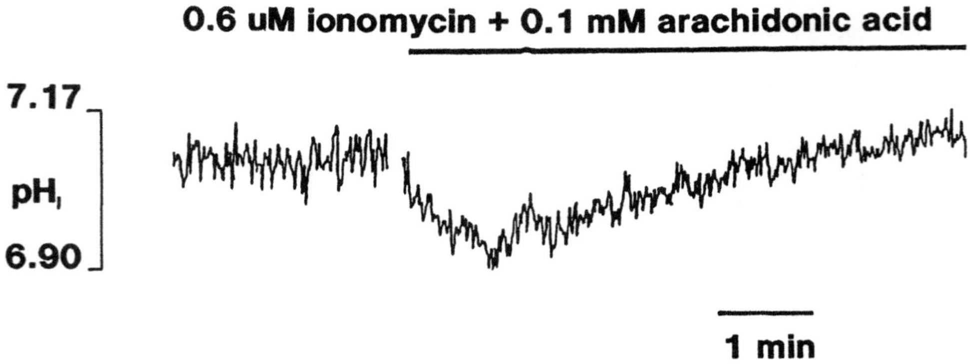

Fig. 4 The effect of arachidonic acid and ionomycin on intracellular pH.

Fig. 4 illustrates the effect of adding 100 uM arachidonic acid simultaneously with 0.6 uM ionomycin. A clear acidosis followed by a recovery is observed. The size of the acidosis is comparable (though not identical) to that observed with ACh (0.07 pH units compared with 0.09 pH units for ACh). 100 uM arachidonic acid alone does not have any marked effect on pH_i and so it seems that activation of the arachidonic acid pathway is a further pre-requisite for channel activation in these acini. However, neither the lipoxygenase inhibitor nordihydroguaiaretic acid, nor the cyclooxygenase inhibitor indomethacin have any effect on this arachidonic acid-induced acidosis, suggesting that the effects of arachidonic acid are not brought about through the synthesis of leukotrienes or prostaglandins. Instead, the data suggest that it is arachidonic acid itself that is important as a mediating factor in channel activation in these acini.

Conclusion

Our work, so far, clearly indicates that $Ca^{2+}{}_i$ mobilisation alone is an insufficient mediator of channel activation in these acini. The more obvious co-factors that could be involved (i.e. diacyl glycerols and cylic nucleotides) appear to be ruled out by our data.

We have not yet studied the possibility that it is the shape of the initial $Ca^{2+}{}_i$ response that is important in the signalling process. Studies using fluorescence microscopy show that there is a very rapid initial `spike' of $Ca^{2+}{}_i$ which we are unable to reproduce using ionomycin (Berrie & Elliott, 1990). This `spike' may be important in the signalling process.

We are able to mimic partially the receptor activated response by a combination of ionomycin and arachidonic acid. Exactly what role arachidonic acid plays is unclear, but it does not appear to involve synthesis of leukotrienes or prostaglandins so that arachidonic acid itself may be the important co-factor required for channel activation. If so, then this would appear to be an unusual example of intracellular signalling.

References

Berrie CP & Elliott AC (1990) unpublished observations.

Bonner TI, (1989) The molecular basis of muscarinic receptor diversity. Trends Neurosci 12:148-151

Brown PD, Elliott AC & Lau KR (1989) Indirect evidence for the presence of non-specific anion channels in rabbit mandibular salivary gland acinar cells. J Physiol (Lond) 414:415-431

Case RM, Howorth AJ & Padfield PJ (1988) The influence of acetylcholine, isoprenaline and forskolin on the secretion of fluid, electrolytes and individual proteins by rabbit mandibular salivary gland. J Physiol (Lond) 406:411-430

Lau KR, Elliott AC & Brown PD (1989) Acetylcholine-induced intracellular acidosis in rabbit salivary gland acinar cells. Am J Physiol 256:C288-C295

Warhurst G, Higgs NB, Lees M, Tonge A & Turnberg LA (1988) C-kinase modulation of cAMP production and electrical responses to prostaglandin E_2 in a colonic epithelial cell line. Am J Physiol 255:G27-G32

FREE CALCIUM IN RED BLOOD CELLS OF HUMAN HYPERTENSIVES IS ELEVATED: HOW CAN THIS BE?

Frank F. Vincenzi, Thomas R. Hinds and Armando Lindner
Departments Pharmacology and Medicine and VA Medical Center
University of Washington
Seattle, WA 98195
USA

INTRODUCTION

Abnormal regulation of calcium transport has been proposed as one mechanism by which increased levels of intracellular free Ca^{2+} may occur and could be a factor in essential hypertension (Bohr and Webb, 1984). Since vascular smooth muscle cells from human subjects are not readily available for studies of ion transport, circulating blood cells have frequently been used as a model for such measurements in hypertension (Erne et al., 1984a; Lindner et al., 1987; Postnov et al., 1977). While several abnormal Na^{+} transport pathways have been described in human red blood cells (RBCs) in hypertension (Postnov et al., 1977; Diez et al., 1987), little is known concerning intracellular free Ca^{2+} in these cells.

Because of the substantial concentration of hemoglobin in RBCs, and the resulting interferences, intracellular fluorescent Ca^{2+} probes have not been widely applied in RBCs. Very recently, one group of investigators reported on the use of fura-2 for measurements of Ca^{2+} in human RBCs (David-Dufilho et al., 1988). However, the characteristic absorption and emission wavelengths of fura-2 do not lend themselves to accurate measurement in the presence of a substantial concentration of hemoglobin. The present study was undertaken to investigate the application of a the fluorescent Ca^{2+} probe, fluo-3 (Minta et al., 1989), to the measurement of intracellular free Ca^{2+} in RBCs of normal and hypertensive humans. Fluo-3 has spectral characteristics (excitation/emission at longer wavelengths) which are more desirable than those of fura-2 for measurements of Ca^{2+} in the presence of a large concentration of hemoglobin. We recently presented preliminary data on the feasibility of using fluo-3 to measure free intracellular Ca^{2+} in RBCs of normal humans (Hinds et al., 1989).

NATO ASI Series, Vol. H 48
Calcium Transport and
Intracellular Calcium Homeostasis
Edited by D. Pansu and F. Bronner

METHODS AND MATERIALS

Fresh blood was collected from healthy, normotensive human volunteers, and from patients with essential hypertension. Blood was drawn in evacuated tubes containing sodium citrate, citric acid and glucose, or heparin coated tubes. The blood was centrifuged at 700 x g for 10 min and the plasma and buffy coat were carefully removed . The packed cells were washed 3 more times with cold saline at 700 x g for 5 min. Five µl of packed RBCs were suspended in 10 ml of RBC-HEPES which contained: NaCl, 145 mM; KCl, 5 mM; HEPES, pH 7.4, 10 mM; glucose, 5 mM; Na_2HPO_4, 0.5 mM; $MgSO_4$, 1 mM; and $CaCl_2$, 1 mM. This resulted in approximately 5 million cells/ml.

Details of the method will be presented elsewhere. The general features of the method will be outlined briefly here. RBCs were loaded with the acetoxymethyl ester of fluo-3 (fluo-3 AM). Fluo-3-AM (100 µM in dry DMSO) was titrated into the stirred cell suspension over 30 minutes, to a final concentration of 1 µM at 37 °C in the dark, and incubated for an additional 30 min in the dark at 37 °C. The loaded cells were centrifuged in a conical 15 ml centrifuge tube for 10 minutes at 500 x g, through a 1.0 ml pad of 5% bovine serum albumin (BSA) in RBC-HEPES without Ca^{2+}. This step separated unincorporated fluorescent probe from the loaded cells and left the cells pelleted under the BSA pad with the loading medium above. The loading medium and the BSA pad were carefully removed then the RBC pellet was resuspended in 2 ml of the RBC-HEPES and kept on ice. The final concentration of cells was approximately 24 million cells/ml; which was a reflection of a 4% loss of cells. For fluorescence measurements, 100 µl of the cell suspension was added to 1.4 ml of RBC-HEPES in a stirred, quartz cuvette, thermostatted to 37 °C. The final concentration of packed RBCs was 0.25 µl/1.5 ml. RBC volume was measured using standard techniques in a Coulter Counter Model Z_{BI} with a channel analyzer. Intracellular free Ca^{2+} was measured by fluorescence at single excitation and emission wavelengths.

Fluorescence was quantified with a Shimadzu RF 5000U spectrofluorophotometer. Excitation wavelength was 506 nm and emission wavelength was 526 nm, with respective slit widths of 5 and 10 nm. Under these conditions there was no measurable light scattering or autofluorescence produced by unloaded cells. Following a stable signal of fluo-3 fluorescence in loaded cells (F), the fluorescence maximum in RBCs was estimated by the addition of 3.8 µM of the ionophore, A23187 (Fmax). The total fluorescence signal was measured after cell membranes were permeabilized by addition of digitonin to a concentration of 33 µM. Minimum fluorescence was obtained by the addition of EGTA at 6.7 mM and then the subsequent addition of $CoCl_2$ at 20 mM which reduced the signal to Fmin, which was defined as zero. Correction for any extracellular fluo-3 was performed for each sample. Fluorescence from loaded RBCs was measured by adding 200 µl (2 x the usual concentration) of the cell suspension to the cuvette. Then, EGTA was added to a final concentration of 6.7 mM. The decrease in signal was assumed to be due to removal of Ca^{2+} from extracellular fluo-3. Finally, addition of $CoCl_2$ (20 mM, final) reduced the signal to zero. The fraction of the fluorescence signal due to extracellular fluo-3 was:

$$\text{fraction outside fluorescence} = \frac{(E)}{(T)}$$

where E = drop in signal with EGTA and T = difference between the fluorescence of the cells and that of the Fmin after $CoCl_2$ addition.

Measured fluorescence signals were corrected as:

$$F^* = F\,\frac{(T\text{-}E)}{(T)}$$

Intracellular free Ca^{2+} was calculated as:

$$Ca^{2+} = Kd\,\frac{(F^*)}{(F^*max - F^*)}$$

where: * denotes the corrected fluorescence signal and Kd = 450 nm.

Fresh blood samples were obtained from a total of 12 fasting subjects. Hypertension was diagnosed when multiple readings on more than two visits averaged 150/90 mm Hg or higher, when sitting after a 5 minute rest. An average of three blood pressure readings, one minute apart, was used. Systolic, diastolic (measured by Korotkoff 5), and mean BP (diastolic plus 1/3 the pulse pressure) were monitored. Group A (n = 7) included normotensive volunteers. These included 4 men and 3 women, whose age averaged 42 ± 10 (S.D.) years. Group B (n = 5) included untreated hypertensive patients. These included 4 men and 1 woman, aged 62 ± 12 (S.D.) years with a diagnosis of essential hypertension. These patients had not received any antihypertensive drugs for a minimum of two weeks before testing, and some had never been treated.

A model of the RBC Ca pump/leak system was created using STELLA (High Performance Systems, Inc., Lyme, NH), operating on a Macintosh Plus microcomputer. It was assumed that all influx of Ca^{2+} into the RBC is passive and directly proportional to the concentration gradient and the permeability. It was further assumed that all of the efflux of Ca^{2+} from the RBC is active, is mediated by the Ca pump ATPase and is the sum of that mediated by the basal and CaM activated forms of the enzyme. The model was based on published values for the passive permeability (Ferreira and Lew, 1977) and Ca pump kinetics (Niggli et al., 1981) of the normal human RBC. It was assumed that the Ca pump ATPase exists in two states: a basal (CaM free) state, and a CaM activated state. It was further assumed that the binding to (and activation of) the Ca pump ATPase by CaM is reversible and dependent on the concentration of free intracellular Ca^{2+} and shows positive cooperativity, as observed for the Ca pump ATPase (Vincenzi and Larsen, 1980) as well as other CaM activated enzymes (Dedman et al., 1977). The cooperativity of CaM activation was assumed to be 4 (Dedman et al., 1977). The cooperativity of Ca binding to the Ca pump ATPase, either in the basal of CaM activated state was assumed to be 1.5 (Vincenzi and Larsen, 1980; Niggli et al., 1981). The Kd Ca for the basal state was assumed to be 14 μM; that of the CaM activated state was assumed to be 1 μM (Niggli et al., 1981). The respective Vmax values for the two states of the Ca pump ATPase were assumed to be 100 and 300 μmol/min/liter (Niggli et al., 1981; Vincenzi, 1990). Steady state values are presented in all of our results.

RESULTS

Application of the fluo-3 method for the determination of intracellular free Ca^{2+} in RBCs of normotensive humans resulted in a value of intracellular free Ca^{2+} in RBCs of normotensive humans of 195 nM (± 31 nM, S.E.M., n = 7). By contrast , as shown in Fig. 1, the average value found in the current study in RBCs of hypertensive humans was 367 nM (± 50 nM, S.E. M., n = 5). The difference between the mean values was significant ($P < 0.01$). Although there was a significant difference between the average age of the normotensive and hypertensive subjects, there was no apparent relationship between age and intracellular free Ca^{2+} (not shown). On the other hand, intracellular free Ca^{2+} and mean arterial pressure were positively correlated (not shown).

The model of the Ca pump/leak system of the human RBC predicted a steady state in vitro level of free intracellular Ca^{2+} of 180 nM. The level of intracellular Ca^{2+} predicted by the model was achieved at a steady state transport rate of 0.167 μmole/min/liter. This is within with the range of resting influx values for Ca^{2+} in normal human RBCs in vitro (Vincenzi, 1990). Of the total efflux, 87% was mediated by the basal and 13% was mediated by the CaM activated Ca pump ATPase. This ratio of effluxes was predicted even while only 0.1% of the total Ca pump

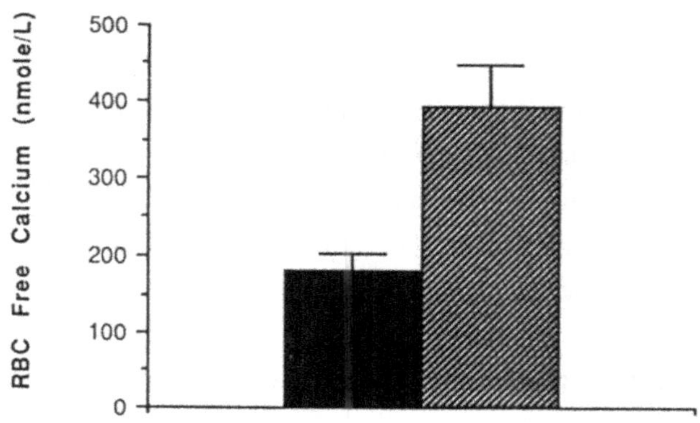

Fig. 1. Intracellular free calcium in human RBCs. Columns indicate the levels of free calcium (± S.E.M.) in RBCs from 7 normotensive subjects (solid) and 5 hypertensive patients (cross hatched). The difference between the means was statistically significant ($P < 0.01$). Intracellular free Ca^{2+} was determined as described in Methods and Materials.

ATPase would be in the CaM activated form under this condition. Thus, while it is qualitatively more or less correct to assume that the Ca pump 'shuts off' at very low intracellular Ca^{2+} by dissociation of CaM from the ATPase, a significant fraction of the actual transport is nevertheless maintained by the presence of CaM. We suggest that this observation may be of significance to interpretation of results obtained in RBCs from humans with certain diseases.

Perhaps the most important perspective is that the rate of the Ca pump ATPase 'at rest' (i.e., in vitro) is only 0.06% of the CaM activated Vmax. This re-emphasizes a point we have made before, i.e., that the Ca pump ATPase capacity far exceeds the apparent needs of the normal human RBC (Vincenzi and Hinds, 1980; Vincenzi, 1990), at least as it is observed it in vitro. Is it suggested that such 'overdesign' is important to regulation of intracellular free Ca^{2+} within narrow limits and is probably important to rapid recovery from transient perturbations.

We attempted to create a model based upon known chemical properties of the system, rather than by curve fitting. This allowed variations in assumptions of such familiar variables as Vmax, Kd, etc. Our model does bear functional similarity (reversible CaM binding/activation) to that created by Scharff et al. (1983) who used a fitting approach. In the current study we did not specifically model the kinetics of the CaM activation/deactivation process and so we can not comment on possible hysteretic behavior of the system as emphasized by Scharff et al. (1983). However, it is clear that the system, when subjected to transient changes in the load of intracellular free Ca^{2+} does exhibit some hysteretic behavior as it adjusts to new steady states. It has been recently pointed out that the more rapid kinetics of fluo-3 compared to fura-2 make it potentially more useful for studying oscillations of free intracellular Ca^{2+} (Eberhard and Erne, 1989). It remains to be determined whether fluo-3 can be used to follow Ca^{2+} transients in RBCs (and whether our model may fit such data).

Some other features of the model system are also instructive and may help one to understand what might account for the elevated intracellular free Ca^{2+} in human RBCs in hypertension. Thus, the predicted intracellular free Ca^{2+} was rather insensitive to the value of passive permeability. Over a range of passive permeabilities of greater than 1000 fold,the predicted intracellular free Ca^{2+} was maintained between 100 and 1000 nM. This is an outstanding expression of the system to 'regulate' intracellular Ca^{2+}. Thus, it required the assumption of an increase of passive permeability of approximately 15-fold to increase the predicted intracellular free Ca^{2+} from 180 to 360 nM. Likewise, the predicted intracellular free Ca^{2+} was rather insensitive to changes in the Vmax of the Ca pump. In order to increase the predicted level of intracellular Ca^{2+} from 180 to 360 nM, Vmax would have to change by an amount corresponding to over 90% inhibition of the enzyme. If a change in the Kd Ca of the basal and CaM activated Ca pump ATPase forms is assumed, then a 7-fold increase in Kd Ca would result in doubling of predicted intracellular Ca^{2+}. This level of change in Kd Ca should be readily detectable in an appropriate assay system.

DISCUSSION

In the current study we found that intracellular free Ca^{2+} in normal human RBCs was 195 nM. Our value is somewhat higher than the 78 nM value reported by David-Dufilho et al. (1988) who used fura-2. We found that intracellular free Ca^{2+} in RBCs from hypertensives was 367 nM. The present results, although obviously preliminary because they are based on a small number of subjects, are consistent with a number of studies which have suggested that some underlying abnormality of plasma membrane Ca transport may exist in essential hypertension. While it may seem easy to accept that elevated intracellular free Ca^{2+} exists in RBCs, as it does in platelets for example (Erne et al., 1984a; Erne et al., 1984b), the mechanism is not obvious.

There is consensus from a number of laboratories that human RBCs do not exhibit Na/Ca exchange (Ferreira and Lew, 1977). Therefore, the hypothesis of Blaustein and Hamlyn

(Blaustein, 1977; Blaustein, Hamlyn, 1984) can not account for the present data. Of course, the consensus on Na/Ca exchange in human RBCs is based on normal RBCs. One could speculate that RBCs of hypertensive humans exhibit Na/Ca exchange while those of normal human do not. If so, one would have to further speculate that this exchanger places a sufficient load on the Ca pump to double intracellular free Ca^{2+}. Based on modeling, such a load of inward Ca flux would have to be equivalent to a 15-fold increase in membrane permeability over normal. Thus, increased inward passive permeability, whether mediated by exchange or by 'leak' seems unlikely as a basis for the increased intracellular free Ca^{2+} associated with human essential hypertension. For purposes of simplification, alterations in modeled variables (such as passive permeability) were assumed to occur independently. It is likely that alterations in more than one variable may occur in a given patient, and that different kinds of alterations may define subgroups of what is currently lumped as 'essential hypertension'.

In earlier studies we concluded that there was no inherent defect in the Ca pump ATPase of RBCs from hypertensive patients (Vincenzi et al., 1988). However, it must be admitted that we, like many others doing similar analyses, measured the enzymatic activities in question under more or less optimal (Vmax) conditions. Thus, it would be more correct to conclude that we found no evidence for an inherent defect in the Vmax of the Ca pump ATPase of RBCs from hypertensive patients. In modeling the Ca pump/leak system of the human RBC, it was found that substantial changes in the Vmax of the enzyme would be needed to predict a doubling of intracellular free Ca^{2+}. By inference, these do not exist. By contrast, a modest increase in the Km Ca of the Ca pump ATPase would produce a doubling of intracellular Ca^{2+}. Increased Km Na of the sodium pump has been reported in RBCs from hypertensive patients (Diez et al., 1987). Increased Km Ca of the Ca pump has not been rigorously eliminated. Therefore, we suggest that essential hypertension is associated with increased Km Ca of the plasma membrane Ca pump of the RBC membrane with or without a change in Vmax. If increased Km Ca were a generalized phenomenon, then it could be a causal influence of increased intracellular free Ca^{2+} in vascular smooth muscle and, thus, of increased vascular resistance.

The mechanism of the hypothetical change in Km Ca remains to be elucidated. Whatever its basis, it is apparently not readily reversed upon removal of RBCs from the in vivo environment. The present results can not rule out some irreversible effect of the hypertensive state or elevated shear stress on the regulation of the Ca pump. It is anticipated that, if such an irreversible change does occur, then the RBC offers a useful cell in which to determine its mechanism.

It is conceivable that elevated intracellular free Ca^{2+} in RBCs (or platelets, (Erne et al., 1984a; Erne et al., 1984b)) is a result of, and not an expression of, the cause of hypertension. As suggested by Larsen et al. (1981) and by Vincenzi and Cambareri (1985), shear stress in vivo may cause increased passive permeability to Ca. And shear stress may be elevated by elevated blood pressure. However, effects of shear stress occurring in vivo would be rapidly reversed in vitro following removal of blood from the body. Even if intracellular free Ca^{2+} were abnormally elevated in the hypertensive circulation, the relatively high capacity Ca pump would be expected to rapidly return intracellular free Ca^{2+} to 'normal' in the face of low shear stress outside the body; unless the regulation of the pump were changed in some irreversible, or slowly reversible manner. Thus, although increased shear stress may occur in vivo in hypertension, it would not be expected to exert a significant effect on the steady state intracellular free Ca^{2+} in RBCs which have been maintained in vitro for over an hour. It may be that chronically elevated intracellular Ca^{2+} in vivo causes some irreversible change in RBC Ca pump ATPase regulation; via the action of a Ca^{2+} activated protease, for example. Calpain seems a likely candidate because of its presence and sensitivity. However, calpain initially results in increased activity of the Ca pump ATPase (Wang et al., 1988a; Wang et al., 1988b; Wang et al., 1989) so a relationship to the current data is not readily apparent .

We consider the hypothesis of Blaustein and Hamlyn (1984) to be a very attractive explanation for the increased intracellular free Ca^{2+} which is widely believed to play a role in essential hypertension. However, we believe that the Na/Ca exchange mechanism can not account for the current data. We suggest that elevated intracellular free Ca^{2+} in RBCs associated with hypertension is indicative a generalized alteration in regulation of the plasma membrane Ca pump. We further suggest that such an alteration probably involves an increase in the Km Ca of the pump. The alteration in the regulation of the plasma membrane Ca pump is sufficiently long lasting so as to be detected after at least an hour of in vitro incubation of RBCs. The alteration may be inherited or acquired, the latter via 'damage' or via some circulating 'hypertensive factor(s)' (Lindner et al., 1987); at least some of which have been characterized as digitalis-like (Abbott, 1988). However, it has long been known that the plasma membrane Ca pump ATPase is not inhibited by compounds such as ouabain (Schatzmann and Vincenzi, 1969) and it was recently reported that one endogenous digitalis-like factor exerts no effect on the plasma membrane Ca pump ATPase (Hamlyn et al., 1989).

SUMMARY AND CONCLUSIONS

Intracellular free Ca^{2+} in RBCs in vitro was found to be 195 nM in RBCs from 7 normal human subjects and 367 nM in RBCs from 5 hypertensive human subjects. Because human RBCs lack a Na/Ca exchange mechanism, altered Na/Ca exchange can not account for these data. Based on modeling, it is suggested that intracellular free Ca^{2+} in RBCs is caused by an alteration in the regulation of the Ca pump which has the effect of reducing the sensitivity of the pump to intracellular Ca^{2+} (increased Km Ca) with or without a change in the maximum capacity of the pump. Whatever, the mechanism of altered Ca pump regulation, its influence on the Ca pump does not appear to be readily reversed in vitro. The results are compatible with the interpretation that a generalized alteration in plasma membrane Ca transport exists in human hypertension. The results are also compatible with, but do not prove, that the alteration is caused by one or more circulating 'hypertensive factors' (Lindner et al., 1987; Hamlyn et al., 1985).

ACKNOWLEDGEMENTS

This work was supported in part by research funds from the Veteran's Administration. Ms. Carol Carrillo provided excellent technical assistance.

REFERENCES

Abbott A (1988) Interrelationship between Na^+ and Ca^{2+} metabolism in hypertension. TIPS 9:111-113

Blaustein MP (1977) Sodium ions, calcium ions, blood pressure regulation and hypertension: a reassessment and a hypothesis. Am J Physiol 232:C165-C173

Blaustein MP, Hamlyn JM (1984) Sodium transport inhibition, cell calcium, and hypertension. The natriuretic hormone/Na^+-$Ca2^+$ exchange/hypertension hypothesis. Am J Med 77:45-59

Bohr DF, Webb RC (1984) Vascular smooth muscle function and its changes in hypertension. Am J Med 77 Suppl 4A:3-16

David-Dufilho M, Montenay-Garestier T, Devynck M-A (1988) Fluorescence measurements of free Ca^{2+} concentration in human erythrocytes using the Ca^{2+}-indicator fura-2. Cell Calcium 9:167-179

Dedman JR, Potter JD, Jackson RL, Johnson JD, Means AR (1977) Physicochemical properties of rat testis Ca^{2+}-dependent regulator protein of cyclic nucleotide phosphodiesterase. Relationship of Ca^{2+}-binding, conformational changes and phosphodiesterase activity. J Biol Chem 252:8415-8422

Diez J, Hannaert P, Garay RP (1987) Kinetic study of Na^+-K^+ pump in erythrocytes from essential hypertensive patients. Am J Physiol 252:H1-H6

Eberhard M, Erne P (1989) Kinetics of calcium binding to fluo-3 determined by stopped-flow fluorescence. Biochem Biophys Res Commun 163:309-314

Erne P, Bolli P, Bürgisser E, Bühler F (1984a) Correlation of platelet calcium with blood pressure. N Engl J Med 310:1084-1088

Erne P, Bürgisser E, Bolli P, Ji B-H, Bühler F (1984b) Free calcium concentration in platelets closely relates to blood pressure in normal and essentially hypertensive subjects. Hypertension 6 (Suppl I):I-166-I-169

Ferreira HG, Lew VL (1977) Passive Ca transport and cytoplasmic buffering in intact red cells. In Membrane Transport in Red Cells. Ellory JC, Lew VL (eds) Academic Press New York, pp. 53-91

Hamlyn JM, Levinson PD, Ringel R, Levin PA, Hamilton BP, Blaustein MP, Kowarski AA (1985) Relationships among endogenous digitalis-like factors in essential hypertension. Fed Proc 44:2782-2788

Hamlyn JM, Harris DW, Ludens JH (1989) Digitalis-like activity in human plasma. Purification, affinity and mechanism. J Biol Chem 264:7395-7404

Hinds TR, Lindner A, Vincenzi FF (1989) Estimation of intracellular free calcium in red blood cells using the fluorescent probe FLUO-3. J Cell Biol 107:75a (Abstract)

Larsen FL, Katz S, Roufogalis BD, Brooks DE (1981) Physiological shear stresses enhance the Ca^{2+} permeability of human erythrocytes. Nature 294:667-668

Lindner A, Kenny M, Meacham AJ (1987) Effects of a circulating factor in patients with essential hypertension on intracellular free calcium in normal platelets. New Engl J Med 316:509-513

Minta A, Kao JPY, Tsien RY (1989) Fluorescent indicators for cytosolic calcium based on rhodamine and fluorescein chromophores. J Biol Chem 264:8171-8178

Niggli V, Adunyah ES, Penniston JT, Carafoli E (1981) Purified (Ca^{2+}-Mg^{2+})-ATPase of the erythrocyte membrane. Reconstitution and effect of calmodulin and phospholipids. J Biol Chem 256:395-401

Postnov YV, Orlov SN, Shevchenko A, Adler AM (1977) Altered sodium permeability, calcium binding, and Na, K-ATPase activity in the red cell membrane in essential hypertension. Pflugers Arch 371:263-269

Scharff O, Foder B, Skibsted U (1983) Hysteretic activation of the Ca^{2+} pump revealed by calcium transients in human red cells. Biochim Biophys Acta 730:295-305

Schatzmann HJ, Vincenzi FF (1969) Calcium movements across the membrane of human red cells. J Physiol (Lond) 201:369-385

Vincenzi FF, Cambareri JJ (1985) Apparent ionophoric effects of red blood cell deformation. In Cellular and Molecular Aspects of Aging: The Red Cell as a Model. Eaton JW, Konzen DK, White JG (eds) Alan R. Liss New York, pp. 213-222

Vincenzi FF (1990) Regulation of the plasma membrane Ca^{2+}-pump. In The Red Cell Membrane. Raess BU, Tunnicliff G (eds) Humana Press Clifton, NJ, pp. 123-142

Vincenzi FF, DiJulio D, Morris CD, McCarron D (1988) Measurements on the activity of the plasma membrane Ca pump ATPase in human hypertension. In Cellular Calcium and Phosphate Transport in Health and Disease. Bronner F, Peterlik M (eds) Alan R. Liss New York, pp. 379-383

Vincenzi FF, Hinds TR (1980) Calmodulin and plasma membrane calcium transport. In Calcium and Cell Function, Vol. I. Cheung WY (ed) Academic Press New York, pp. 127-165

Vincenzi FF, Larsen FL (1980) The plasma membrane calcium pump: Regulation by a soluble Ca^{2+} binding protein. Fed Proc 39:2427-2431

Wang KKW, Villalobo A, Roufogalis BD (1988a) Activation of the Ca^{2+}-ATPase of human erythrocyte membrane by an endogenous Ca^{2+}-dependent neutral protease. Arch Biochem Biophys 260:696-704

Wang KKW, Roufogalis BD, Villalobo A (1988b) Further characterization of calpain-mediated proteolysis of the human erythrocyte plasma membrane Ca^{2+}-ATPase. Arch Biochem Biophys 267:317-327

Wang KKW, Roufogalis BD, Villalobo A (1989) Calpain I activates Ca^{2+} transport by the reconstituted erythrocyte Ca^{2+} pump. J Membr Biol 112:233-245

SECTION V

EPITHELIAL CALCIUM TRANSPORT

TRANSEPITHELIAL CALCIUM TRANSPORT IN GUT AND KIDNEY

Felix Bronner
Department of BioStructure and Function
University of Connecticut Health Center
Farmington, CT 06032, USA

INTRODUCTION

From the earliest times, cells had to struggle to keep calcium out in order to prevent the precipitation of phosphate as calcium phosphate. Utilization of phosphate bond energy and of phosphorylation reactions had been an early metabolic decision by the ur-cells from which life evolved (26). Three general strategies to minimize the rise in cytosolic calcium are to restrict entry, to develop ways to extrude the ion from the cell, and to store it by binding. Indeed all three of these strategies have been used--calcium channels restrict calcium entry, calcium pumps extrude calcium, and organelles bind and store calcium.

When organisms left the ocean or remained behind as the waters receded, they had to develop external or internal skeletons to sustain themselves. Moreover, they no longer lived in a homogeneous environment, nor in one with a constant concentration of calcium. They developed an internal fluid environment in which the calcium concentration was constant and closely regulated. They evolved mechanisms for ingesting calcium-containing foods, extracting the ion from those foods and ways of transporting calcium across the intestinal epithelia. They also evolved ways of extracting and returning calcium from the body fluids as these circulated and were renewed.

NATO ASI Series, Vol. H 48
Calcium Transport and
Intracellular Calcium Homeostasis
Edited by D. Pansu and F. Bronner

It is not surprising that the two processes, intestinal calcium absorption and renal calcium reabsorption, have utilized quite similar mechanisms. The challenge was to transform defense mechanisms that were designed to keep calcium out into instruments to help move calcium across cells. As it turned out, the transformation does not appear to have modified entry, buffering and extrusion mechanisms. Rather intestinal and renal cells involved in calcium transport evolved a ferrying mechanism that speeded up the rate of calcium diffusion that otherwise would have been too slow. This was accomplished by utilizing a soluble calcium binding protein for the ferrying role.

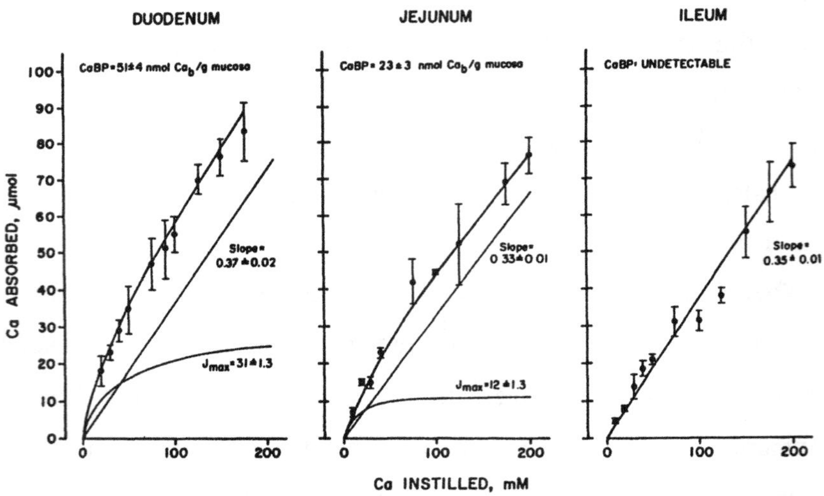

Figure 1. Calcium absorption in three intestinal segments (10 cm) of rats, proximal duodenum, proximal jejunum and distal ileum. Seventy male weanling rats were placed on a high calcium diet (1.5% Ca, 1.5% P) for 10 days. When their body weights averaged 77 $\pm$ 1.2 (SE) g and duodenal calcium-binding protein (CaBP, calbindin D9K) was 25.5 $\pm$ 2 (SE) nmol CaBP/g mucosa, calcium absorption was analyzed by an in situ loop procedure. J_{max} (V_m) is expressed in μmol Ca/g/2.5 h; the slope represents the fraction absorbed by the non-saturable route per 2.5 h and CaBP is expressed as nmol Ca_{bound}/g mucosa (2 nmol Ca_{bound} = 1 nmol CaBP)

Reproduced by permission from (38)

But before the details of this process are described, something needs to be said about the second process of transepithelial calcium transport, namely the paracellular route.

THE PARACELLULAR PATHWAY

Fig. 1 shows the results of a typical in situ loop experiment in which intestinal loops were prepared from duodenum, jejunum and ileum of rats that had been fed a high calcium diet that contained adequate amounts of vitamin D. The expressions describing the amount absorbed in 2.5 h, ordinate, as a function of the amount of calcium in the lumen, the amount instilled, abscissa, can be analyzed as the sum of a saturable and a non-saturable function in the duodenum and jejunum, but in the ileum the function is strictly linear, i.e. non-saturable. Moreover, the slope of the linear functions is essentially the same throughout the intestine and equals about 16% per hour. In other words, about 16% of the calcium present in the lumen crosses the epithelium by this linear route. It can be shown (2) that this fraction is the same at all concentrations and is essentially unchanged under a variety of physiological and nutritional conditions. It is also unchanged with age (2,37). By measuring calcium loss from intestinal sacs under conditions when active transport was inhibited and comparing it with the rate of loss of phenol red (Fig. 2), it was possible to infer that transfer occurred paracellularly. Theoretical considerations and kinetic experiments have further supported this conclusion (7).

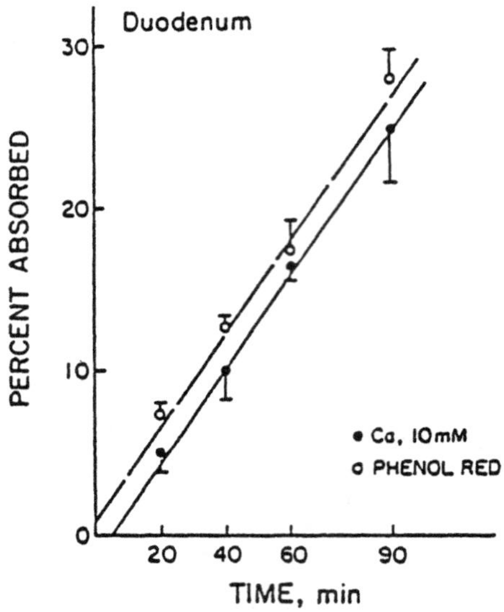

Figure 2. Rates of calcium and phenol red absorption from duodenal sacs (~0.5 g, 10 cm in length) were filled with 0.4 ml of a buffer solution (10 mM $CaCl_2$, 200 μM trifluoperazine, TFP, 10 mM phenol red in 0.9% NaCl, pH 7.2) and bathed in the buffer containing 0.1 mM $CaCl_2$ and no phenol red. ^{45}Ca was added to both inside and outside solutions to obtain the same specific activity. TFP inhibits active transport (44). The slopes of the two time-functions were 0.28/min for Ca and 0.29/min for phenol red, i.e., 16.8%/h for Ca and 17.2%/h for phenol red.

Adapted from (8)

Transepithelial calcium transport in the proximal tubule of the kidney is largely passive (4). Very little calcium is reabsorbed in the loop of Henle. In the thick ascending limb of the nephron, in the connecting and collecting ducts, calcium moves across the epithelium by a passive, paracellular route (4), with only calcium reabsorption in the distal convoluted tubule being effected by an active mechanism (see below).

Relatively little is known about regulation of the paracellular route. Patency of the tight junctions clearly affects calcium absorption. Thus when the intestine becomes distended, as following the instillation of hyperosmolar solutions (40), the rate of passive calcium movement can be doubled. Systemic and local regulators of tight junctions are therefore likely also to modulate passive calcium absorption, but systematic studies concerning possible regulation of the paracellular route of calcium absorption have not been published.

Yet the passive paracellular movement of calcium across intestinal and renal epithelia is of great functional importance. A general equation, adapted from Wasserman and Taylor (55), that describes intestinal calcium transport, is

$$v_a = \frac{V_m \ [Ca]}{K_m + [Ca]} + b \ [Ca] \qquad (1)$$

Where v_a = rate of calcium absorption

V_m = maximum rate of saturable calcium absorption

[Ca] = luminal calcium concentration

K_m = luminal calcium concentration at which $V_m/2$ is attained

b = rate of absorption by the non-saturable route

V_m varies inversely with calcium intake (36) and age (37) and term 1 of Eq. 1 can therefore become quite small, whereas term 2, representing the passive, non-saturating step of calcium transport, is relatively age-invariant (37) and can therefore constitute a very significant proportion of v_a.

Before concluding the discussion of paracellular transport, it may be interesting to speculate why this route exists. Two possible reasons come to mind: a) the need to provide a constant input of calcium so as to be able readily to maintain a constant plasma calcium level; b) the need for redundancy if the regulatable and more complicated active transport system were to become inoperative. As an example of the former, it has been argued (9) from experimental evidence (25) that passive calcium reabsorption in the kidney suffices if the total plasma calcium is near 1.3 mM, i.e. half of normal, but that when plasma calcium is 2.5 mM, i.e. normal, active calcium reabsorption is needed in the distal convoluted tubule to prevent undue calcium losses. The need for redundancy and system preservation can be illustrated by pointing out that even at low calcium intakes, when the active transport system is functioning at its maximum, it contributes only about half the amount of calcium that is absorbed (7).

TABLE 1

TRANSCELLULAR CALCIUM MOVEMENT IN INTESTINE

occurs largely in proximal intestine, especially duodenum
energy-dependent
oxygen-dependent
subject to regulation
primary regulator: vitamin D
down-regulated by high calcium intake, in old age
up-regulated by low calcium intake, in pregnancy, during growth
V_m = 7.3 fmol/min/cell

TRANSCELLULAR CALCIUM TRANSPORT

Table 1 summarizes significant aspects of transcellular calcium movement in the intestine. The three steps in transcellular calcium movement--entry, intracellular movement and extrusion--are sequential. As a result the total transport rate cannot exceed the rate of the slowest component. In the rat, the V_m of active transport is 22 μmol Ca/h/g duodenal mucosa (7), which translates to 7.3 fmol/min/cell. In the kidney, the equivalent rate in the distal convoluted tubule (DCT), the segment of the kidney where the major portion of active calcium transport takes place, is 10 pmol/min per mm tubule (9). Fig. 3 illustrates the developmental course of the intestinal V_m in the rat.

CALCIUM ENTRY

Brush-border membrane vesicles have served as the principal tool for the study of the entry process. The vesicles are right-side out (22), take up calcium spontaneously in a concentration-dependent

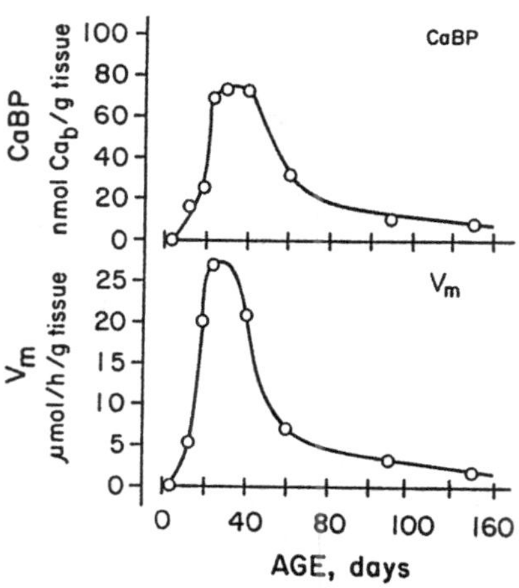

Figure 3. Developmental time course of tissue CaBP content and active calcium transport (V_m) in rat duodenum.

Adapted from (37)

fashion (Fig. 4), with the calcium that has crossed the membrane becoming bound to the inner aspect (30,31). The K_m of the process in the rat is 1.1 mM Ca (30). Thus calcium uptake by the vesicles --and by extension by intestinal or renal cells--does not require metabolic energy and proceeds down an electrochemical gradient. The inside of the cell is electronegative with respect to the lumen and its free calcium ion concentration is typically less than 0.5 μM. This contrasts with a calcium concentration of some 1 to 1.5 mM in the body fluids. In the intestinal lumen the calcium concentration can be well above 10 mM.

The route by which calcium crosses the brush border is not yet known. It seems reasonable to think that calcium channels, the

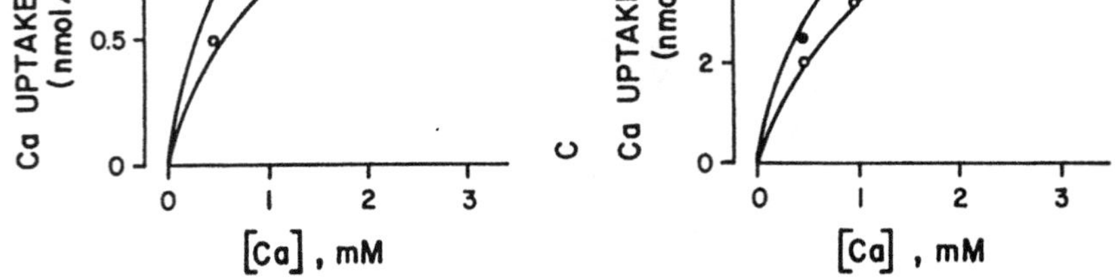

Figure 4. Calcium uptake by brush-border membrane vesicles, BBMV, from vitamin D-deficient rats, (○), or deficient rats that 16 h previously had received 0.5 μg of 1,25-$(OH)_2$-D_3 by intraperitoneal injection, (●). Panel A refers to initial (0.5 min) uptake rates; panel B refers to 15 min rates, when uptake has plateaued.

Reproduced from (30) by permission

widely utilized mechanism for calcium entry, also exist and are made use of in transporting gut and kidney cells. No report of calcium channels in the duodenum has as yet appeared[1]; evidence for the existence of calcium channels in renal cells has been presented by Saunders and Isaacson (46).

Quantitative evaluation of the rate of calcium uptake by brush border vesicles has yielded values that are an order of magnitude below the experimental V_m for transcellular transport (7). The reason for this discrepancy is not readily apparent; better kinetic approaches than those utilized in the early experiments have brought the uptake V_m nearer the transport V_m (Murer, personal communication), but a significant discrepancy still remains.

It has often been suggested that the entry step is rate-limiting for transcellular calcium transport. Two arguments can be advanced against this logical possibility. Unless the experimental V_m of calcium uptake by brush-border membranes can be shown to equal the experimentally evaluated V_m for total transport, there is no basis for attributing to it a rate-limiting role, since no component step can be smaller than the process as a whole. A second perhaps more potent argument stems from the fact that the rate of calcium entry is depressed only about 30 percent in brush-border membrane vesicles harvested from vitamin D-deficient rats (Fig. 4) or chicks (42), whereas transcellular calcium transport is totally suppressed in vitamin D deficiency (38).

---

1) Homaidan et al (23) have reported evidence for the existence of calcium channels in the basolateral membrane of rabbit ileal epithelial cells.

CALCIUM EXTRUSION

The generalized cell has been shown to be equipped with two calcium extrusion pumps, the plasma membrane Ca-ATPase and the Na/Ca exchanger. The relative role and importance of these two systems have been widely debated. There seems to be a growing consensus that the exchanger is probably of little functional importance in the intestinal cell (32), as far as calcium transport is concerned. In the renal cell, the exchanger may be of greater importance, although it has been suggested, on the basis of experimental studies, that the primary function of the exchanger in kidney cells is to help regulate sodium channel activity and intracellular pH, rather than extrude calcium as part of intracellular calcium regulation (4,57).

The plasma membrane Ca ATPase has been analyzed in detail (53) and its overall structure and function are beginning to be understood (11). The pump molecule consists of transmembrane elements, a calmodulin-binding domain near the carboxyl terminus, two domains that are rich in serine and threonine, one of which matches sequences found in the protein kinase substrates that are cAMP-dependent. Near the N-terminus are two sequences that on the basis of their resemblance to E-F hands may bind calcium. What appear to be the calmodulin-sensitive, calcium-binding and phosphoprotein domains are thought to be located on the cytoplasmic side, with calcium then being expelled through a channel-like opening formed by the transmembrane elements. For this to occur, phosphorylation is thought to have effected a conformational change in the transmembrane elements such that calcium, having become bound to the ATPase, could then pass through that opening. It is important to state that this picture of calcium extrusion has as yet no experimental basis and for two reasons: Functional studies with the plasma membrane Ca ATPAse are far from complete. Moreover, the Ca ATPase from neither intestine nor kidney has as yet been isolated or found to exhibit the properties attributed to the plasma membrane pump isolated from a human teratoma library (53). Indeed the renal Ca APTase may differ importantly from the plasma membrane Ca ATPase. Thus Brunette (personal communication) has found that vanadate has only a limited effect on the renal enzyme.

Whatever the precise nature of the renal and intestinal Ca ATPases, their extrusion capacity must be such as to satisfy the quantitative requirements of the transport process. Ghijsen et al (20), on the basis of transport studies with isolated basolateral membrane vesicles, have estimated a V_m for the enzyme of ~25 nmol Ca/min per mg protein, a value similar to that estimated by Carafoli for the plasma membrane enzyme (10). Active calcium transport in the intestine is characterized by a V_m of 18 nmol Ca/min/mg protein (7), a value sufficiently smaller than the Vm of the enzyme to make it unlikely that extrusion is a limiting rate. Data for the kidney ATPase are less certain, ranging from some 50 to some 300 nmol Ca/min per mg protein (4). It is difficult to transform the data to the transport V_m, usually expressed per mm DCT tubule. However, it seems probable that the calcium extrusion process in the kidney also is not the limiting rate.

INTRACELLULAR CALCIUM MOVEMENT

If neither calcium entry nor calcium extrusion is limiting the rate of transcellular calcium transport, there remains only the movement from apical to serosal pole as the category that can impose a limitation on transcellular movement. How can that be?

The general equation for flux or self-diffusion of calcium in an aqueous medium at 37C is given by the relationship (7):

$$F = A/L \times D_{Ca} \times \Delta[Ca] \qquad (2)$$

where F = flux (units: $M \times T^{-1} \times cell^{-1}$)

A = area per cell (units: L^2)

L = length of diffusion path between apical and serosal poles (units: L)

D_{Ca} = the diffusion constant of Ca^{2+}, 1.8×10^{-5} cm^2/s

$\Delta[Ca]$ = calcium concentration difference between apical and serosal poles, assumed to be 200 nM

Only $\Delta[Ca]$ is uncertain in Eq. 2. Since however, the free calcium ion concentration of cells is now generally agreed to be in the neighborhood of 100 nM, a gradient of 200 nM seems reasonable. Appropriate substitution in Eq. 2 yields a value of 0.96×10^{-16} mol/min per cell for the intestine and 0.13×10^{-9} mol/min per mm DCT (5). Preliminary experiments (4) on the rate of self-diffusion of Ca^{2+} in non-transporting segments of the nephron have yielded a value of 0.14 ± 0.02 (SE, n = 13) pmol/min per mm.

If the diffusion values are compared with the experimental transport values of $V_m = 0.72 \times 10^{-14}$ mol/min per duodenal cell and of $V_m = 10 \times 10^{-9}$ mol/min per mm DCT, it can be seen that the experimental transport values exceed the self-diffusion rates of calcium by some 70-fold in both intestine and DCT.

How can the intestinal or renal cell effect the needed augmentation so as to be able to transport calcium at a rate some 70 times greater than if calcium would self-diffuse? Kretsinger et al (27) have proposed and Feher (13-15) has experimentally verified that if a calcium-binding protein is present in the cytosol at near millimolar quantities, the protein can act to ferry the calcium across the cell by a mechanism of facilitated diffusion. Indeed this process is similar to the manner in which hemoglobin functions as a carrier of oxygen.

To calculate how CaBP augments the intracellular diffusion of calcium, one rewrites Eq. 2 as follows:

$$F_{Ca/CaBP} = D_{CaBP} \times \frac{A[(Ca_1/CaBP_1)-(Ca_2/CaBP_2)]}{L} \tag{3}$$

Where $F_{Ca/CaBP}$ = the flux of calcium bound to CaBP (per unit time per cell or per mm DCT)

D_{CaBP} = diffusion constant of CaBP through the cytosol

$Ca_1/CaBP_1$ and $Ca_2/CaBP_2$ are the concentrations of calcium bound to CaBP at the apical and serosal poles of the cell or DCT. L and A are defined as in Eq. 2.

The concentration of calcium bound to CaBP can be found with the aid of the dissociation constant, K_D, for calcium bound to the protein. Substitution in Eq. 3 will yield an augmentation factor for calcium diffusion as follows:

$$\text{Augmentation} = \frac{\dfrac{D_{CaBP}}{D_{Ca}} \times K_D \times [CaBP]}{(K_D + [Ca_1])\ (K_D + [Ca_2])} \tag{4}$$

where K_D is the dissociation constant for CaBP, 0.3 μM for CaBP from the intestine and 2 μM for CaBP from the kidney (5). If $K_D >> [Ca_1]$ and $[Ca_2]$, Eq. 4 would simplify to:

$$\text{Augmentation} \simeq \frac{D_{CaBP}}{D_{Ca}} \times \frac{[CaBP]}{K_D} \tag{5}$$

The value of Eq. 5 is that it illustrates more sharply that augmentation of self-diffusion of calcium is approximately proportional to the concentration of CaBP, even though diffusion of calcium bound to the binding protein is slower than self-diffusion of calcium.

For example, if the CaBP concentration is near millimolar (50), the K_D near micromolar, and the ratio of the diffusion constants is 1/6 (15), CaBP would provide a fifty-fold augmentation according to Eq. 5.

Experimental data of CaBP and V_m for the rat intestine are shown in Fig. 5. The relationship is linear throughout in both series of experiments, steady-state, and induction of transport and CaBP. Thus active transcellular calcium transport is a direct linear function of the tissue content of CaBP. Further support for this finding comes from recent work (39) that has shown that when calcium-binding by CaBP is interfered with, active calcium transport is impaired in direct proportion to the binding inhibition.

While the relationship between CaBP and active transport seems firm and causal, early experiments (1,49,51) suggested that increased calcium entry into cells from vitamin D-deficient animals preceded the appearance of ribosomes that contain the messenger RNA for CaBP. Later studies failed to dissociate CaBP appearance from increased

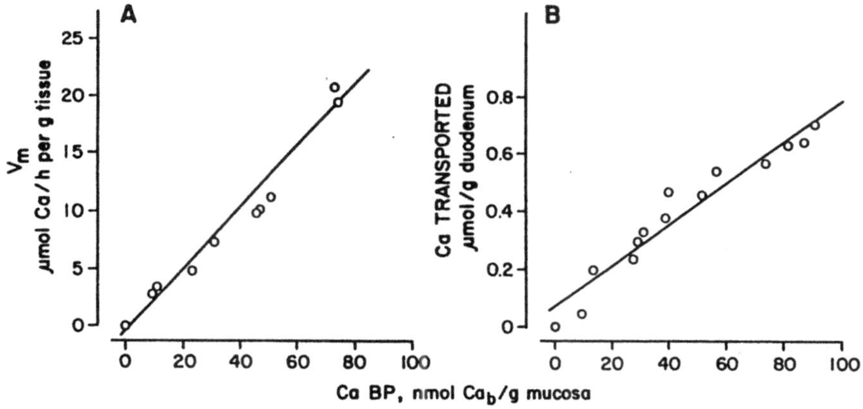

Figure 5. The relationship between intestinal calcium transport and CaBP content. A. V_m, calculated from in situ duodenal, jejunal and ileal loop experiments (36-38), shown as a function of CaBP content. The relationship is highly significant (r = 0.98). B. Active calcium transport, evaluated from everted duodenal sac experiments, shown as a function of CaBP content (44). The animals in the experiments for Panel B were either vitamin D-deficient or vitamin D replete, some of which had been treated with varying doses of 1,25-$(OH)_2$-D_3 either 4h (+D) or 9h (-D) before the experiment (44). The relationship is highly significant (r = 0.97).

Reproduced from (7).

calcium transport (6,18). Since then the total dependence of CaBP biosynthesis on 1,25-$(OH)_2$-D_3 has been established for both kidney (24) and intestinal (12) cells, but careful kinetic studies correlating calcium transport with CaBP expression have not been reported.

It has also been suggested (33,34) that lysosomes function in intracellular transport, perhaps constituting the ferry that transports calcium from one cell pole to the other. Lysosomes do move intracellularly, bind calcium and may even contain some CaBP (33). The notion that calcium traverses the cell in some packaged from, thereby not compromising intracellular calcium homeostasis, is not new (45,54). However, available data on the amount of calcium carried by lysosomes and the probable number of these per transporting cell make it unlikely that lysosomal calcium transport contributes significantly to transcellular transport.

Since most organelles have calcium-binding sites, one might think they contribute to calcium transport. Some of these sites are indeed vitamin D-stimulated (53). However, fixed organelles can only impede, not contribute to calcium transport. Their contribution is more likely that of added buffering capacity when calcium flux increases, as after vitamin D stimulation (7).

REGULATION

Inasmuch as general aspects of the regulation of calcium transport have been reviewed widely (3,28,41), discussion here will emphasize regulation of active transport, especially of factors that play on CaBP expression. These can be classified into a) factors acting directly on CaBP, b) factors influencing 1,25-$(OH)_2$-D_3 biosynthetic activity, c) factors influencing vitamin D metabolism, d) factors modifying calcium entry or extrusion. Table 2 lists these by way of illustration, but is not intended to be exhaustive.

TABLE 2

MODULATION OF ACTIVE CA TRANSPORT

CATEGORY	EXAMPLE	REFERENCE
Direct action on CaBP	Theophylline (inhibits Ca-binding by CaBP & inhibits Ca transport)	39
Biosynthetic action of 1,25-$(OH)_2$-D_3	Glucocorticoids (depress CaBP synthesis and Ca absorption)	16
Alteration of vitamin D metabolism	Parathyroidectomy (leads to diminution of circulating 1,25-$(OH)_2$-D_3 and decreased Ca absorption)	51
Modification of Ca entry	Verapamil (decrease in active Ca transport)	17
Modification of Ca Extrusion	Trifluoperazine (diminishes CaATPase activity and active Ca transport)	44

Most studies on regulation have dealt with various aspects of vitamin D and its metabolism; in general, active transport, CaBP levels and 1,25-$(OH)_2$-D_3 plasma levels vary in parallel. However, one can imagine effects at the subcellular level. Theophylline, for example, diminishes the capacity of CaBP to bind calcium and also inhibits active calcium transport (39). One might also imagine some modulation of transcription or translation that might be specific to CaBP biosynthesis. For example, Singh and Bronner (47,48) have reported that the post-transcriptional activation of CaBP synthesis that can be effected in vitro with the aid of duodenal cells derived from animals on a high-calcium diet can be inhibited by high calcium in the medium.

Parathyroid hormone is known to enhance the enzymatic conversion of 25-OH_2-D_3 to 1,25-$(OH)_2$-D_3. Parathyroidectomy leads to lower levels of 1,25-$(OH)_2$-D_3, lower amounts of CaBP and lower levels of calcium absorption (Table 2).

On the other hand, glucocorticoids seem to depress CaBP biosynthesis and active calcium transport (16), yet 1,25-$(OH)_2$-D_3 plasma levels under these circumstances seem unchanged, at least in some studies (29). It is unclear but probable that this effect of glucocorticoids is general, as far as protein biosynthesis is concerned. One might also imagine drugs or situations where transcription or translation leading to CaBP are interfered with specifically.

Finally, it is interesting but not surprising that inhibition of calcium channel activity (17) or of calcium extrusion, with the aid of trifluoperazine, TFP, added to an everted intestinal sac preparation or the basolateral membrane vesicle preparation (44), interfered with calcium availability or extrusion and thus blocked transport. Another substance that inhibits calcium extrusion and therefore blocks transport is the vasoactive intestinal peptide, VIP (43). How TFP and VIP act to block the pump enzyme is not known.

The effect of varying calcium intake on calcium transport has already been referred to above. Lowering the calcium density of the food stimulates the renal production of 1,25-$(OH)_2$-D_3 and thus increases CaBP levels in the duodenum. This in turn up-regulates calcium transport. Raising calcium intake reverses this process. In addition, however, the increased calcium input into the enterocyte may directly down-regulate CaBP biosynthesis (47,48), presumably at some post-transcriptional step. As pointed out above, the effect of varying calcium intake on the non-saturable, paracellular calcium transport component is the opposite, i.e. the amount of calcium absorbed by the paracellular path varies in direct proportion with the luminal calcium concentration.

CONCLUDING REMARKS

Even though the basic aspects of calcium transport are now well established, much needs to be done to clarify and extend current knowledge. Thus the nature of paracellular calcium movement and its regulation are largely unknown. As far as transcellular calcium transport is concerned, the functional and molecular nature of calcium channels in calcium-transporting epithelial need to become known. Similarly, the genes specifying the Ca ATPase(s) of intestine and kidney need to be identified and their expressions studied in detail. The Na/Ca exchangers of these epithelia also need to be identified, cloned and studied. Finally, the role of CaBP as a calcium ferry would be elegantly demonstrated in transfection experiments aimed at transforming a non-transporting cell, i.e. one that does not express CaBP, into one that does so. The tools and techniques to carry out the experiments outlined above already exist. Undoubtedly, however, future work will not only establish what has been predicted, but will yield surprises, unpredicted findings, and intellectual excitement.

ACKNOWLEDGMENTS

The concepts outlined in this essay are the results of the experimental and theoretical efforts of many. I am particularly grateful to my colleagues, Dr. D. Pansu and Dr. W.D. Stein, for their stimulating friendship and the many hours we have spent in discussing and formulating ideas that led to experiment and synthesis.

REFERENCES

1. Bikle DD, Zolock DT, Morissey RL, and Herman RH (1978) Independence of 1,25-dihydroxyvitamin D_3-mediated calcium transport from de novo RNA and protein synthesis. J. Biol. Chem. 253: 484-488.
2. Bronner F. (1987) Calcium absorption. In: Johnson, LR, Christensen, J., Jacobson, ED, Jackson, MJ, Walsh, JH (eds) Physiology of the gastrointestinal tract. 2nd ed. Raven, New York, pp 1419-1435.
3. Bronner F. (1988) Gastrointestinal absorption of calcium. In: Calcium in human biology. B.E.C. Nordin, ed. Springer, London, pp. 93-123.
4. Bronner F (1989) Renal calcium transport: mechanisms and regulation--an overview. Am. J. Physiol. 257 (Renal Fluid Electrolyte Physiol. 26): F707-F711.
5. Bronner F. (1990) Transcellular calcium transport. In: Intracellular calcium regulation, F. Bronner, ed. Wiley-Liss, New York. pp. 415-437.
6. Bronner F, Lipton J, Pansu D, Buckley M, Singh R and Miller A III (1982) Molecular and transport effects of 1,25-dihydroxyvitamin D_3 in rat duodenum. Fed. Proc., 41: 61-65.
7. Bronner F, Pansu D, and Stein WD (1986) An analysis of intestinal calcium transport across the rat intestine. Am. J.Physiol. 250(Gastrointest. Liver Physiol. 13): G561-G569.
8. Bronner F, and Spence K (1988) Non-saturable Ca transport in the rat intestine is via the paracellular pathway. In: Bronner, F., Peterlik, M. (eds). Cellular calcium and phosphate transport in health and disease. Alan R. Liss, New York, p. 277-284.
9. Bronner F and Stein WD (1988) $CaBP_r$ facilitates intracellular diffusion for Ca pumping in distal convoluted tubule. Am. J. Physiol. 255(Renal Fluid and Electrolyte Physiol. 24): F558-F562.
10. Carafoli E (1988) Membrane transport of calcium: an overview. In: Methods of Enzymology. 157(Q): 3-11.

11. Carafoli E, James P and Strehler EE (1990) Structure-function relationships in the calcium pump of plasma membranes. In: Molecular and cellular regulation of calcium and phosphate metabolism. M. Peterlik, F. Bronner (eds). Wiley-Liss, New York, pp. 818-193.
12. Dupret JM, Brun P, Lomri N, Thomasset M and Cuisinier-Gleizes, P. (1987) Transcriptional and post-transcriptional regulation of vitamin D-dependent calcium-binding protein gene expression in the rat duodenum by 1,25-dihydroxy-cholecalciferol. J. Biol. Chem. 262:16553-16557.
13. Feher JJ (1983) Facilitated calcium diffusion by intestinal calcium-binding protein. Am. J. Physiol. 244(Cell Physiol. 13): C303-C307.
14. Feher JJ (1984) Measurement of facilitated calcium diffusion by a soluble calcium-binding protein. Biochim. Biophys. Acta. 773:91-98.
15. Feher JJ and Fullmer CS (1988) Facilitated diffusion of calcium by calcium-binding protein: its role in intestinal calcium absorption. In: Cellular calcium and phosphate transport in health and disease. F. Bronner, M. Peterlik (eds) Alan R. Liss, New York. p. 121-126.
16. Feher JJ and Wasserman RH (1979) Intestinal calcium-binding protein and calicum absorption in cortisol-treated chicks: effects of vitamin D_3 and 1,25-dihydroxyvitamin D_3. Endocrinology. 104: 547-551.
17. Fox J and Green DT (1986) Direct effects of calcium channel blockers on duodenal calcium transport in vivo. Europ. J. Pharmacol. 129: 159-164.
18. Franceschi RT and DeLuca HF (1981) The effect of inhibitors of protein and RNA synthesis on 1α,25-dihydroxyvitamin D_3-dependent calcium uptake in cultured embryonic chick duodenum. J. Biol. Chem. 256: 3848-3852.
19. Freund T, and Bronner F (1975) Regulation of intestinal calcium-binding protein by calcium intake in the rat. Am. J. Physiol. 228: 861-869.
20. Ghijsen WEJM, DeJong MD, and Van Os CH (1982) ATP-dependent calcium transport and its correlation with Ca^{2+}-ATPase activity in basolateral plasma membranes of rat duodenum. Biochem. Biophys. Acta. 689: 327-336.
21. Hurwitz S, Stacey RE and Bronner F (1969) Role of vitamin D in plasma calcium regulation. Am. J. Physiol. 216: 254-262.

22. Haase W, Schafer A, Murer H, and Kinne R (1978) Studies on the orientation of brush-border membrane vesicles. Biochem. J. 172: 57-62.
23. Homaidan FR, Donowitz M, Weiland GA, and Sharp GWG (1989) Two calcium channels in basolateral membranes of rabbit ileal epithelial cells. Am J Physiol 257(Gastrointest. Liver Physiol 20):G86-G93.
24. Huang YC and Christakos S (1988) Modulation of rat calbindin-D28 gene expression by 1,25-dihydroxyvitamin D3 and dietary alteration. Molecular Endocrinology 2: 928-935.
25. Hurwitz S, Stacey RE and Bronner F (1969) Role of vitamin D in plasma calcium regulation. Am. J. Physiol. 216: 254-262.
26. Kretsinger RH (1990) Why cells must export calcium. In: Intracellular calcium regulation. F. Bronner, ed. Wiley-Liss, New York, pp. 439-457.
27. Kretsinger RH, Mann JE and Simmons JG (1982) Model of facilitated diffusion of calcium by the intestinal calcium binding protein. In: Vitamin D chemical, biochemical and clinical endocrinology of calcium metabolism. A.W. Norman, K. Schaefer, D.V. Herrath, and H.-G. Gringoleit (eds)., 233-246, DeGruyter, Berlin and New York, pp. 233-246.
28. Levine BS, Walling MW, and Coburn JW (1982) Intestinal absorption of calcium: its assessment, normal physiology, and alterations in various disease states. In: Disorders of mineral metabolism. F. Bronner and J.W. Coburn (eds). Academic Press, New York, Vol. 2, pp. 103-188.
29. Lukert BP, Stanbury SW and Mawer EB (1973) Vitamin D and intestinal transport of calcium: effects of prednisolone. Endocrinology 93: 718-722.
30. Miller A III and Bronner F (1981) Calcium uptake in isolated brush-border vesicles from rat small intestine. Biochem. J. 196: 391-401.
31. Miller A III, Li ST and Bronner F (1982) Characterization of calcium binding to brush border membranes from rat duodenum. Biochem. J. 208: 773-782.
32. Nellans HN and Popovitch JR (1984) Role of sodium in intestinal calcium transport. In: Epithelial calcium and phosphate transport: Molecular and cellular aspects. F. Bronner and M. Peterlik (eds), Alan R. Liss, New York, 1984, pp. 301-306.

33. Nemere I (1990) Organelles that bind calcium. In: Intracellular calcium regulation, F. Bronner, ed. Wiley-Liss, New York, pp. 163-179.

34. Nemere I, Leathers V and Norman AW (1986) 1,25-dihydroxyvitamin D_3-mediated intestinal calcium transport. Biochemical identification of lysosomes containing calcium and calcium-binding protein (calbindin-D28K). J. Biol. Chem. 261:16106-16114.

35. Palmer LG and Frindt G (1987) Effects of cell Ca and pH on Na channels from rat cortical collecting tubules. Am. J. Physiol. 253 (Renal Fluid Electrolyte Physiol. 22): F333-F339.

36. Pansu D, Bellaton C and Bronner F (1981) The effect of calcium intake on the saturable and non-saturable components of duodenal calcium transport. Am. J. Physiol. 240(Gastrointest. Liver Physiol. 3): G32-G37.

37. Pansu D, Bellaton C and Bronner F (1983) Developmental changes in the mechanisms of duodenal calcium transport in the rat. Am. J. Physiol. 244(Gastrointest. Liver Physiol. 7): G20-G26.

38. Pansu D, Bellaton C, Roche C and Bronner F (1983) Duodenal and ileal calcium absorption in the rat and effects of vitamin D. Am. J. Physiol. 244(Gastrointest. Liver Physiol. 7): G695-G700.

39. Pansu D, Bellaton C, Roche C, and Bronner F (1989) Theophylline inhibits transcellular Ca transport in intestine and Ca-binding by CaBP. Am. J. Physiol. 257(Gastroint. Liver Physiol. 20): G935-G943.

40. Pansu D, Chapuy MC, Milani M and Bellaton C (1976) Transepithelial calcium transport enhanced by xylose and glucose in the rat jejunal ligated loop. Calcif. Tiss. Res. 21: 45-52.

41. Peacock M (1988) Renal excretion of calcium. In: Calcium in human biology. B.E.C. Nordin, ed. Springer, London. pp. 125-169.

42. Rasmussen H, Fontaine O, Max EE and Goodman DP (1979) The effect of 1(α)-hydroxyvitamin D_3 administration on calcium transport in chick intestine brush border membrane vesicles. J. Biol. Chem. 254: 2993-2999.

43. Roche D, Bellaton C, Pansu D and Bronner F (1985) Vasoactive intestinal peptide (VIP) decreases active duodenal Ca transport directly. In: Abstracts, 7th annual scientific meeting American Society for Bone and Mineral Research, Washington, DC.

44. Roche C, Bellaton C, Pansu D, Miller A III and Bronner F (1986) Localization of vitamin D-dependent active Ca2+ transport in rat duodenum and relation to CaBP. Am. J. Physiol. 251(Gastrointest. Liver Physiol. 14): G314-G320.

45. Rubinoff MJ, and Nellans HN (1985) Active calcium sequestration by intestinal microsomes. Stimulation by increased calcium load. J. Biol. Chem. 260: 7824-7828.

46. Saunders JCJ and Isaacson LC (1989) Non-selective cation and Ba-permeable apical channels in rabbit renal tubules. Abstracts. Intl. Congress Physiol. Sci. Helsinki.

47. Singh RP and Bronner F (1980) Vitamin D acts post-transcriptionally. In vitro studies with the vitamin D-dependent calcium-binding protein of rat duodenum. In: Calcium binding proteins: Structure and function. F.L. Siegel, E. Carafoli, R.H. Kretsinger, D.H. MacLenan and R.H. Wasserman (eds) Elsevier-North Holland, New York, pp. 379-383.

48. Singh RP and Bronner F (1982) Duodenal calcium binding protein: induction by 1,25-dihydroxyvitamin D in vivo and in vitro. Ind. J. Exp. Biol. 20: 107-111.

49. Spencer R, Chapman M, Wilson P and Lawson E (1976) Vitamin D-stimulated intestinal calcium absorption may not involve calcium-binding protein directly. Nature. 263: 161-163.

50. Thomasset M, Cuisinier-Gleizes P, and Mathieu H (1979) 1,25-dihydroxycholecalciferol: Dynamics of the stimulation of duodenal calcium-binding protein, calcium transport and bone calcium mobilization in vitamin D- and calcium-deficient rats. FEBS Letters. 107: 91-94.

51. Thomasset M, Cuisinier-Gleizes P, Mathieu H, Golub EE and Bronner F (1979) Regulation of intestinal calcium binding protein in rats: role of parathyroid hormone. Calcif. Tissue. Intl. 29: 141-145.

52. Thomasset, M., Parkes, C.O., Cuisinier-Gleizes, P. (1982) Rat calcium-binding proteins: distribution, development and vitamin D-dependence. Am. J. Physiol. 243(Endocrin. Metab. 6): E483-E488.

53. Verma AK, Filoteo AG, Stanford DR, Wieben ED, Penniston JT, Strehler EE, Fischer R, Heim R, Vogel G, Mathews S, Strehler-Page MA, Vorherr T, Krebs J and Carafoli E (1988) Complete primary structure of the human plasma membrane Ca^{2+} pump. J. Biol. Chem. 263: 14152-14159.

54. Warner RR, and Coleman JR (1975) Electron probe analysis of calcium transport by small intestine. J. Cell Biol. 64: 54-74.

55. Wasserman RH, and Taylor AN (1969) Some aspects of the intestinal absorption of calcium, with special reference of vitamin D. In: Mineral Metabolism--An advanced treatise. C.L. Comar and F. Bronner (eds), Academic Press, New York and London, pp. 321-403.

56. Weiser MM, Bloor JH, Dasmahapatra A, Freedman RA and MacLaughlin JA (1981) Vitamin D-dependent rat intestinal Ca^{2+} transport. Ca^{2+} uptake by Golgi membranes and early nuclear events. In: Calcium and phosphate transport across biomembranes. F. Bronner and M. Peterlik (eds.) Academic Press, New York, pp. 264-273.

57. Yang JM, Lee CO, and Windhager EE (1988) Regulation of cytosolic free calcium in isolated perfused proximal tubules of Necturus. Am. J. Physiol. 255 (Renal Fluid Electrolyte Physiol. 24): F787-F799.

INTESTINAL CALCIUM ABSORPTION: THE CALBINDINS, THE VISUALIZATION OF TRANSPORTED CALCIUM AND A NEW RAPID ACTION OF 1,25-DIHYDROXYVITAMIN D_3

R.H. Wasserman, C.S. Fullmer, S. Chandra*, H. Mykkanen+, N. Tolosa de Talamoni#, and G. Morrison*
Department of Physiology
Department of Chemistry*
Cornell University
Ithaca, NY USA, 14853

The calbindins, calbindin-D_{28K} and calbindin-D_{9K}, were discovered about 25 years ago as a vitamin D-inducible calcium-binding protein in chick and rat intestinal mucosa, respectively. The calbindins have now been identified in a wide variety of different types of tissues. Despite detailed information on their biochemical and physical properties, the exact function of the calbindins in epithelial transport and their role in non-epithelial tissues have not yet been resolved.

Investigations on the localization of calcium during intestinal absorption were undertaken which might bear on the role of the calbindins in the absorptive process. These studies utilized ion microscopy, cryosectioning procedures and a stable isotope of calcium (Ca-44). Intestinal calcium is first sequestered in the brush border region of the enterocyte. Only in the vitamin D-treated animal is calcium rapidly released from this region to complete the absorption process. Calbindin, acting as a diffusional facilitator, could account for these observations. Other models of calcium transport could also account for these findings.

Studies on isolated brush borders membrane vesicles (BBMV) showed that the intravenously injection of $1,25(OH)_2D_3$ into rachitic chicks results in a rapid increase in the reactive sulfhydryl groups of proteins associated with those membranes. This effect occurs at low physiological doses of the vitamin D hormone (0.1 ng/chick), and with a short lag period (within 10 min). Estradiol and testosterone elicit a similar response but are less potent. Adaptation of chicks to calcium or phosphorus deficient diets also results in a significant increase in BBMV sulfhydryl groups. The relationship of these changes in BBMV sulfhydryl groups to the early localization of absorbed calcium in the intestinal microvillar region and to overall calcium absorption is unknown at this time.

+ Present address: Department of Nutrition, University of Kuopio, Finland.
# Permanent address: Department of Biochemistry, Faculty of Medicine, University of Cordoba, Argentina.

NATO ASI Series, Vol. H 48
Calcium Transport and
Intracellular Calcium Homeostasis
Edited by D. Pansu and F. Bronner

Significant events in the field of vitamin D and calcium metabolism occurred some 20-25 years ago. Within that time frame, 25-hydroxyvitamin D_3 ($25(OH)D_3$) was identified by DeLuca and colleagues (5). Shortly thereafter the vitamin D hormone, 1,25-dihydroxyvitamin D_3 ($1,25(OH)_2D_3$), was identified and its existence established through the efforts of the Cambridge, Wisconsin and Riverside groups (cf. ref. 43 for an early review). Much before this time period, Lindquist (24) from physiological studies noted that a time lag of several hours was required before vitamin D enhanced Ca absorption in rachitic animals. This time lag was considerably shortened when $1,25(OH)_2D_3$ instead of vitamin D_3 was administered to vitamin D deficient animals (23,33). The usual explanation invoked to account for a delay of several hours between hormone administration and a biological response is a requirement for hormone-dependent _de novo_ protein synthesis. This concept was strongly supported when the synthesis of the calbindins in chick intestine was shown to be vitamin D dependent (45) and these proteins are considered to be a prominent product of the interaction of the $1,25(OH)_2D_3$-receptor complex with the gene coding for calbindin mRNA synthesis.

The early experiments on the intestinal calbindins, done at Cornell and several other laboratories, demonstrated the presence of the calbindins in a number of tissues in addition to the intestine (cf. ref. 10 and 34 for recent summaries). In some organs, such as the brain, the calbindins appear to be constitutively expressed in the post-embryonic state and not dependent on the vitamin D status of the animal (37,38). However, the CaBP content of brain tissue is subject to modification by various conditions, such as in animals made epileptic by the kindling procedure (3) and in certain neurodegenerative diseases of man (11).

Physiological studies of various sorts revealed a high correlation between intestinal CaBP and calcium absorption, although discrepancies in this relationship were revealed in the early studies. For example, the initial increase in calcium absorption in response to a single dose of $1,25(OH)_2D_3$, at least in our hands, corresponds in time with the appearance of CaBP (42). However, at later times, when calcium absorption decreased to near baseline, CaBP was still present in the intestinal mucosa (44). Further, if $1,25(OH)_2D_3$ is given to a normal animal or a rachitic animal partially repleted with vitamin D, an increase in calcium absorption occurs without a substantial lag time and without a corresponding significant increase in CaBP, signalling a response not dependent on gene transcription (42). The conclusion was inescapable; $1,25(OH)_2D_3$ elicits more than one effect on the intestinal calcium absorptive system, one dependent and others not dependent on _de novo_ protein synthesis.

Assume that CaBP is involved in vitamin D-dependent calcium absorption. How might it function in this capacity? The theoretical analysis of the transport system by Kretsinger et al (22) and the in vitro model studies of Feher (16) and Feher et al (17), suggested that CaBP acts as a diffusional facilitator, increasing the rate of diffusion of Ca^{2+} from the apical region to the basal region of the intestinal cell where the ATP-dependent calcium pump is located. The analysis of the intestinal system by Bronner et al (8) and van Os (39), and analysis of the renal system by Bronner et al (9) and Borke et al (6), indicated that CaBP could very well have an important role in the transcellular movement of calcium.

Bearing on these questions has been a lack of precise information on the locality of Ca^{2+} during the course of its transfer across the intestinal epithelium. Previous studies suggested that transported Ca^{2+} is transiently sequestered by mitochondria (28) or transferred in packets (21,28,41). Lysosomes containing CaBP have been implicated as part of a vesicular transport mechanism (32).

Recent studies in our laboratory, in collaboration with Drs. S. Chandra and G. Morrison of the Department of Chemistry at Cornell, were undertaken to examine this problem. Advantage was taken of the unique ion microscopy system developed by Dr. Morrison and colleagues which allows one to directly image the locality of specific elements in tissues.

Duodenal segments derived from rachitic and vitamin D-repleted chicks were cryosectioned and the localization of various elements in the tissue was determined. A solution containing the stable isotope of calcium, Ca-44, was injected into the duodenal lumen in situ at different times before the duodenal tissue was taken for analysis. The stable isotope was used in order to distinguish absorbed calcium (Ca-44) from residual calcium (Ca-40).

In duodenum from rachitic chicks, Ca-44 was sequestered in the microvillar-terminal web region of the enterocyte and remained concentrated in this region over a 20 min absorption period. In contrast, absorbed Ca-44 in the vitamin D-replete tissue, although initially concentrating in this same microvillar-terminal web region, rapidly moved through and out of the intestinal cell into the circulatory system. It is clear that the prior administration of vitamin D_3 in some fashion caused the release of Ca^{2+} from its binding sites in the apical region and allowed transcellular transport to proceed considerably more rapidly than in the absence of vitamin D_3. Certainly these observations can be explained by the presence of CaBP in the vitamin D_3-treated tissue, CaBP acting as the suggested diffusional facilitator. CaBP, with its relatively high binding affinity for calcium, could theoretically remove Ca^{2+} from the apical binding sites and accelerate its transcellular transfer to the basal region of the cell. The uptake of Ca^{2+} by CaBP-containing vesicles and the intracellular transfer of Ca^{2+} encapsulated in vesicles (34) could also account for these observations. Unfortunately, limitations in the sensitivity and resolution of the procedure does not allow one to distinguish between these two processes but perhaps this can be done with further developments in ion microscopy technology.

The localization data are consistent with previous observations showing that isolated brush border membrane vesicles from chicks (46) and rats (26) have the capacity to bind calcium with relative high affinities, of the order of $5x10^5$ M^{-1}. A set of lower affinity sites with a k_a of about 3×10^2 M^{-1} were also identified (26). Each group (46,26) also noted that the binding of calcium to the isolated brush border membranes was increased by vitamin D treatment of deficient animals. Calmodulin, with an affinity similar to that of the isolated brush border membranes, is a known component of the microvillus complex, bound to a 110 kD cytoskeletal protein (4,18). Calmodulin binding to the 110 kD protein is not dependent on the presence of Ca^{2+} (4,18) but is increased by vitamin D (18). Interestingly, Bikle and Munson (18) reported a direct relationship between the calmodulin content of, and Ca^{2+} uptake by, brush border membrane vesicles isolated from chicks. The calmodulin bound to the 110 kD protein could account for a considerable amount of the calcium binding reaction, assuming that the bound calmodulin binds Ca^{2+} to the same extent as free calmodulin (4 sites per molecule).

Miller et al (27) also identified a vitamin D-responsive, calcium-binding protein in rat intestinal brush border membrane vesicles with a molecular weight similar to calmodulin.

The calmodulin-110 kD protein complex, as suggested by Bikle and Munson (18), might have a functional role in controlling calcium transport across the microvillus membrane, in addition to merely sequestering Ca^{2+}. The ion microscope visualization of calcium localization in the microvillus-terminal web region does suggest some control on the entry of Ca^{2+} into the cytosol compartment of the enterocyte. A possible feed-back mechanism might be in place in which saturation of binding sites in this region by calcium, in some fashion, determines the rate of Ca^{2+} entry into cell. Such a mechanism would prevent the uncontrolled, rapid entry of Ca^{2+} into the enterocyte that might produce untoward, toxic effects.

Now, what of the other action or actions of $1,25(OH)_2D_3$ on the intestine that might be translated into an increase in calcium absorption? Fluidity changes of the plasma membrane have been documented (7,35), increases in cyclic nucleotide (cAMP, cGMP) synthesis have been reported (12,25,40) and, more recently, the activation of Ca channels was noted (13).

We have been examining another effect of 1,25(OH)$_2$D$_3$, a rapid, apparently non-genomic increase in available protein-associated sulfhydryl groups of isolated brush border membrane vesicles (BBMV). The studies on sulfhydryl groups were prompted by the observation that vitamin D$_3$ and 1,25(OH)$_2$D$_3$ stimulated the uptake and binding of selenite, a form of the essential trace element selenium, by BBMV (29). Recognizing that selenite tightly binds to sulfhydryl groups, a number of experiments revealed a direct correspondence between vitamin D-dependent increases in selenite binding and the vitamin D-dependent increase in BBMV sulfhydryl groups. Both the total concentration of sulfhydryl groups, as determined by the Ellman (15) reaction in the presence of SDS, and the readily reactive sulfhydryl groups as assessed by the use of a fluorescent probe DACM (N-7-dimethylamino-4-methylcoumarin-3-yl maleimide), are affected by vitamin D. DACM only becomes fluorescent after binding to sulfhydryl groups. The enhancement of BBMV sulfhydryl groups occurs within 10 min. after the intravenous injection of 1,25(OH)$_2$D$_3$ into rachitic chicks and occurs at low physiological doses (0.1 ng/chick) of the vitamin D hormone. In the intestine, this effect is relatively specific for the brush border membranes, with little or no change occurring in the sulfhydryl groups associated with isolated mitochondria, microsomes or basal-lateral membranes. The localization of a high concentration of sulfhydryl groups in the intestinal brush border region was also verified by the use of fluorescence histochemistry. The brush border region was intensely stained by DACM in contrast to other parts of the intestinal cell.

The BBMV sulfhydryl groups from chick kidney are similarly responsive to 1,25(OH)$_2$D$_3$. It was also observed that other steroids, like estradiol, testosterone and aldosterone, elicited similar increases in intestinal BBMV sulfhydryl groups. This brought into question the exact relationship of these changes to a specific effect on the calcium transport system. There is evidence that each of these steroid hormones does exert a biological response in intestine. For example, aldosterone increases Mg-HCO_3^--ATPase and carbonic hydrase activity (36); a synthetic estrogen (ethinylestradiol) modulates the Na^+-H^+ exchange reaction in colonic brush borders (14); and testosterone (17 alpha-methyltestosterone) enhances intestinal L-leucine transport, possibly by increasing the activity of Na^+, K^+-ATPase (19). It has been observed, however, that 1,25(OH)$_2$D$_3$ is more potent than estrogen and testosterone (aldosterone not yet compared) in affecting BBMV sulfhydryl groups. In any event, these results suggest that the intestine rapidly responds in a general fashion to different steroid hormones which might operate in concert with their more specific actions. In regard to vitamin D, earlier studies by Adams and Norman (2) and

Holdsworth (20) showed that vitamin D-dependent calcium absorption is inhibited by sulfhydryl blocking reagents whereas these reagents were without effect on vitamin D-independent intestinal calcium transport. It is also of interest to note that the activity of intestinal alkaline phosphatase, an enzyme often implicated in the transport of calcium, is dependent on the status of its sulfhydryl groups (1,31). And further, it was recently reported that the activity of this enzyme in rat duodenum is increased within 10 min. after $1,25(OH)_2D_3$ administration to vitamin D-deficient animals (30), the same time frame in which an increase in BBMV sulfhydryl groups occurs.

In the above described studies on BBMV sulfhydryl groups, the effects of exogenous steroids were examined. It was important to know if similar changes would occur under more physiological circumstances that influence calcium absorption. To this end, chicks were adapted to dietary calcium and dietary phosphorus deficiencies. Previous data indicated that these two deficiencies increase the absorption of calcium, accompanied by an increase in the synthesis of mRNA for calbindin-D_{28K} and calbindin-D_{28K} per se. Analysis of the BBMV isolated from the adapted chicks showed a 2-3 fold increase in total and available sulfhydryl groups. The latter response could certainly be due to the elevated synthesis of $1,25(OH)_2D_3$ although an effect of other factors associated with the adaptation process cannot be eliminated.

The possible role of the modification of the status of protein-associated sulfhydryl groups on the physiological actions of vitamin D and other steroids, and the biochemical and molecular basis of this phenomenon, continues to be investigated.

(Supported by NIH DK04652, NIH ES04072 and NIH GM 24314).

References

1) Abdobrazaghi Z, Butterworth PJ (1983) Reactive thiol groups in calf-intestinal alkaline phosphatase. Enzyme 30:12-20.
2) Adams TH, and Norman AW (1970) Studies on the mechanism of action of calciferol. Basic parameters of vitamin D-mediated calcium transport. J Biol Chem 245:4421-4431.
3) Baimbridge, KG, Mody, I, Miller, JJ (1985) Reduction of rat hippocampal calcium-binding protein following commissural, amygdala, septal, perforant path, and olfactory bulb kindling. Epilepsia 26: 460-465.
4) Bikle DD, Munson S (1986) The villus gradient of brush border membrane calmodulin and the calcium-independent calmodulin-binding protein parallels that of calcium-accumulating ability. Endocrinology 118:727-732.
5) Blunt, JW, DeLuca HF, Schnoes HK (1968) 25-Hydroxycholecalciferol. A biologically active metabolite of vitamin D_3. Biochemistry 7:3317-3322.
6) Borke JL, Caride A, Verma AK, Penniston JT, Kumar R (1989) Plasma

membrane calcium pump and 28-kDa calcium binding protein in cells of rat kidney distal tubules. Amer J Physiol 257:F842-F849.

7) Brasitus TA, Dudeja PK, Eby B, Lau K (1986) Correction by 1,25-dihydroxycholecalciferol of the abnormal fluidity and lipid composition of enterocyte brush border membranes in vitamin D-deprived rats. J Biol Chem 261:16404-16409.

8) Bronner F, Pansu D, Stein WE (1986) An analysis of intestinal calcium transport across the rat intestine. Am J Physiol 250:-G561-G569.

9) Bronner F, Stein WD (1988) CaBPr facilitates intracellular diffusion for Ca pumping in distal convoluted tubule. Am J Physiol 255:F558-F562.

10) Christakos, S (1989) Vitamin D-dependent calcium binding proteins: chemistry, distribution, functional considerations and molecular biology. Endocrine Reviews 10:3-26.

11) Christakos S, Iacopino I, Li H, Lee S, Gill W (1990) Regulation of calbindin-D_{28K} gene expression (this volume).

12) Corradino RA (1974) Embryonic chick intestine in organ culture: interaction of adenylate cyclase system and vitamin D_3-mediated calcium absorptive mechanism. Endocrinology 94:1607-1614.

13) de Boland AR, Nemere I, Norman AW (1990) Ca^{2+}-channel agonist BAY K8644 mimics $1,25(OH)_2$-vitamin D_3 rapid enhancement of Ca^{2+} transport in chick perfused duodenum. Biochem. Biophys Res. Comm. 166:217-222.

14) Dudja PK, Foster ES, Dahiya R, Brasitus TA (1987) Modulation of Na^+-H^+ exchange by ethinyl estradiol in rat colonic brush-border membrane vesicles. Biochim Biophys Acta 899:222-228.

15) Ellman GL (1959) Tissue sulfhydryl groups. Arch Biochem Biophys 22:70-77.

16) Feher JJ (1983) Facilitated calcium diffusion by intestinal calcium-binding protein. Am J Physiol 244:303-307.

17) Feher JJ, Fullmer CS and Fritzsch GK (1989) Comparison of the enhanced steady-state diffusion of calcium by calbindin-D9K and calmodulin: possible importance in intestinal calcium absorption. Cell Calcium 19:189-203.

18) Glenney, JR Jr, Weber K (1980) Calmodulin-binding proteins of the microfilaments present in isolated brush borders and microvilli of intestinal epithelial cells. J Biol Chem 255:10551-10554.

19) Habibi HR, Ince BW (1984) A study of androgen-stimulated L-leucine transport by the intestine of rainbow trout (Salmo gairdneri Richardson) in vitro. Comp Biochem Physiol 79(1):143-149.

20) Holdsworth ED (1965) Vitamin D_3 and calcium absorption in the chick. Biochem J 96:475-483.

21) Jones RG, Davis WL, Hagler HK (1979) Calcium containing lysosomes in the normal chick duodenum: a histochemical and analytical electron microscopic study. Tissue and Cell 11:127-138.

22) Kretsinger RH, Mann JE, Simmons JB (1982) Model of the facilitated diffusion of calcium by the intestinal calcium binding protein. In: Norman AW, Schaefer K, Herrath DV, Gregoleit H-G (eds), Vitamin D: Chemical, Biochemical and Clinical Endocrinology of Calcium Metabolism, Berlin, de Gruyter, pp 233-248.

23) Lawson DEM (1980) Metabolism of vitamin D. In: Norman AW (ed). Vitamin D. Molecular Biology and Clinical Nutrition. Marcel Dekker, Inc., New York, p. 93-126.

24) Lindquist, B (1952) Effect of vitamin D on the metabolism of radiocalcium in rachitic rats. Acta Paediat. 41: Suppl. 86.

25) Long RG, Bikle DD, Munson SJ (1986) Stimulation by 1,25-dihydroxyvitamin D_3 of adenylate cyclase along the villus of chick duodenum. Endocrinology 119:2568-2573.

26) Miller A 3d, Li ST, Bronner F (1982) Characterization of calcium binding to brush-border membranes from rat duodenum. Biochem J 208:773-781.

27) Miller A 3rd, Ueng T-H, Bronner F (1979) Isolation of a vitamin D-dependent,

calcium-binding protein from brush borders of rat duodenal mucosa. FEBS Letters 103:319-322.

28) Morrissey RL, Zolock DT, Mellick PW, Bikle DD (1980) Influence of cycloheximide and 1,25-dihydroxyvitamin D on mitochondrial and vesicle mineralization in the intestine. Cell Calcium 1:69-79.

29) Mykkanen H, Wasserman RH (1989) Uptake of ^{75}Se-selenite by brush border membrane vesicles from chick duodenum stimulated by vitamin D. J. Nutr 119:242-247.

30) Nasr LB, Monet J-D, Lucas PA (1988) Rapid (10 minute) stimulation of rat duodenal alkaline phosphatase activity by 1,25-dihydroxyvitamin D_3. Endocrinology 123:1778-1782.

31) Navaratnam N, Stinson RA (1986) Modulation of activity of human alkaline phosphatases by Mg^{2+} and thiol compounds. Biochim Biophys Acta 869:99-105.

32) Nemere I, Leathers W, Norman AW (1986) 1,25-Dihydroxyvitamin D_3-mediated intestinal calcium transport. Biochemical identification of lysosomes containing calcium and calcium-binding protein (calbindin-D_{28K}). J Biol Chem 261:16106-16114.

33) Norman AW (1980) 1,25$(OH)_2$-D_3 as a steroid hormone. In: Norman AW (ed) Vitamin D. Molecular Biology and Clinical Nutrition. Marcel Dekker, Inc., New York, p. 197-250.

34) Norman AW (1987) Studies on the vitamin D endocrine system in the avian. J. Nutr. 117: 797-807.

35) Rasmussen H, Matsumoto T, Fontaine O, Goodman DB (1982) Role of changes in membrane lipid structure in the action of 1,25-dihydroxyvitamin D_3. Fed Proc 41:72-77.

36) Suzuki S, Ren LH, Chen H (1989) Further studies on the effect of aldosterone on Mg^{2+}-HCO^3(-)-ATPase and carbonic anhydrase from rat intestinal mucosa. J Steroid Biochem 33:89-99.

37) Taylor AN (1974) Chick brain calcium binding protein: comparison with vitamin D-induced calcium binding protein. Arch. Biochem. Biophys. 161: 100-108.

38) Taylor AN (1977) Chick brain calcium binding protein: response to cholecalciferol and some developmental aspects. J. Nutr. 107: 480-486.

39) Van Os CH (1987) Transcellular calcium transport in intestinal and renal epithelial cells. Biochim Biochem Acta 906:195-222.

40) Vesely DL, Juan D (1984) Cation-dependent vitamin D activation of human renal cortical guanylate cyclase. Am J Physiol 246:E115-E120.

41) Warner RR, Coleman JR (1975) Electron probe analysis of calcium transport by small intestine J Cell Biol 64:54-74.

42) Wasserman RH, Brindak ME, Meyer SA, Fullmer CS (1982) Evidence for multiple effects of vitamin D_3 on calcium absorption: response of rachitic chicks, with or without partial vitamin D_3 repletion, to 1,25-dihydroxyvitamin D_3. Proc Natl Acad Sci USA 79: 7939-7943.

43) Wasserman RH, Corradino, RA (1971) Metabolic roles of vitamins A and D. Ann. Rev. Biochemistry 40:501-532.

44) Wasserman RH, Corradino RA, Feher JJ, Armbrecht HJ (1977) Temporal patterns of response of the intestinal calcium absorptive system and related parameters to 1,25-dihydroxycholecalciferol. In: Norman AW, Schaefer K, Coburn JW, DeLuca HF, Fraser D, Grigoleit HG, v. Herrath D(eds) Vitamin D: Biochemical, Chemical and Clinical Aspects Related to Calcium Metabolism. de Gruyter, Berlin, p. 331-340.

45) Wasserman RH, Taylor AN (1966) Vitamin D_3-induced calcium-binding protein in chick intestinal mucosa. Science 152: 791-793.

46) Wilson PW, Lawson DEM (1979) Calcium binding activity by chick intestinal brush-border membrane vesicles. Pflugers Arch 389:69-74.

INTESTINAL CALCIUM TRANSPORT: VESICULAR CARRIERS, NONCYTOPLASMIC CALBINDIN D 28K, AND NON-NUCLEAR EFFECTS OF 1,25-DIHYDROXYVITAMIN D_3

I. Nemere
Department of Biochemistry
University of California
Riverside, CA 92521 USA

The mechanisms of 1,25-dihydroxyvitamin D_3 [1,25-$(OH)_2$-D_3] actions in intestinal epithelium and the components of the calcium transport pathway have been studied from many aspects: Uptake at the brush border membrane, movement of the divalent cation across the cell, and extrusion at the basal lateral membrane. Over the years it has become apparent that the system is complex, mingling elements of intracellular calcium homeostasis present in nearly all cell types, such as phosphoinositide metabolism (Lieberherr et al. 1989) and activation of calcium channels (de Boland et al. 1990), with vesicular trafficking systems involved in net transport (Nemere et al. 1986, 1988, 1989). The present communication will attempt to integrate past and present observations, as well as aspects of intracellular calcium homeostasis involved in signalling the transport pathway.

Both ultrastructural (Jande and Brewer, 1974: Bikle et al. 1979; Davis et al. 1979) and biochemical evidence (Nemere and Szego, 1981; Nemere et al. 1986, 1988, 1989) support a role for vesicular calcium carriers: Uptake of calcium at the brush border by endocytic vesicles, transfer of contents to lysosomes and exocytosis at the basal lateral membrane, with directional transport provided by movement along microtubules.

Biochemically, endocytic vesicles and lysosomes fill the requirements for calcium carriers on the basis of time-course and dose-response studies. In these experiments, vitamin D-deficient chicks were treated with vehicle or seco-steroid hormone prior to

NATO ASI Series, Vol. H 48
Calcium Transport and
Intracellular Calcium Homeostasis
Edited by D. Pansu and F. Bronner

absorption of ^{45}Ca in vivo from ligated duodenal loops. The mucosae from such loops were collected, homogenized and fractionated by differential- and Percoll gradient centrifugation. In the time course analyses (Nemere and Norman 1988), the 1,25-$(OH)_2$-D_3-mediated increase in lysosomal ^{45}Ca levels correlated quite well with stimulated net intstinal calcium absorption, judged by ^{45}Ca in serum. By comparison, seco-steroid hormone dependent augmentation of radionuclide content in endocytic vesicles was maximal at 5 h after 1,25-$(OH)_2$-D_3, a time substantially earlier than the maximum for enhanced transport. It was postulated that hormone-induced alterations in intestinal epithelial microtubules might be required before delivery of endocytic vesicles to lysosomes, and subsequent transport steps could occur. Thus, while while uptake was fully stimulated 5 h after 1,25-$(OH)_2$-D_3, net absorption was not affected in parallel fashion because the correct cytoplasmic "tracks" (microtubules) were not present.

A similar explanation was invoked to account for the following observation in the dose-response studies: Lysosomal ^{45}Ca accumulation was maximal after 10 pmoles of 1,25-$(OH)_2$-D_3, whereas net transport was fully stimulated between 52-260 pmoles of hormone.

Initial evidence for the influence of vitamin D status on microtubule isotypes was obtained at the level of alpha-tubulin mRNA content and by the [^3H]-colchicine binding assay for tubulin protein (Nemere et al. 1987). The latter analyses indicated an absence of [^3H]-colchicine binding in high speed supernatants of intestinal epithelium from vitamin D-deficient chicks, and a progressive increase of binding with time after 1,25-$(OH)_2$-D_3 treatment. Lack of[^3H]-colchicine binding could be due to the absence of tubulin, which was not the case, or the presence of a tubulin isotype that did not bind colchicine.

It is now known that the subunits of microtubules, alpha- and beta-tubulin, each have multiple isotypes (Cleveland et al., 1985). Some of these isotypes arise at the level of transcription, while others are due to post translational modifications such as phosphorylation, tyrosylation, or acetylation. Because of the divergent sources of isotype variation, the influence of vitamin D status on tubulin microheterogeneity was analysed at the protein level. Vitamin D-deficient chicks were dosed with vehicle or 1,25-$(OH)_2$-D_3 prior to the removal of duodena (four per group) under ether anesthesia. The scraped mucosae were homogenized and taxol-stabilized microtubules (Vallee, 1982) prepared from high speed supernatants. Separation of microtubule isotypes was accomplished on isoelectric focusing (IEF) gels following the procedure of Fields et al. (1984).

Figure 1 depicts the results of a partial time course study.

Fig. 1. Effect of 1,25-$(OH)_2$-D_3 on tubulin isotypes of chick intestinal epithelium.

Microtubules were prepared from the intestinal epithelium of vitamin D deficient (-D) chicks treated with vehicle or 1.3 nmoles of 1,25-$(OH)_2$-D_3 15 h or 43 h prior to sacrifice. The major difference between tubulin from rachitic and treated chicks is the appearance of two bands in the bottom (acidic) region of the gel in samples from 1,25-$(OH)_2$-D_3 dosed chicks. Based on electrophoretic migration, the hormone-induced bands correspond to alpha tubulin isotypes, which are usually more acidic than beta tubulins. Similar analyses at earlier times after 1,25-$(OH)_2$-D_3 indicated that the bands appeared 5 h after hormone, and increased in intensity 10 h after dosing, a time corresponding to maximal calcium transport (data not shown). Forty-three h after 1,25-$(OH)_2$-D_3 both lysosomal ^{45}Ca and intestinal calcium absorption return towards basal levels (Nemere and Norman 1988), while CaBP 28K levels are elevated, and no decreases in hormone-mediated tubulin isotypes are evident (Fig. 1). Thus, cessation of augmented absorption may be due to the absence of 1,25-$(OH)_2$-D_3 acting at a non-nuclear level (see below), rather than due to the absence of a transport pathway element such as an induced protein.

The seemingly pivotal role of microtubules in augmented transport suggested that "soluble" CaBP 28K might actually be associated with cytoskeletal elements (Nemere et al., manuscript in preparation). Although an enriched specific immunoreactivity has been reported in endocytic vesicle and lysosomal fractions (Nemere et al. 1986), a substantial proportion of CaBP 28K remains soluble. Microtubules become depolymerized in the cold, a condition employed in biochemical preparations, and thus are also "soluble" under circumstances that yield soluble CaBP 28K. Several approaches have been employed to determine whether CaBP 28K is associated with cytoskeletal elements: Isolation of microtubules or tubulin by a variety of methods resulted in associated immunoreactive CaBP 28K, and double indirect immunofluorescent labelling of fixed, forzen intestinal sections has revealed colocalization of tubulin and calbindin (Nemere and Norman, 1988b). More recently, microtubule components were separated on SDS-polyacrylamide gels and overlayed with $[^{125}I]$-CaBP 28K. The results shown in Fig 2 were reproduced in two independent experiments.

Overlays conducted in the presence of 1 mM $CaCl_2$ exhibited reduced binding of iodinated CaBP 28K relative to gels incubated in the presence of 1 mM EGTA. Two protein bands (indicated by arrows) exhibited a seco-steroid hormone induced increase in binding of calbindin: The smaller protein, which corresponds to tubulin by immunoblot analysis, exhibited a +D/-D binding ratio of 2.40 and 4.88 in the first and second experiments, respectively, while the larger protein (MW 187,000) revealed a binding ratio of 1.35 and 4.23 in the two separate experiments.

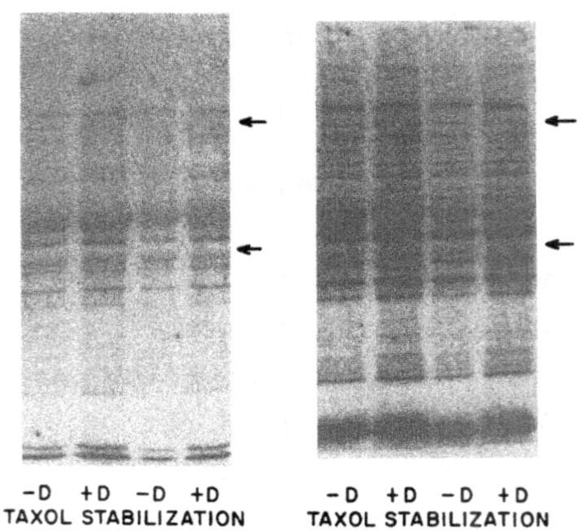

Fig. 2. Calcium-dependent association of iodinated CaBP 28K with tubulin and microtubule associated proteins on SDS-polyacrylamide gels. Arrows indicate two protein bands that exhibit reproducible 1,25-$(OH)_2$-D_3- mediated increases in CaBP binding.

The apparent overabundance of ^{125}I-CaBP acceptor sites suggested a lack of specificity and prompted an attempt to competitively displace labelled CaBP binding with cold calbindin. There was a marked failure to displace radioactive binding, corroborating the findings of others (Marian R. Walters, personal communication). However, it is conceivable that the total calbindin added (iodinated and cold) was insufficient to saturate all available binding sites. The possibility is strengthened by recent work at the ultrastructural level that indicates CaBP 28K immunoreactivity is associated with intestinal epithelial cell microtubules, in addition to abundant representation in vesicles and lysosomes. Despite the predominant association with discrete cell structures, CaBP 28K also reveals a cytoplasmic localization in thin sections prepared from vitamin D_3 treated chick intestine. How the various subcellular distributions of calbindin interrelate and contribute to calcium transport remains to be determined. It is tempting to speculate that the CaBP in vesicles and lysosomes acts to sequester calcium during transport, while that in the cytoplasm might act in a similar fashion as proposed by Bronner et al. (1986) and Feher et al. (1989). Cytoplasmic calbindin might also be the source of cytoskeletal CaBP 28K and contribute to the structure and function of microtubules and filaments.

Given that these various elements of the calcium transport pathway are present but quiescent in the intestinal epithelium of normal, vitamin D-replete chicks, how does the introduction of $1,25\text{-(OH)}_2\text{-D}_3$ to the basal lateral membrane activate the rapid, hormonal stimulation of calcium transport (transcaltachia)? From the early observation that parathyroid hormone parallels the actions of $1,25\text{-(OH)}_2\text{-D}_3$ in isolated rat intestinal cells (Nemere and Szego, 1981), and the more recent finding that PTH stimulates transport in normal chick duodena (Nemere and Norman, 1986), it is likely that phosphorylation events play a role in the initiation of transcaltachia, either through activation of protein kinase A or C. Current work with the perfusion system tends to support the likelihood (de Boland and Norman, submitted). In addition, activation of calcium channels at the basal lateral membrane by $1,25\text{-(OH)}_2\text{-D}_3$ appears to be intimately involved (de Boland et al., 1990). Figure 3 summarizes these steps.

Hormonal activation of protein kinase(s) and calcium channels (and possibly mobilization of intracellular calcium stores) results in a situation that resembles stimulus-secretion coupling: redistribution of ionic calcium, from either intracellular or extracellular sources stimulates the movement of vesicles/lysosomes along microtubules and/or exocytosis of the contents. Thus, a number of pathways routinely occuring in other cell types appear to have been adapted to the specialized function of calcium transport in intestinal epithelial cells. A major puzzle remaining in this model of hormone mediated intestinal calcium transport is regulation at the brush border membrane: What is the calcium "receptor" and how do events at

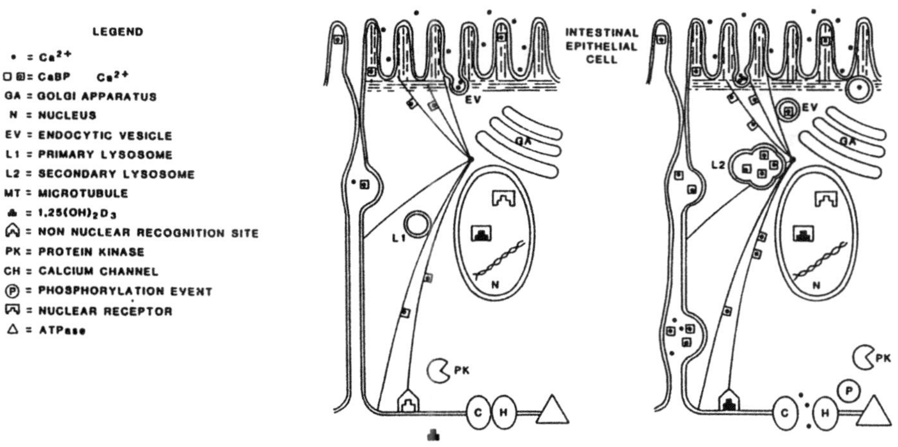

Fig. 3. Schematic diagram of vesicular calcium transport and potential steps involved in activation by 1,25-$(OH)_2$-D_3.

the basal lateral membrane modulate events at the brush border membrane? What determines unidirectional flow of transport vesicles, and how are they regenerated for subsequent demands? These and other questions should provide ample new and continuing directions for study.

ACKNOWLEDGMENTS The author wishes to thank Dr. Anthony W. Norman for grant support that allowed her to experimentally test her own intellectual contributions to the field, namely nongenomic effects, vesicular carriers, subcellular localization of CaBP, and microtubule isotypes. Thanks also to Dr. William Fletcher for his expertise in morphology and unflagging good humor.

References

Bikle DD, Morrissey RL, Zolock DT (1979) The mechanism of action of vitamin D in the intestine. Am J Clin Nutr 32: 2322-2338

Boland A de, Nemere I, Norman AW (1990) Ca^{2+}-Channel agonist bay K8644 mimics 1,25$(OH)_2$-vitamin D_3 rapid enhancement of Ca^{2+} transport in chick perfused duodenum. Biochem Biophys Res Commun 166: 217-222.

Bronner F, Pansu D, Stein WD (1986) An analysis of intestinal calcium transport across the rat intestine. Am J Physiol 250: G561-G569

Cleveland DW, Sullivan KF (1985) Molecular biology and genetics of tubulin. Ann Rev Biochem 54: 331-365

Davis WL, Jones RG, Hgler HK (1979) Calcium containing lysosomes in the normal chick duodenum: A histochemical and analytical electron microscopic study. Tiss Cell 11: 127-138

Feher JJ, Fullmer CS, Fritzsch GK (1989) Comparison of the enhanced steady-state diffusion of calcium by calbindin-D9K and calmodulin: Possible importance in intestinal calcium absorption. Cell Calcium 10: 189-203

Field DJ, Collins RA, Lee JC (1984) Heterogeneity of vertebrate brain tubulins Proc Natl Acad Sci USA 81: 4041-4045

Jande SS, Brewer LM (1974) Effects of vitamin D_3 on duodenal absorptive cells of chick. Z Anat Entwickl-Gesch 144: 249-265

Lieberherr M, Grosse B, Duchambon P, Drucke T (1989) A functional cell surface type receptor is required for the early action of 1,25-dihydroxyvitamin D_3 on the phosphoinositide metabolism in rat enterocyte. J Biol Chem 264: 20403-20406

Nemere I, Leathers VL, Norman AW (1986) 1,25-Dihydroxyvitamin D_3-mediated calcium transport across the intestine: Biochemical identification of lysosomes containing calcium and the calcium binding protein (calbindin-D_{28K}). J Biol Chem 261: 16106-16114

Nemere I, Norman AW (1986) Parathyroid hormone stimulates calcium transport in perfused duodena of normal chicks: Comparison with the rapid effect (transcaltachia) of 1,25-dihydroxy-vitamin D_3. Endocrinology 119: 1406-1408

Nemere I, Norman AW (1988) 1,25-Dihydroxyvitamin D_3-mediated vesicular transport of calcium in intestine: Time course studies. Endocrinology 122: 2962-2969

Nemere I, Norman AW (1988b) Transcaltachia, vesicular calcium transport, and microtubule-associated calbindin-D_{28K}. In: Norman AW et al. (eds), Vitamin D: Molecular, cellular and Clinical endocrinology. Walter de Gruyter, Berlin, p 549-557.

Nemere I, Norman AW (1989) 1,25-Dihydroxyvitamin D_3-mediated vesicular transport of calcium in intestine: Dose-response studies. Molec Cell Endocrinol 67: 47-53

Nemere I, Szego CM (1981) Early actions of parathyroid hormone and 1,25-dihydroxycholecalciferol on isolated epithelial cells from rat intestine: I. Limited lysosomal enzyme release and calcium uptake. Endocrinology 108: 1450-1462.

Nemere I, Theofan G, Norman AW (1987) 1,25-Dihydroxyvitamin D_3 regulates tubulin expression in chick intestinal epithelium. Biochem Biophys Res Commun 148: 1270-1276

Vallee RB (1982) Microtubule isolation by the taxol method. J Cell Biol 92: 435-442

CALCIUM TRANSPORT ACROSS THE COLON

U. Karbach
Medical Clinic Innenstadt
University of Munich
Ziemssenstr. 1
8000 München 2
Fed. Rep. of Germany

In 1969 Harrison and Harrison were the first who found Ca absorption in the rat colon in vitro. For a long time, however, it has been suggested that bulk of dietary Ca absorption occurs across the small intestine with little contribution from the large bowel. Perfusion studies confirmed that the rat colon is capable to absorb Ca at physiological luminal concentrations (Petith and Schedl 1976). Furthermore, there is clinical evidence that in patients suffering from malabsorption in the small intestine the preservative function of the colon plays an important role in maintaining a sufficient Ca absorption(Hylander et al. 1980). Therefore it seemed of interest to study Ca transport across different segments of the colon in order to compare Ca transport across the large bowel with the transport pattern of Ca across the small intestine.

Methods: Unidirectional mucosa to serosa (ms) and sm ^{45}Ca flux (nmol/cm^2·h ± SEM) simultaneously with the flux of the paracellular marker ^{3}H-mannitol in the absence of electrochemical gradients (SCC, Ussing chamber) and by the voltage clamp (VC) technique in vitro was studied across different segments of the rat small and large intestine. The VC method allows to discriminate between the cellular and paracellular compenent of the unidirectional Ca transport. Fluxes across clamped preparations were evaluated using the equation $J_i = {}_oJ_d \cdot \xi^{-1/2} + J_m$ ($\xi = z \cdot F \cdot PD/RT$; ${}_oJ_d$ = SCC Ca flux; J_i = Ca flux at any clamp PD). Accordingly, the PD-independent cellular Ca flux is given by the intercept, whereas the slope of the line represents the PD-dependent, i.e. paracellular fraction of the unidirectional Ca transport (Frizzell and Schultz 1972). Ca transport was measured in controls and in rats pretreated with 1,25-$(HO)_2$-D_3 (250ng/d ·kg sc for 3d), dexamethasone (dexa, 1mg/kg·d sc for 3d), or diphosphonate (EHDP, 40 mg/kg·d sc for 7d). EHDP is known to inhibit the synthesis of 1,25-$(OH)_2$-D_3 by blocking the 1a-hydroxylase in the kidney (Gasser1972).

Results: Fluxes across the short-circuited tissue demonstrate that Ca

NATO ASI Series, Vol. H 48
Calcium Transport and
Intracellular Calcium Homeostasis
Edited by D. Pansu and F. Bronner

is only absorbed in the duodenum and colon but it is secreted in the jejunum and ileum (Tab.1).

	CONTROLS			1,25-$(OH)_2$-D_3		
	Jms	Jsm	Jnet	Jms	Jsm	Jnet
Duodenum	50±9	36±4	14±5	74±4°	51±5°	23±2°
Jejunum	67±7	91±7	-24±7	103±9°	104±8	- 1±9°
Ileum	62±5	74±7	-12±3	85±5°	103±7°	-18±3
Cecum	186±9	24±2	162±9	167±8	35±5	132±7
C.ascend.	101±4	28±3	73±1	109±8	33±2	76±3
C.descend.	31±2	24±3	7±2	64±3°	19±2	45±3°

Tab.1
Unidirectional mucosa to serosa (Jms), serosa to mucosa (Jsm) and net Ca flux (Jnet = Jms - Jsm) across different segments of the small and large intestine in controls and in rats pretreated with 1,25-$(OH)_2$-D_3. Negative sign indicates secretion; ° significant difference when compared with the corresponding control flux.

In the C. descendens Ca absorption is comparable to that in the duodenum. Ca absorption in ascending colon is five times and in the cecum even ten times higher than in the descending colon or duodenum. 1,25-$(OH)_2$-D_3 has only a small effect on Ca absorption in the duodenum but abolishes Ca secretion in the jejunum. In the C.descendens, however, ms Ca flux is increased by 100% and Ca absorption is six times higher than in controls. On the other hand 1,25-$(OH)_2$-D_3 has no effect on net Ca flux across the ileum (Karbach and Rummel 1987a), cecum or ascending colon (Tab.1). In 1,25-$(OH)_2$-D_3-depleted rats after pretreatment with EHDP, however, ms flux and Ca absorption in the cecum and ascending colon is decreased by 60% and Ca absorption is completely abolished in the C.descendens (Fig.1). After the simultaneous application of EHDP and 1,25-$(OH)_2$-D_3 Ca absorption is turned to control values in all colonic segments (Fig.1). The finding suggests that in the C.descendens Ca absorption is purely 1,25-$(OH)_2$-D_3-dependent, whereas in the proximal colon in addition a 1,25-$(OH)_2$-D_3-independent absorptive mechanism exists. Furthermore, the result demonstrates that the 1,25-$(OH)_2$-D_3-sensitive transporter in the proximal colon under normal nutritional conditions already works at the maximal level. The segmental heterogeneity of the Ca transport between the proximal and distal colon becomes also evident when comparing the kinetics of the unidirectional Ca fluxes (Fig.2). In the ascending colon in the Ca transport in both directions a saturable component is involved.

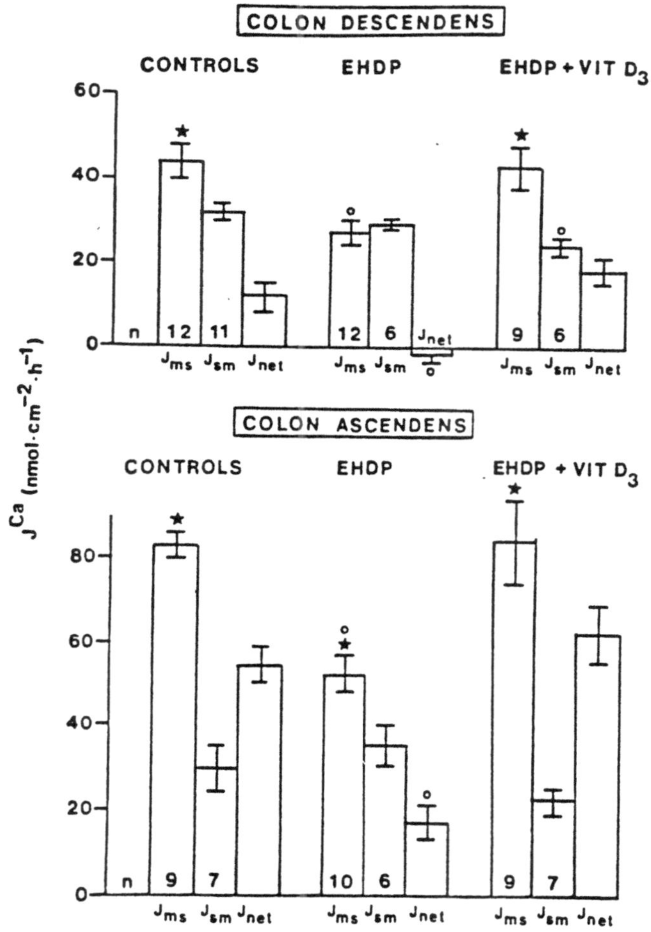

Fig.1
Unidirectional mucosa to serosa (ms), serosa to mucosa (sm) and net Ca transport across the ascending and descending colon of controls and rats pretreated with EHDP or after the simultaneous application of EHDP and 1,25-$(OH)_2$-D_3. *indicates significant difference when compared with the oppositely directed flux; ^{o} significantly different from the corresponding control flux.

The maximal transport capacity (Vmax = 238$\pm$7 vs 154$\pm$12nmol/cm^2·h) and the affinity of the carrier (K_t = 1.8$\pm$0.2 vs 5.4$\pm$0.8mmol/l) for the ms is remarkably higher than for the sm flux so that Ca is absorbed at all concentrations. In contrast, in the distal colon only in the ms flux a saturable fraction is involved (Vmax = 52$\pm$15nmol/cm^2·h; K_t = 0.7$\pm$0.3mmol/l). The sm flux in the distal colon, however, is a linear function of the Ca concentration. At low concentrations Ca is absorbed, at Ca concentration higher than 2.5mmol/l Ca is secreted (Karbach et al. 1986). This secretion is

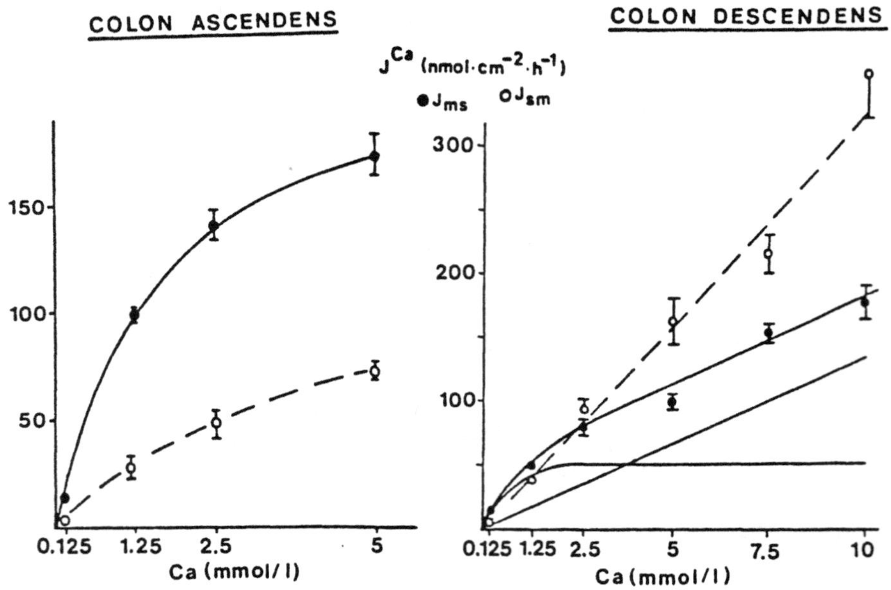

Fig.2
Concentration dependence of unidirectional mucosa to serosa (points) and serosa to mucosa (circles) Ca flux across the short-circuited ascending and descending colon.

not an active process but is the consequence of a prevalence of the purely diffusive sm Ca flux, which is about 2.5 times higher than the diffusive fraction of ms transport (33±2· Ca vs 13±2· Ca nmol/cm^2·h). The voltage dependence (Fig.3) agree well with the concentration dependence (Fig.2) of the Ca fluxes across both colonic segments. In the proximal colon in both directions a voltage-independent cellular mechanism is involved. The cellular ms transport is three times higher than the cellular sm flux (30±3 vs 9.2±2nmol/cm^2·h). Furthermore, the paracellular ms flux in the ascending colon is two times higher than the paracellular sm flux (46±2 vs 23±3·$\xi^{-1/2}$). From the VC experiments it can be calculated that only 38% of the ms flux measured across the short-circuited ascending colon is cellular whereas 62% is paracellular. In the distal colon, however, only in the ms Ca flux a cellular fraction is involved (13±2nmol/cm^2·h), which is about 60% smaller than the cellular ms Ca flux in the ascending colon. In the distal colon only 46% of the ms flux is cellular whereas 54% is paracellular. With other words, still in the relatively tight colonic mucosa Ca flux across the paracellular route plays an important role. The sm flux across the descending colon is purely paracel-

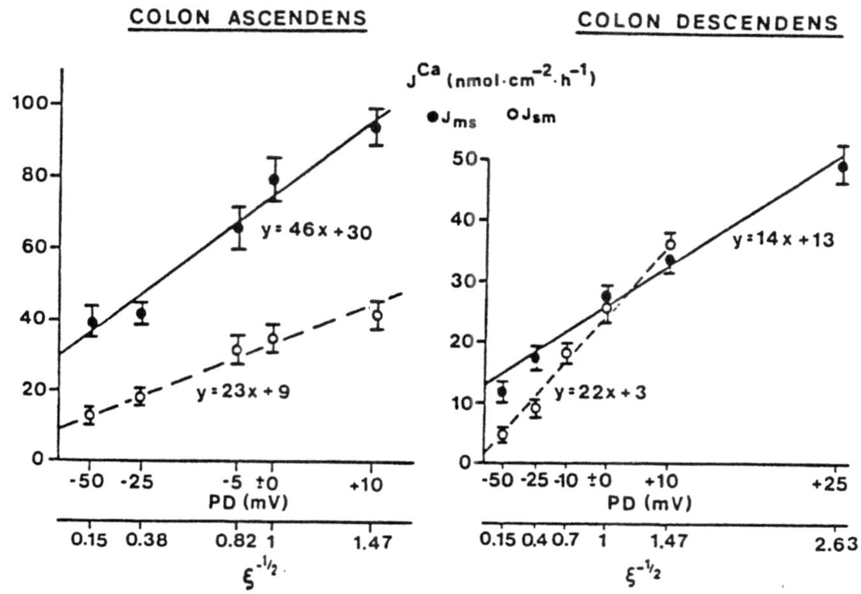

Fig.3
Voltage dependence of unidirectional mucosa to serosa (points) and serosa to mucosa (circles) Ca flux across the ascending and descending colon. The sign of PD is given with reference to that side of the tissue flux is measured from. The intercept of the line represents the voltage-independent, i.e. cellular flux. The voltage-dependent paracellular fraction of the flux is given by the slope of the line.

lular. But in contrast to the proximal colon in the distal colon there is an asymmetry of the paracellular flux with a prevalence of the sm flux over that in ms direction (22±3 vs 14±1·$\xi^{-1/2}$). It is hypothesized that the preferential paracellular sm Ca flux induces Ca secretion across the short-circuited descending colon at high Ca levels when the carrier is saturated (Fig.2).

	J^{Ca}						J^{man}					
	CONTROLS			DEXA			CONTROLS			DEXA		
	ms	sm	net	ms	sm	net	ms	sm	net	ms	sm	net
CA	64±5	25±5	39±4	83±6°	51±5°	32±2	54±4	39±2	15±5	89±4°	82±7°	7±5°
CD	26±2	15±1	11±1	26+2	22±2°	-4±1°	30±2	29±2	1±1	32±4	42±3°	-10±2°

Tab.2
Influence of dexamethasone on unidirectional mucosa to serosa (ms), serosa to mucosa (sm) and net flux of Ca and the simultaneously measured paracellular marker mannitol (man). ^{o} indicates significant difference when compared with the corresponding control flux.

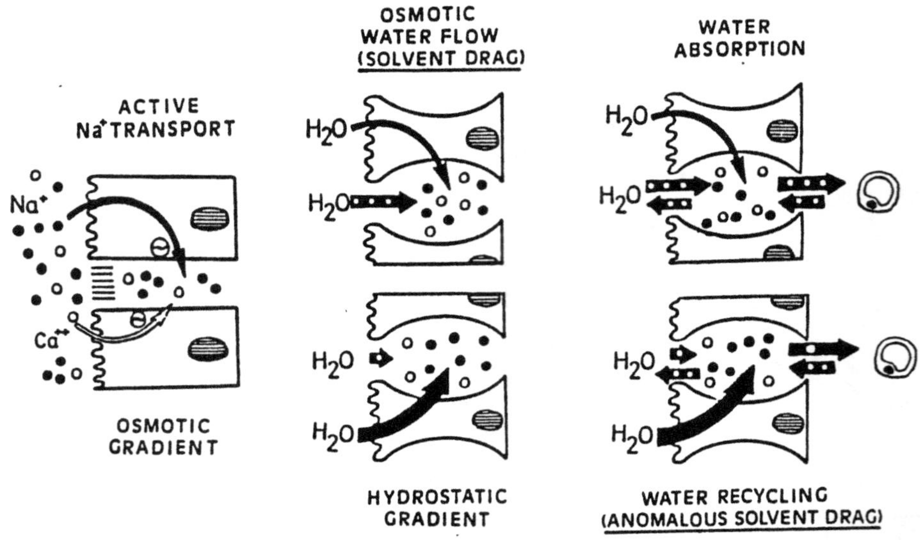

Fig.4
Model of the epithelium to explain the oppositely directed asymmetry of the paracellular Ca flux across the proximal and distal colon as the consequence of an asymmetric transepithelia water flow across the paracellular shunt pathway.

Driving force for solute stream along the paracellular pathway is transepithelial water flow.Therefore Ca flux under the influence of dexa was measured in order to characterize the effect of transepithelial water flow on paracellular Ca flux (Tab.2). Dexa is known to increase Na and hence water absorption by an activation of the Na, K-ATPase. In the ascending colon dexa increase Ca and mannitol flux in both directions to the same degree and has no essential effect on the net flux of both substances (Karbach and Rummel 1987b). In the distal colon, however, dexa only stimulates sm Ca and mannitol flux and hereby induces secretion of both substances. Such a preferential paracellular solute transport in the opposite direction to water flow has been names "anomalous solvent drag effect" (Ussing and Johansen 1969). It is described for the sm calcium and mannitol flux across the rat ileum (Nellans and Kimberg 1979). The results demonstrate that Ca transport across the colon is remarkably determined by the paracellular Ca movement as the consequence of transepithelial water flow. Furthermore, the data confirm that a segmental heterogeneity not only for the cellular but also with respect to the paracellular Ca flux between the proximal and distal colon exists (Karbach 1989) (Fig.4).

<u>Conclusions:</u> The results demonstrate that Ca is only absorbed in the duodenum and colon. Passage through the duodenum is too fast to allow the absorption of substantial amounts of Ca. Ca is secreted in the jejunum and ileum, consequently the luminal concentration of available Ca in the proximal colon (pH<5) might amount to 40mmol/l (Hill 1982). Ca absorption in the proximal colon, which is the highest all along the intestine, therefore might be of physiological importance in the regulation of Ca absorption: Under normal nutritional conditions the carrier for the ms Ca flux operates at the optimal level and still in the absence of $1,25\text{-}(OH)_2\text{-}D_3$ in the proximal colon Ca is absorbed by solvent drag. Ca absorption in the distal colon, similar to the value found in the duodenum, can be stimulated by $1,25\text{-}(OH)_2\text{-}D_3$ and therefore might preserve Ca under certain pathophysiological conditions, i.e. malabsorption in the proximal intestine. The qualitative data are consistent with the possibility, previously grossly underestimated, that the colon may play a major role in the regulation of intestinal Ca homeostasis.

<u>References:</u>

Frizzell RA, Schultz SG (1972) Ionic conductances of extracellular shunt pathway in rabbit ileum. J Gen Physiol 59:318-346

Gasser AB, Morgan DB, Fleisch A, Richelle J (1972) The influence of two diphosphonates on calcium metabolism in the rat. Clin Sci 43:31-45

Harrison HC, Harrison HE (1969) Calcium transport by rat colon in vitro. Am J Physiol 217:121-125

Hill GL (1982) Metabolic complications of ileostomy. Clin Gastroenterol 11/2:260-267

Hylander E, Ladefoged K, Jarnum S (1980) The importance of the colon in calcium absorption following small bowel resection. Scand J Gastroenterol 15:55-60

Karbach U (1989) Intestinal calcium transport in health and disease. Dig Dis 7:1-18

Karbach U, Bridges RJ, Rummel W (1986) The role of the paracellular pathway in the net transport of calcium across the colonic mucosa. Naunyn-Schmiedeberg's Arch Pharmacol 334:525-530

Karbach U, Rummel W (1987a) Cellular and paracellular calcium transport in the rat ileum and the influence of 1,25-dihydroxyvitamin D_3 and dexamethasone. Naunyn-Schmiedeberg's Arch Pharmacol 336:117-124

Karbach U, Rummel W (1987b) Calcium transport across the mucosa of the colon ascendens and the influence of 1,25-dihydroxyvitamin D_3 and dexamethasone. Eur J Clin Invest 17:368-374

Nellans HN, Kimberg DV (1979) Anomalous calcium secretion in rat ileum: role of paracellular pathway. Am J Physiol 264:E473-481

Petith MM, Schedl HP (1976) Intestinal adaptation to dietary calcium restriction: in vivo cecal and colonic calcium transport in the rat. Gastroenterology 71:1039-1042

Ussing HH, Johansen B (1969) Anomalous transport of sucrose and urea in toad skin. Nephron 6:317-328

CALCIUM ABSORPTION. VITAMIN D EFFECTS ON RAT INTESTINAL CALCIUM TRANSPORT SYSTEMS DEPEND ON THE STATE OF ENTEROCYTE DIFFERENTIATION ALONG THE CRYPT-VILLUS AXIS

Milton M. Weiser
Jay Zelinski
Division of Gastroenterology and Nutrition
Department of Medicine, SUNY Buffalo and
Buffalo General Hospital
100 High Street
Buffalo, New York 14203
United States of America

Introduction. Calcium absorption by the rat small intestine is most efficient in the duodenum and least efficient in the jejunum (Kimberg et al., 1961). Both active and passive transport processes contribute to calcium absorption with active transport accounting for over 75 percent of the duodenum's ability to maintain the highest serosal-to-mucosal ratios (Bronner et al., 1986) and the intestinal response to 1,25$(OH)_2$cholecalciferol (Halloran and Deluca). The response of the vitamin D-deficient rat intestine to 1,25$(OH)_2$cholecalciferol appears to be biphasic with an initial restoration in calcium absorption by 6 hours, a subsequent loss by 8-12 hours followed by a more constant restoration by 24-48 hours (Maclaughlin et al., 1980 and Halloran and Deluca, 1981) (Fig. 1). Our laboratory has provided evidence to suggest that a partial explanation for this biphasic response is related to the state of enterocyte differentiation. The enterocyte has a rapid turnover, 48-72 hours, as it undergoes differentiation in the crypt and lower villus until it is finally expelled into the lumen in the upper villus tip. Our data suggest that the upper villus cells are responsible for active calcium absorption and that these cells respond differently to 1,25$(OH)_2$cholecalciferol than do the undifferentiated and differentiating enterocyte. In this article,

NATO ASI Series, Vol. H 48
Calcium Transport and
Intracellular Calcium Homeostasis
Edited by D. Pansu and F. Bronner

this data will be summarized and questions arising from these findings presented.

<u>Methods.</u> The method used separates enterocytes as isolated cell fractions representative of the crypt-villus axis of differentiation (Weiser, 1973). Confirmation of this axis of differentiation is determined enzymatically by a gradient of alkaline phosphatase and/or sucrase activity, these activities being highest in the upper villus cells, and by thymidine kinase activity, it being highest in the lower crypt cells. More recently, we have demonstrated reciprical gradients of actin mRNA (compatible with the increase in the villus cell's microvillus actin content) and histone H2B mRNA (increased in the mitotically active lower crypt cell zone) (Weiser, et al.,1990). The major advantages of isolated enterocytes over the traditional intestinal scrapings is the decreased contamination with intestinal mucin and pancreatic hydrolytic enzyme activities. Rat intestinal mucin has long been a troublesome problem in purification of enzymes from the intestine and in our hands mucin significantly lowers the yield of RNA. Demonstration of an ATP-dependent calcium uptake by membrane vesicles prepared from the lateral domain of the enterocyte was apparent only in preparations derived from isolated cells (Walters, et al., 1986). The possible disadvantages of the isolated cell preparation are the increased time of preparation (up to 90 min for lower crypt cells), potential loss of intracellular material, particularly small molecules such as ATP, intracellular changes and/or degradation, and alterations in plasma mambrane topography. Despite these potential problems membrane vesicles have been isolated, as defined by Na^+,K^+-ATPase activity enriched membranes, felt to represent the lateral domain of the enterocyte plasma membrane.

<u>Results.</u> When plasma membrane vesicles were prepared from isolated cells derived from a crypt-villus gradient of rat enterocytes, an ATP- vitamin D-dependent calcium uptake was demonstrated (Walters and Weiser, 1987). With membrane vesicles prepared from duodenal

villus cells, this calcium uptake was shown to be inhibited by vanadate, not functional in the presence of the calcium ionophore A23187, and decreased in vitamin D-deficient animals. In normal, vitamin D-sufficient animals calcium uptake by lateral membrane vesicles was approximately two-fold greater in vesicles from upper villus as compared to those from crypt cells. Analysis of the kinetics of uptake suggested that the difference was not due to a change in the affinity of the pump for calcium (both showed a K_m=0.3uM), but rather due to a difference in the number of functional pumps within the membrane (Walters and Weiser, 1987). Essentially, in the vitamin D-deficient state the rate of calcium uptake by lateral plasma membrane vesicles of villus cells is reduced to that observed with vesicles from crypt cells whether these crypt cells are derived from vitamin D-deficient or vitamin D-sufficient rats. Figure 2 schematically summerizes these data but the repletion data are mostly hypothetical although partly supported by data with Golgi membranes (Maclaughlin et al., 1980). In experiments with lateral membranes, the effect of 1,25$(OH)_2$cholecalciferol repletion on calcium uptake was only determined for one time point, six hours after injection (Walters and Weiser, 1987) (Fig. 2), a time when the first restoration peak of calcium absorption was observed (Fig. 1). It appears that six hours after repletion, calcium uptake by duodenal villus lateral plasma membranes was significantly increased, but only in the mid and lower villus fractions with no increase seen in the villus tip cell membrane vesicles. The data implied that there was a calcium pump (Ca-pump) in the lateral domain of villus cells that was responsive to the vitamin D status of the animal, or that the Ca-pump, present in both crypt and villus cells, could be increased by a vitamin D responsive system active only in the mature villus cell. Did the villus cell have an additional Ca-pump not present in the crypt cells or was it the same pump, whose content could be increased over a threshold by vitamin D-responsive elements? Was the relatively small increase in Ca-pump activity observed in villus cells below the tip sufficient to account for the near total restoration of calcium absorption at six hours as measured in gut sacks? To answer these questions we tried to purify the Ca-pump(s) to further characterize its properties, compare villus with crypt and produce antibodies to better study its synthesis.

Purification proved difficult because of its relatively low abundance. Initial attempts to monitor Ca-pump activity during purification by measuring Ca-ATPase activity was thwarted by the finding that the lateral domain of the enterocyte plasma membrane also contained an abundant nucleotide phosphatase activity that was not specific for ATP (Moy, et al., 1986). This appeared not to be due to alkaline phosphatase or 5'-nucleotidase activities but was an activity similar to that reported by Lin, 1988, as a non-specific Ca-ATPase activity on liver cell membranes, an ecto-nucleotide phosphatase activity separable from the Ca-ATPase activity attributable to the Ca-pump. Subsequently, we were able to purify the villus enterocyte Ca-pump by formation of a ^{32}P-labeled intermediate, and purification by calmodulin affinity chromatography (Wajsman, et al., 1988). Its estimated molecular size was 130kDa, compatible with other plasma membrane Ca-pumps. Its properties suggested an acylphosphate bond, preferentially phosphorylated by ATP, a phosphorylated intermediate stabilized by the presence of Ca^{2+} and La^{3+} and inhibited by vanadate. Because of the very low quantities present in the lateral membrane, and developments by other investigators, further investigation concentrated on detecting the transcript for the enterocyte plasma membrane Ca-pump.

A number of cDNAs for plasma membrane Ca-pumps have been described. Shull has isolated cDNA probes from a rat brain library that define a number of isoforms of the Ca-pump. He has used a 3' restriction fragment of one isoform, PMCA1, to survey the presence of the Ca-pump transcript in various tissues (Greeb and Shull, et al., 1988). RNA extracted from full thickness rat intestine showed the presence of Ca-pump transcripts. Our laboratory used a similar probe of the 3'untranslated region (Gift of Gary Shull, University of Cincinnati) to study the steady state content of the rat enterocyte

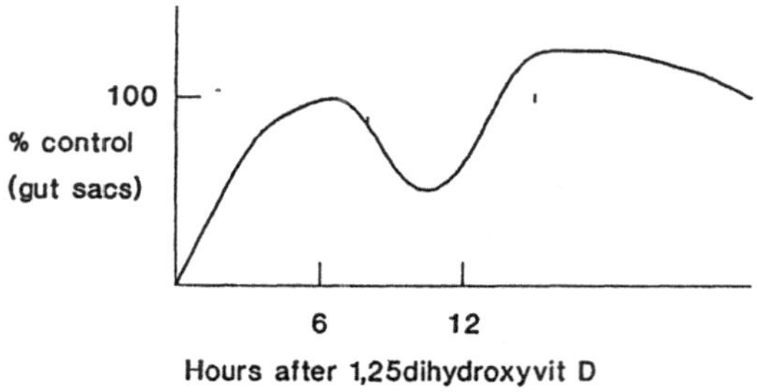

Fig.1. Restoration of Calcium Absorption Relative to Vit D-sufficient Rat

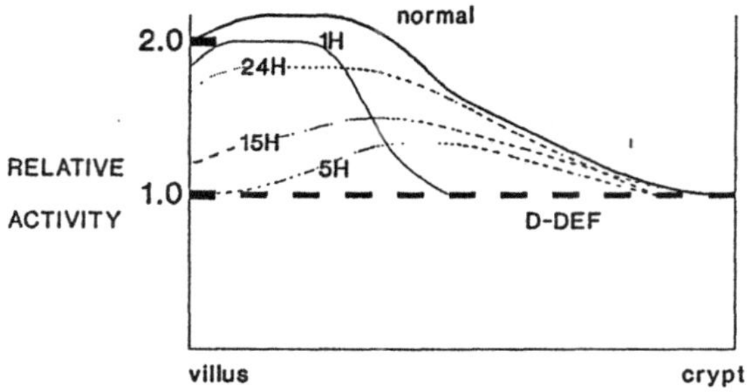

Fig. 2. Relative Activity of Ca-pump along Crypt-Villus Gradient of Differentiation Effect of 1,25dihydroxyvitD Repletion

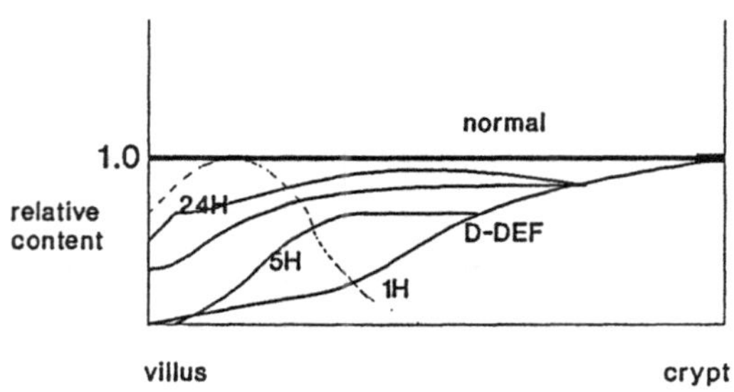

Fig. 3. Ca-Pump mRNA Steady-state Content Villus-Crypt Gradient of Differentiation Effect of 1,25dihydroxyvitD Repletion

plasma membrane Ca-pump mRNA as related to the crypt-villus axis of enterocyte differentiation, and to the effect of vitamin D deficiency and 1,25$(OH)_2$cholecalciferol repletion on the steady state content of this transcript. Figure 3 summarizes our results schematically. Actually, four different sized transcripts were detected on Northerns, one of which was more represented in the crypt cell, an 8kB transcript. Each of the four acted differently in response to repletion but there was a general pattern for each that is shown in Fig. 3. The major change in the D-deficient state was the marked decrease in steady state content of Ca-pump mRNA in upper villus cells. But it was in these cells that one observed the most rapid and fleeting response; by one hour after repletion, the earliest time tested, there was a full recovery of Ca-pump mRNA, levels close to D-sufficient controls, but by 3 hours content was back to untreated D-deficient levels only to slowly rise to approach D-sufficient levels at 24 hours or later. In the lower villus, upper and lower crypt zones, recovery to D-sufficient levels appeared to be first detected at 5 hours, slightly dip and then slowly increase up to the 24 hours tested.

Discussion. These results indicate that the action of 1,25$(OH)_2$cholecalciferol may differ depending on the state of cell differentiation. Enterocytes are not at one state of differentiation but are constantly dividing and differentiating at a rate exceeded only by the hematopoietic cells of the bone marrow. The results above do not suggest a direct effect of vitamin D on enterocyte differentiation; this was not evaluated. It does suggest that the effects of vitamin D on promoting Ca-pump synthesis may involve different control mechanisms depending on the state of enterocyte differentiation. More time points may prove it to be even more complicated but the present evidence suggests that the mature enterocyte has a system, presumably nuclear, which is maximally primed to receive 1,25$(OH)_2$cholecalciferol and rapidly synthesize Ca-pump mRNA. Its fairly rapid disappearance over 2 hours can be partly attributed to turnover of the upper villus cells, some being shed into the lumen. However, unless

$1,25(OH)_2$cholecalciferol repletion also increases cell turnover and loss, another explanation must be proposed for the subsequent decrease in Ca-pump mRNA content observed at 3 hours in these upper villus cells. A simple interpretation is that these transcripts have completed their task of synthesizing more Ca-pumps and are subsequently destroyed along with inhibition of Ca-pump mRNA synthesis. Space limits discussion of other possibilities. The response of the lower villus and upper crypt cells is compatible with a *normal* reaction of differentiating cells responding to a fairly constant level of $1,25(OH)_2$cholecalciferol; these cells would move up the villus in increasing numbers as differentiated cells expressing their allotment of Ca-pump mRNA needed to maintain synthesis of the required amount of Ca-pumps for enhancement of active calcium absorption. It is not clear from our data whether the response of Ca-pump mRNA to vitamin D is a direct action of a vitamin D responsive element close to the Ca-pump genome or a reaction to effects unrelated genetically. It is our working hypothesis that the genes for a number of vitamin D dependent proteins may be closely interlinked with inhibitors of transcription, specific ribonucleases, and elements of alternative splicing mechanisms in a way that would lead to exquisitely sensitive controls for maintaining intracellular [Ca^{2+}] and controlling the rate of Ca^{2+} transit through the enterocyte.

REFERENCES

Greeb J, Shull GE (1989) Molecular cloning of a third isoform of the calmodulin sensitive plasma membrane calcium transporting ATPase that is expressed prodominantly in brain and skeletal muscle. J Biol Chem 264:18569-18576

Halloran BP, DeLuca HF (1981) Intestinal calcium transport: evidence for two distinct mechanisms of action of 1,25-dihydroxy vitamin D_3. Arch Biochem Biophys 208:477-486

Kimberg DV, Schachter D, Schenker H (1961) Active transport of calcium by intestine: effects of dietary calcium. Amer J Physiol 200:1256-1262

Lin S-H, Russel WE (1988) Two Ca^{2+}-dependent ATPases in rat liver plasma membrane. The previously purified (Ca^{2+}-Mg^{2+})-ATPase is not a Ca^{2+}-pump but an ecto-ATPase. J Biol Chem 263:12253-12258

MacLaughlin JA, Weiser MM, Freedman RA (1980) Biphasic recovery of vitamin D-dependent Ca^{2+} uptake by rat intestinal Golgi membranes. Gastroenterology 78:325-332

Moy TC, Walters JRF, Weiser MM (1986) Intestinal basolateral membrane Ca-ATPase activity with properties distinct from those of the Ca-pump. Biochem Biophys Res Commun 141:979-985

Wajsman R, Walter JRF, Weiser MM (1988) Identification and isolation of the phosphorylated intermediate of the calcium pump in rat intestinal basolateral membranes. Biochem J 256:593-598

Walters JRF, Horvath PJ, Weiser MM (1986) Preparation of subcellular membranes from rat intestinal scrapings or isolated cells: Different Ca^{2+}-binding, non-esterified fatty acid levels and lipolytic activity. Gastroenterology 91:34-40

Walters JRF, Weiser MM (1987) Calcium transport by rat duodenal villus and crypt basolateral membranes. Am J Physiol 252:G170-G177

Weiser MM (1973) Intestinal epithelial cell surface membrane glycoprotein synthesis: 1. An indicator of cellular differentiation. J Biol Chem 248:2536-2541

Weiser MM, Sykes DE, Killen PD (1990) Rat intestinal basement membrane synthesis, epithelial vs non-epithelial contribution. Lab Invest 62:325-330

OSCILLATIONS IN INTESTINAL CALCIUM ABSORPTION AS OBSERVED AND AS PREDICTED BY COMPUTER SIMULATION

S. Hurwitz, and A. Bar
Institute of Animal Science
The Volcani Center
Bet Dagan, Israel

The recognition of oscillatory behavior as a characteristic of various biological systems which have been believed to be maintained at a constant level, may be considered to be a departure from the classical cybernetic approach to homeostasis. The deviation from a constant level due to oscillation may even be an essential component of the control-system (Rapp, 1987) rather than an "error", as conventionally termed. For example, the oscillatory behavior of intracellular calcium may provide the mode for its participation in cellular information transfer, since hormonal stimulation results in an increased frequency of oscillations (Cuthberston and Cobbold, 1985). Extracellular calcium concentration has been shown to be under homeostatic control and rated as one of nature's constants (McLean and Hastings, 1935). The cybernetic representation of calcium metabolism (Aubert and Bronner, 1965) considered plasma calcium concentration to be regulated against a single reference value. The deviation from this value, the error, was taken as the driving force for the control systems, mainly bone. In recent years considerable evidence has accumulated on the oscillatory behavior of plasma calcium and some of its controlling systems such as bone formation-resorption and intestinal absorption, and of the calcium-regulating hormones such as parathyroid hormone, calcitonin and 1,25 dihydroxycholecalciferol [$1,25(OH)_2D_3$], as reviewed by Hurwitz (1989) and Staub *et al.* (1989).

Oscillations in calcium absorption due to reproductive activity

Rhythmicity in intestinal calcium absorption in the chicken (*Gallus domesticus*), induced by the the process of egg shell deposition, was observed experimentally over two decades ago (Hurwitz and Bar, 1965; Hurwitz *et al.*, 1973). Within a single ovulatory cycle of 24-27 h in the chicken, a period of 15-17 h is devoted to calcium deposition into the forming egg shell. Measurements of absorption *in vivo*, with the aid of an isotopic non-absorbed reference substance (^{91}Y), showed that during shell calcification the rate of calcium absorption increased to a level 2-4 times higher than during shell gland inactivity (Table 1). This change occurred mainly in the duodenum and jejunum, and was detectable as soon as 2 h after the start of egg shell calcification (Hurwitz *et al.*, 1973).

NATO ASI Series, Vol. H 48
Calcium Transport and
Intracellular Calcium Homeostasis
Edited by D. Pansu and F. Bronner

Table 1. *Variables of calcium and vitamin D metabolism during the reproductive cycle in the mature chicken (± SE)*

Variable	Egg shell calcification	
	Active	Inactive
Ca absorption, total, % of intake	81.5	20.8
Ca absorption, jejunum, % of intake	45.7±5.8	17.7±5.8
^{45}Ca absorption, jejunal loops, % dose	44.0±7.0	49.0±3.2
25(OH)D_3-1-hydroxylase, pmol/15 min	22.6±2.3	15.7±2.3
Plasma 1,25$(OH)_2D_3$, pg/ml	245.0±25.0	321.0±54.0
Duodenal calbindin, mg/g	2.16±0.19	2.62±0.13
Duodenal calbindin mRNA, pmol/g	1.29±0.15	1.13±0.05
Uterine calbindin, mg/g	1.14±0.08	1.20±0.01

In vivo calcium absorption, total and jejunal, was estimated from the decrease in the ratio of calcium to ^{91}Y added to the diet as a a non-absorbed radiotracer (Hurwitz and Bar, 1965). ^{45}Ca absorption was taken as the disappearance of the tracer from jejunal loops during a 20 min incubation in situ . 25(OH)-1-hydroxylase activity was measured in kidney homogenates in vitro (Bar and Hurwitz, 1979). 1,25$(OH)_2D_3$ *was measured by a competitive binding assay. Calbindin mRNA was determined by a solution-hybridization assay. Calbindin was measured by a radioimmunoassay (Bar and Hurwitz, 1979).*

In chickens with daily oviposition, rhythmicity in calcium absorption occurs with no corresponding changes in intestinal calbindin concentration or even in its mRNA (Table 1). Only a slight change is noted in the 25(OH)-cholecalciferol-1-hydroxylase which is not reflected in any significant change in the concentration of 1,25$(OH)_2D_3$ in blood plasma. Furthermore, ^{45}Ca transport by intestinal loops *in situ* is not modified as a consequence of shell calcification. Results of Wasserman and associates and our studies showed that intestinal ^{45}Ca transport was always an expression of control of absorption by vitamin D. Finally, bypass of the 1,25$(OH)_2D_3$ - intestinal control axis by the external supply of the hormone as the only vitamin D source in the diet, did not interfere with the rhythmicity of calcium absorption associated with egg shell calcification (Bar *et al.*, 1976). This body of evidence suggests that the rhythmicity in calcium absorption is not a consequence of 1,25$(OH)_2D_3$ action. The increase in calcium absorption during shell calcification may be the consequence of an increase in the transmembrane calcium gradient due to a decrease in ionic calcium in plasma and the increase in its concentration in the intestinal lumen, as a result of the metabolic acidosis during shell calcification. However, this possibility remains to be confirmed.

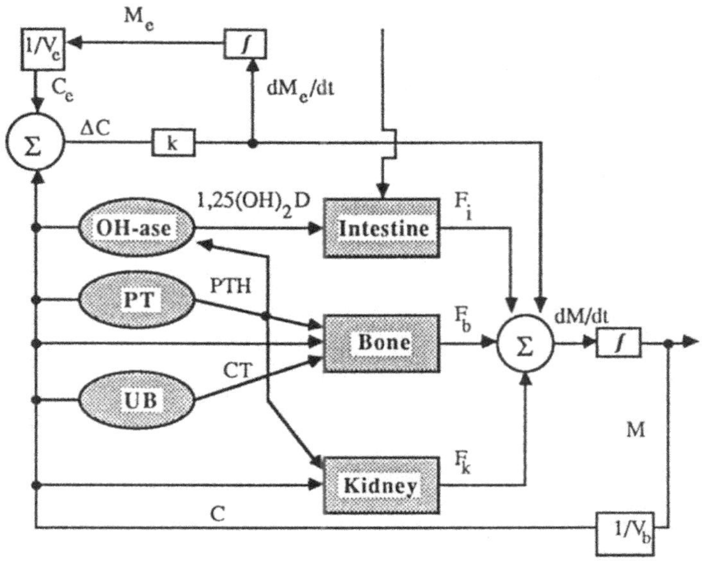

Fig. 1. *Feedback regulation of plasma calcium concentration* [C]. *The sum* (Σ) *of the net calcium flows from intestine* (F_i), *bone* (F_b) *and kidney* (F_k), *is the change in total plasma calcium* (dM/dt) *which, when added* ($\int$) *to existing plasma calcium and divided by the blood volume* (V_b), *yields the plasma calcium concentration* (C). *Plasma calcium then undergoes a rapid exchange with extracellular calcium* (M_e) *which upon division by the extracellular volume* (V_e) *yields extracellular calcium concentration* (C_e). *Plasma calcium concentration determines the rates of urinary calcium excretion* (F_k), *bone calcium flow* (F_b), *calcitonin* (CT) *secretion by the ultimobranchial gland* (UB), *parathyroid hormone* (PTH) *secretion from the parathyroid gland* (PT) *and production of* $1,25(OH)_2D$ *by the kidney 25-hydroxycholecalciferol-1-hydroxylase system (OH-ase). Bone flow and* $1,25\ (OH)_2D$ *production are also controlled by* PTH.

Oscillations in calcium absorption in growing chickens

Another type of oscillatory behavior of plasma calcium and its control systems was predicted for growing birds by a simulation model (Hurwitz *et al.*, 1983, 1987a,b) based on the conceptual model shown in Fig.1. This computerized algorithm integrates numerically the various rate equations describing hormone (PTH and $1,25(OH)_2D_3$) concentrations, uptake of $1,25(OH)_2D_3$ by intestinal mucosa, and the generation of a calcium absorption mechanism. The change in extracellular (including plasma) calcium concentration is given by the sum of the rates of calcium transfer by intestine, bone and kidney.

In the algorithm, growth is described by a Gompertz equation. Feed intake is calculated by summing the energy needs for maintenance and for weight gain, as a function of body fat content. Calcium intake is then calculated for any concentration of dietary calcium. The various pool sizes and rates as well as new bone deposition and calcium intake, change constantly due to growth.

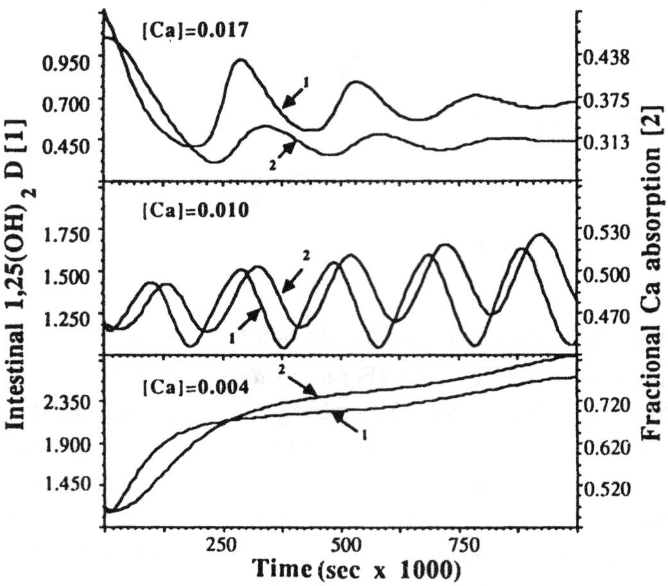

Fig. 2. *Simulated oscillations in intestinal 1,25(OH)$_2$D$_3$ and calcium absorption in growing chickens fed low-Ca (0.004 g/g), normal Ca (0.010 g/g), or high-Ca (0.017 g/g) diets.*

Computer simulation predicts for growing birds an oscillatory behavior of the various calcium control systems including calcium absorption (Fig. 2), without any external rhythmic stimulation. This prediction is supported by results of Miller and Norman (1982). The magnitude and frequency of oscillations in plasma calcium and its various controlling systems are dependent on the high calcium turnover rate imposed by the process of growth (Hurwitz *et al.*, 1987a). No oscillations (simulated) occur when growth rate is made to equal zero. When dietary calcium intake is made to deviate from normal towards either high or very low levels (Fig. 2), an adaptation to the change in dietary calcium is effected through a negative feedback change in $1,25(OH)_2D_3$ production which leads to a corresponding change in the fraction of calcium absorption. This adaptation is accompanied by the disappearance of oscillations in either control system.

Normal vitamin D intake and the integrity of the 25-hydroxycholecalciferol-1-hydroxylase system are essential for generation of oscillations in calcium absorption and other variables of the calcium control system.

When analyzed, oscillations in the various components of the calcium control system exhibit phase shifts according to the known sequence of information transfer from one component to the next. The important determinant of the generation of oscillations appears to be the difference in response-time to PTH, between bone calcium outflow on the one hand, and intestinal calcium absorption - which is affected by PTH through the 1-hydroxylase - $1,25(OH)_2D_3$ pathway - on the other. The events that are responsible for eliciting a single oscillation begin with the process of growth, which removes calcium from the central pool, leading to a reduction in calcium concentration. A rapid increase in PTH secretion rate and its concentration of blood plasma ensues. PTH then stimulates, with a time factor of minutes, bone resorption, which acts to diminish the magnitude of hypocalcemia. Meanwhile, PTH stimulates the 25-hydroxycholecalciferol-1-hydroxylase enzyme responsible for increased production of $1,25(OH)_2D_3$ and an increased uptake of the hormone in intestinal mucosa. The resulting increase in the calcium-absorbing capacity leads to increased calcium absorption several hours after the original PTH stimulus, when plasma calcium concentration has approached normal values through bone resorption. As a result, plasma calcium continues to rise, leading to a suppression of PTH release which results first in a decrease in bone resorption and later in a decrease in calcium absorption, thereby completing the single oscillation.

Conclusions

Periodicity in intestinal calcium absorption appears to be a property of the normal calcium regulatory mechanism in both reproducing and growing birds. Oscillations may disappear when the normal processes of growth or reproduction are interrupted, or when the "normal" dietary calcium supply is disrupted. Oscillations in calcium absorption, triggered by the non-continuous process of egg shell calcification occur in laying hens and do not involve vitamin D control. In growing birds, computer simulation suggests oscillations in calcium absorption as well as in other calcium-regulating systems. The oscillatory behavior depends on the normal dietary intake of calcium and results from the difference in response time to PTH of bone resorption and the $1,25(OH)_2D_3$- intestinal absorption axis.

References

Aubert, J-P, Bronner, F (1965) A symbolic model for the regulation by bone metabolism of the blood calcium level in rats. Biophys J 5: 349-358.

Bar A, Eisner U, Montecuccoli G, Hurwitz S (1976) Regulation of intestinal calcium absorption in the laying quail independent of vitamin D hydroxylation. J Nutr 106: 1332-1338.

Bar A, Hurwitz S (1979) The interaction between dietary calcium and gonadal hormones in their effect on plasma calcium, bone, 25-hydroxycholecalciferol-1-hydroxylase, and duodenal calcium-binding protein, measured by a radioimmunoassay in chicks. Endocrinology 104: 1455-1460.

Cuthberston KS, Cobbold PH (1985) Phorbol esters and sperm activate mouse oocytes by inducing sustained oscillations in cell Ca^{2+}. Nature Lond 316: 541-542.

Hurwitz S (1989) Calcium homeostasis in birds. Vitamins and Hormones, Vol 45 Aurbach GD, McCormick DB (eds), Academic Press New York, p 173-221.

Hurwitz S, Bar A (1965) The absorption of calcium and phosphorus along the gastrointestinal tract of the laying fowl as influenced by dietary calcium and egg shell formation. J Nutr 86: 433-438.

Hurwitz S, Bar A, Cohen I (1973) Regulation of calcium absorption in fowl intestine. Am J Physiol 225: 140-154.

Hurwitz S, Fishman S, Bar A, Pines M, Riesenfeld G, Talpaz H (1983) Simulation of calcium homeostasis: modeling and parameter estimation. Am J Physiol 245: R664-R672.

Hurwitz S, Fishman S, Talpaz H (1987a) Model of plasma calcium regulation: system oscillations induced by growth. Am J Physiol, 252: R1173-R1181.

Hurwitz S, Fishman S, Talpaz H (1987b) Calcium dynamics: a model system approach. J Nutr 117: 791-796.

McLean FC, Hastings AB (1935) Clinical estimation and significance of calcium ion in blood. Am J Med Sci 189: 601-613.

Miller B, Norman AD (1982) Evidence for circadian rhythms in the serum levels of the vitamin D-dependent calcium-binding protein and in the activity of the 25-hydroxyvitamin D_3-1-α-hydroxylase in the chick. FEBS Lett 141: 242-244.

Rapp PE (1987) Why are so many biological systems periodic ? Progress in Neurobiol 29: 261-673.

Staub JF, Traqui P, Lausson S, Milhaud G, Perault-Staub AM (1989) A physiological view of *in vivo* calcium dynamics: the regulation of a nonlinear self-organized system. Bone 10: 77-86.

MECHANISMS AND SITES OF TRANSEPITHELIAL Ca^{2+} TRANSPORT IN KIDNEY CELLS

R.J.M. Bindels, J.A.H. Timmermans, R.J.J.M. Bakens, A. Hartog,
E. van Leeuwen and C.H. van Os
Department of Physiology
University of Nijmegen
P.O.Box 9101
6500 HB Nijmegen
The Netherlands

INTRODUCTION

In epithelial cells, possible mechanisms involved in active transcellular Ca^{2+} transport are: a passive entry step at the apical membrane, diffusion through the cytoplasm, and active extrusion mechanisms located in the basolateral membrane (van Os, 1987). The latter mechanisms, i.e. Ca^{2+}-ATPase and Na^{+}/Ca^{2+} exchange, have been studied extensively in basolateral membranes from the kidney.

In contrast, in epithelial cells little is known about the mechanisms involved in passive entry of calcium from the extracellular space into the cytoplasm. Recently, Ca^{2+} channels have been demonstrated in rabbit ileal epithelial cells (Homaidan et al., 1989). However, in epithelial cells from the kidney there is until now no evidence for the presence of Ca^{2+} channels similar as those found in excitable cells.

The intracellular movement of calcium might be greatly enhanced by facilitated diffusion due to the presence of Ca^{2+}-binding proteins (Bronner, 1989). In the mammalian kidney $1,25(OH)_2D_3$ dependent Ca^{2+}-binding protein, Calbindin-D 28k (CaBP-28k), is localized exclusively in the distal tubule, a site where Ca^{2+} is actively reabsorbed (Taylor et al., 1982).

In this chapter we report studies on two widely used renal cell lines to delineate some features of transepithelial Ca^{2+} transport. In addition, in rat kidney possible sites involved in active Ca^{2+} transport were traced by means of immunohistochemical detection of the calcium-binding proteins: CaBP-28k, CaBP-9k and parvalbumin.

NATO ASI Series, Vol. H 48
Calcium Transport and
Intracellular Calcium Homeostasis
Edited by D. Pansu and F. Bronner

MATERIALS AND METHODS

Cell culture

OK cells (passages 87-93) and MDCK cells (passages 75-85) were routinely maintained at 37 °C in a humidified 5% CO_2 atmosphere in Dulbecco's modified Eagle's medium, supplemented with 5% fetal bovine serum, 100 µg/ml penicillin, 100 µg/ml streptomycin and a mixture of non-essential amino-acids. The cells were fed three times a week and were trypsinized weekly with a PBS solution containing 0.1% (w/v) trypsin and 1 mM EDTA.

Ca^{2+} influx experiments

All experiments were performed with confluent monolayers, grown on 12-wells plates. Immediately before the Ca^{2+} uptake experiments, the cells were washed with buffer containing 140 mM NaCl, 5.4 mM KCl, 1.2 mM $MgSO_4$, 0.7 mM $CaCl_2$, 10 mM HEPES/Tris, pH 7.40 (37 °C). $^{45}Ca^{2+}$ uptake was initiated by adding the same buffer containing 2 µCi $^{45}Ca^{2+}$. The $^{45}Ca^{2+}$ uptake was stopped by addition of ice-cold 150 mM NaCl, 2.5 mM $CaCl_2$, 20 mM HEPES/Tris, pH 7.40 buffer. After several washings with the same stop buffer, an aliquot of the cells was taken and counted for radioactivity.

Transepithelial Ca^{2+} transport experiments

All experiments were performed with confluent monolayers grown on Costar's cell culture chamber inserts. The filter bottom of the inserts was placed in an Ussing-type chamber and bathed in an aerated (95% O_2/5% CO_2) Krebs buffer. $^{45}Ca^{2+}$ fluxes were determined at 37 °C as described previously (Bakker and Groot, 1984).

D888-binding experiments

Basolateral membranes were isolated from rat kidney cortex as described previously (van Heeswijk et al., 1984). Equilibrium binding assays were performed at 37 °C in 125 mM NaCl, 5 mM KCl, 0.1% BSA and 25 mM HEPES/Tris, pH 7.40. Membranes were allowed to equilibrate for 15 min with various concentrations of [^{3}H]-desmethoxyverapamil (D888) (10^{-9} - 10^{-3}M) and membranes were separated from the incubate by centrifugation through ice-cold buffer (100,000xg, 5 min) (Beckmann Airfuge). After another washing the membranes were counted for radioactivity. Non-specific binding was defined as binding in the presence of excess D888 (10^{-3}M).

Immunohistochemical experiments

CaBP-9k was purified from bovine intestine (kindly provided by dr. S. Forsén, Lund, Sweden), CaBP-28k was isolated from chick duodenum (Friedlander and Norman, 1980) and parvalbumin, prepared from rabbit skeletal muscle (Sigma, St.Louis). The purified Ca^{2+}-binding proteins were used for immunization of New Zealand white rabbits. The produced antisera were tested for their specificity and cross-reactivity with an enzyme-linked immunosorbent assay (ELISA). Conventional immunoperoxidase staining was performed on deparaffinized serial sections (5 μm) of PLP-fixed rat kidneys (McLean and Nakane, 1974). Sections of rat kidney were also processed for double-label immunohistochemistry. The antisera against the various CaBP's and monoclonal antibody St.48 against principal cells of collecting ducts (Fejes-Tóth and Nâray-Fejes-Tóth, 1987) were employed and labeled with contrasting fluorescent colours within the same tissue sections. As secondary antibodies were used, goat antirabbit immunoglobulin TRITC-conjugated, respectively, goat antimouse IgG FITC-conjugated.

RESULTS AND DISCUSSION

In OK cells, Ca^{2+} influx was measured at 37 °C and proved to be linear with time up to 40 min of incubation (Fig.1). In the presence of either 2.5 mM $LaCl_3$ or 1 mM methoxyverapamil (D600) the time-dependent uptake was completely inhibited. Hence, calcium entry blocker sensitive Ca^{2+} influx into OK cells could be measured accurately and averaged 21.7 ± 0.3 pmol/min.mg protein. Dose-response curves for several Ca^{2+} entry blockers were determined and are shown in Fig.2. D600, a verapamil analog belonging to the phenylalkylamines; diltiazem, a benzothiazepine, and flunarizine, a phenylpiperazine, inhibited Ca^{2+} uptake in a dose-dependent manner. D600 was the most effective inhibitor of Ca^{2+} uptake ($K_{0.5} \approx 45$ μM), followed by diltiazem and flunarizine (both with $K_{0.5} \approx 100$ μM).

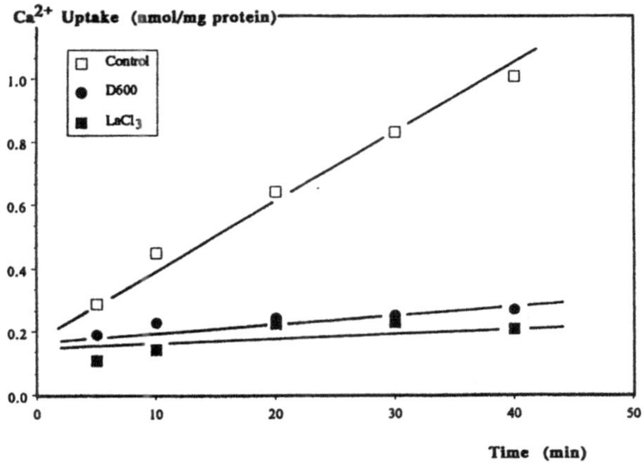

Figure 1.
Time course of $^{45}Ca^{2+}$ uptake in confluent OK cells in the presence of vehicle (□), 1 mM D600 (●) or 2.5 mM $LaCl_3$ (■). Data shown are mean ± SE of 6 experiments.

The half-maximal inhibition concentrations of these compounds turned out to be several orders of magnitude higher than those required to inhibit voltage-dependent Ca^{2+} entry in excitable tissues (Hosey and Lazdunski, 1988), suggesting that we are dealing with receptor- or second messenger-operated Ca^{2+} entry rather than with voltage-operated Ca^{2+} entry (Avodin et al., 1988). Furthermore, felodipine, a blocker belonging to the dihydropyridines and classically known as blocker of the L-type slow Ca^{2+}-channels (Hosey and Lazdunski, 1988), did not affect Ca^{2+} uptake in OK cells

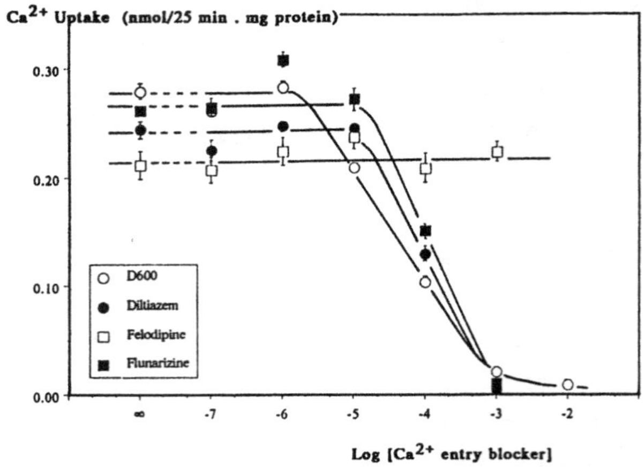

Figure 2. Concentration-dependence of effect of D600 (o), diltiazem (●), felodipine (□) and flunarizine (■) on $^{45}Ca^{2+}$ uptake in OK cells. Data shown are mean ± SE of 6 experiments.

(Fig.2). It should be noted, that at the high concentrations of Ca^{2+} entry blockers used, the action of Ca^{2+}-entry blockers may be unspecific, hence, effects on other cellular processes cannot be excluded.

We also studied directly the equilibrium binding of [^{3}H]-D888 to an enriched preparation of basolateral membranes from rat kidney cortex. The results are shown in Fig.3. Specific binding of D888 could be fitted by a double-site model. The kinetic parameters of the high affinity site were $K_{D1} \approx 3.10^{-8}$M and $B_{max1} \approx 10$ pmol/mg.protein (n=3).

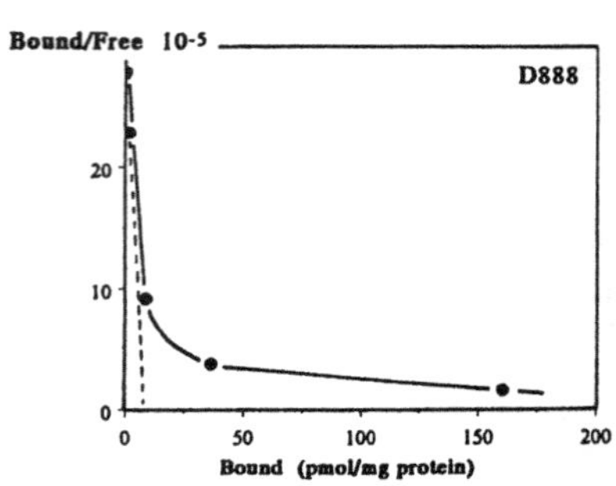

Figure 3. Scatchard plot of specific binding of [^{3}H]-D888 to basolateral membranes of rat kidney cortex. Data shown are mean of 3 experiments.

The discrepancy between the half-maximal $K_{0.5}$ inhibition value for Ca^{2+} influx into OK cells (Fig.2) and the K_{D1} value extracted from binding studies for D888 (Fig.3) is most likely due to the inadequacy of Ca^{2+}-influx measurements in OK cells. Ca^{2+} uptake into OK cells can only be studied by luminal addition of $^{45}Ca^{2+}$ and D600. However, the D888-binding studies with isolated basolateral membranes suggest that Ca^{2+}-influx is via Ca^{2+} channels in the basolateral membrane and not via the luminal membrane. Therefore, the concentration of D600 in the luminal solution, which is used in Fig.2, is not identical to the actual concentration on the other side of the cell layer. Another explanation, albeit less likely, could be an inherent difference between Ca^{2+}-channels in an established renal cell line and those present in proximal tubules.

The effects of bPTH (1-34), forskolin and 12-O-tetradecanoylphorbol 13-acetate (TPA) on Ca^{2+} influx in OK cells have also been studied and the results are shown in Fig.4. bPTH (1-34) (10^{-6}M) reduced the Ca^{2+} influx 24%. Since PTH induces in OK cells a rise in intracellular Ca^{2+} and an increase in intracellular cAMP, it was of interest to study the activation of both pathways independently. Forskolin (10^{-4}M) which increases intracellular cAMP reduced Ca^{2+} uptake by 46%. On the other hand, the phorbol ester TPA (10^{-6}M) decreased the Ca^{2+} influx in confluent OK cells with 50%. Therefore, it is very likely that protein kinase A as well as protein kinase C dependent pathways are involved in inhibition of Ca^{2+} entry in OK cells by PTH. Interestingly, in OK cells a similar regulating mechanism has been demonstrated for PTH action on Na^+/H^+ exchange and Na^+/phosphate cotransport (Helme-Kolb et al., 1990).

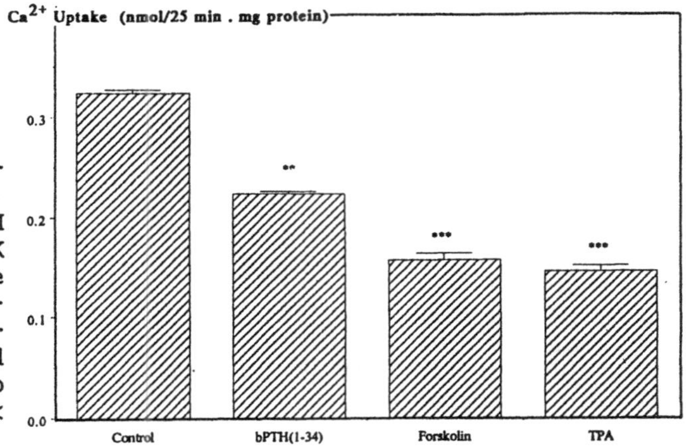

Figure 4.
Effect of vehicle (control), 10^{-6}M bPTH (1-34), 10^{-4}M forskolin and 10^{-6}M TPA on $^{45}Ca^{2+}$ uptake in OK cells. Data shown are mean ± SE of 9 experiments. Statistical significance is as indicated by asteriks. (*, $0.01 < p < 0.05$; **, $0.001 < p < 0.01$; ***, $p < 0.001$).

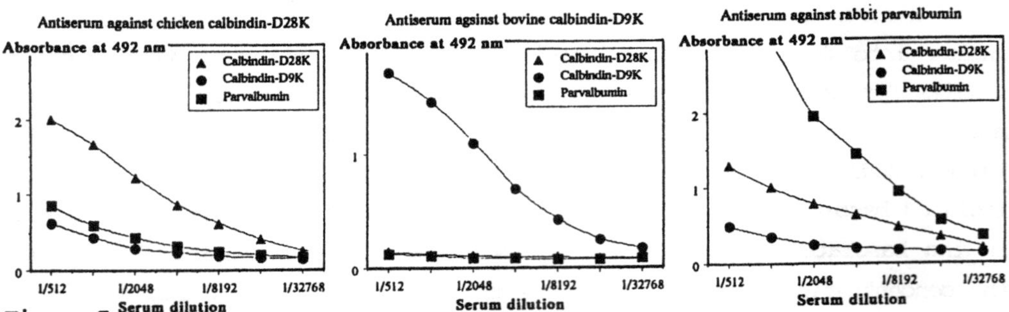

Figure 5.
Concentration-dependent binding of various antisera to bovine CaBP-28k, bovine CaBP-9k and rabbit parvalbumin in an ELISA system.

In MDCK cells, a tight epithelial cell with distal properties, trans-epithelial Ca^{2+} fluxes were estimated in conventional Ussing-type chambers. Only when filters with a large surface area (4.9 cm^2) were used, Ca^{2+} fluxes could be determined, (averaged: J_{MS}= 285 ± 76 and J_{SM}= 135 ± 24 $nmol/h.cm^2$, $p > 0.08$). Accurate determination of net Ca^{2+} fluxes across these cultured kidney cells proved to be very difficult, and no significant net Ca^{2+} fluxes could be determined.

In addition, possible sites in the rat kidney involved in active Ca^{2+} transport were traced by means of immunohistochemical detection of CaBP's. There is strong evidence that CaBP's are involved in transcellular Ca^{2+} transport (Bronner, 1989). Antisera, with high specificity and without cross reactivity for other CaBP's (Fig.5), were raised against CaBP-28k from chicken intestine, CaBP-9k from bovine intestine and parvalbumin from rabbit muscle. In serial sections of rat kidney (Fig.6 A-E), CaBP-28k was only present in the outer part of the cortex, where it was localized in the

Figure 6. (See next page)
A-C) Immunoperoxidase staining of serial sections of rat kidney in transition of thick ascending loop of Henle (cTALH) to distal convoluted tubule (DCT) showing distribution of CaBP-28k (A), CaBP-9k (B) and parvalbumin (C) (x580).
D-E) Immunoperoxidase staining of serial sections of rat kidney showing distribution of CaBP-28k (D) and parvalbumin (E). Connecting tubule arcade (CNT) joining cortical collecting duct (CCD). Notice parvalbumin-containing cells in thick ascending loop of Henle (cTALH) (x280).
F-I) Double immunofluorescent staining of sections of rat kidney showing distribution of principal cells (F,H), CaBP-9k (G) and parvalbumin (I). Cortical collecting duct (CCD) with principal and intercalated cells (ic, arrows) (x420)

Figure 6.

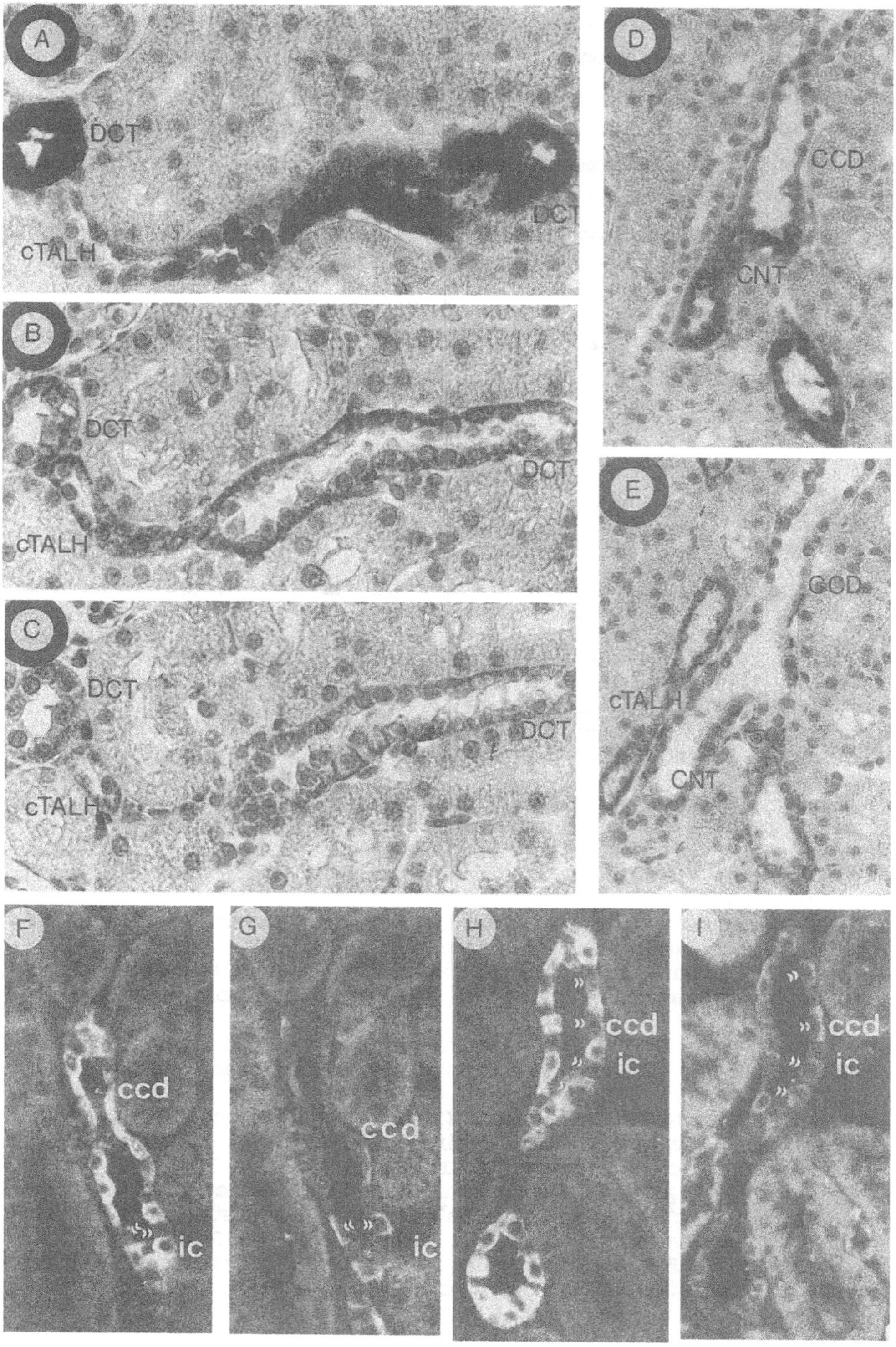
A
DCT
DCT
cTALH
B
DCT
DCT
cTALH
C
DCT
DCT
cTALH
D
CCD
CNT
E
CCD
cTALH
CNT
F
ccd
ic
G
ccd
ic
H
ccd
ic
I
ccd
ic

distal convoluted tubule and in the connecting tubule. In these cells CaBP-28k was evenly distributed through the cytosol. CaBP-28k could not be demonstrated in the loops of Henle or collecting duct. CaBP-9k and parvalbumin were colocalized in both the thin and thick loops of Henle, distal convoluted tubule and connecting tubules (Fig.6 A-E). In the collecting duct their presence was restricted to the intercalated cells (Fig.6 F-I). Interestingly, in all responsive cells parvalbumin and CaBP-9k were present along the basolateral membrane. In distal convoluted tubules and connecting tubules all three tested CaBP's were colocalized in 90% of the cells.

The localization of parvalbumin and CaBP-9k in the loops of Henle and intercalated cells suggests that also these cells may be involved in active Ca^{2+} reabsorption.

REFERENCES

Avodin PV, Menshikor MY, Svitina-Ulitana IV, Tkachuk VA (1988) Blocking of the receptor-stimulated calcium entry into human platelets by verapamil and nicardipine. Thromboses Res 52:587-597

Bakker R, Groot JA (1984) cAMP-mediated effects of ouabain and theophylline on paracellular ion selectivity. Am J Physiol 246:G213-G217

Bronner F (1989) Renal calcium transport: mechanisms and regulation - An overview. Am J Physiol 257:F707-F711

Fejes-Tóth G, Náray-Fejes-Tóth A (1987) Differentiated transport functions in primary cultures of rabbit collecting ducts. Am J Physiol 253:F1302-F1307

Friedlander EJ, Norman AW (1980) Purification of chick intestinal calcium-binding protein. Methods in Enzymology 67:504-508

Helme-Kolb C, Montrose MH, Murer H (1990) Regulation of Na^+/H^+ exchange in opossum kidney cells by parathyroid hormone, cyclic AMP and phorbol esters. Plügers Arch 415:461-470

Hosey MM, Lazdunski M (1988) Calcium channels: Molecular pharmacology, structure amd regulation. J Membrane Biol 104:81-105

Homaidan FR, Donowitz M, Weiland GA, Sharp GWG (1989) Two calcium channels in basolateral membranes of rabbit ileal epithelial cells. Am J Physiol 257:G86-G93

McLean IW, Nakane PK (1974) Periodate-lysine-paraformaldehyde fixative, a new fixative for immunoelectron microscopy. J Histochem Cytochem 22:1077-1083

Taylor AN, McIntosh JE, Bourdeau JE (1982) Immunocytochemical localization of vitamin D-dependent calcium-binding protein in renal tubules of rabbit, rat and chick. Kidney Int 21:765-773

Van Heeswijk MPE, Geertsen JAM, van Os CH (1984) Kinetic properties of the ATP-dependent Ca^{2+} pump and the Na^+/Ca^{2+} exchange system in basolateral membranes from rat kidney cortex. J Membrane Biol 79:19-31

Van Os CH (1987) Transcellular calcium transport in intestinal and renal epithelial cells. Biochim Biophys Acta 906:195-222

MECHANISMS OF H^+/HCO_3^- TRANSPORT IN RAT KIDNEY MEDULLARY THICK ASCENDING LIMB. EFFECTS OF ARGININE VASOPRESSIN

P. Borensztein, M. Delahousse, F. Leviel,M. Paillard, and M. Bichara. Laboratoire de Physiologie et Endocrinologie Rénale, Université Pierre et Marie Curie, Hôpital Broussais, and INSERM CJF 88-07 ; Centre Biomédical des Cordeliers, 15 rue de l'Ecole de Médecine, Paris, France.

The thick ascending limb (TAL) of rat kidney absorbs bicarbonate and substantial rates (Good et al. 1984) that may account for much of the bicarbonate absorbed from Henle's loop in vivo (about 15% of the bicarbonate filtered load, Bichara et al. 1984). Yet, the cellular mechanisms of H^+/HCO_3^- transport by the rat TAL have been studied in only a few studies that have reported in the cortical segment the presence of a luminal $Na^+:H^+$ antiport (Good 1985) and of a basolateral Na^+: $(HCO_3^-)n>1$ symport (Krapf 1988). Also, in the medullary segment (MTAL) of the mouse TAL, a luminal $Na^+:H^+$ antiport and a basolateral $Na^+:(HCO_3^-)$ $n>1$ symport have been described (Kikeri et al. 1990).

The rat TAL is also a target site for arginine vasopressin (AVP), a peptide hormone that exerts major effects on tubular fluid and urine acidification (Bichara 1987) ; particularly, AVP inhibits bicarbonate absorption by Henle's loop in vivo (Bichara 1987) and by the isolated perfused rat MTAL (Good 1990). We summarize here some results we have recently obtained in our laboratory about the cellular mechanisms of H^+/HCO_3^- transport of the rat MTAL and their modulation by AVP.

We used a suspension of fresh rat MTAL fragments prepared as described by Trinh-Trang-Tan et al. (1986). We have monitored the intracellular pH (pHi) and the membrane potential with use of the fluorescent probes BCECF and DIS-C3-5, respectively.

NATO ASI Series, Vol. H 48
Calcium Transport and
Intracellular Calcium Homeostasis
Edited by D. Pansu and F. Bronner

Proton transport : Na^+-H^+ antiport and H^+-ATPase

All experiments were performed in the nominal absence of bicarbonate. To demonstrate a Na^+-H^+ antiport activity, Na-depleted and acidified cells were abruptly exposed to external sodium ; MTAL cells were Na-depleted by preincubation in a Na-free medium (Na replaced with tetramethylammonium) and acidified to pHi ~ 6.90 by exposure to 10 μM nigericin, a K^+-H^+ ionophore that caused an exchange of internal K^+ for external H^+. Addition of NaCl to the medium caused prompt dose-dependent pHi recoveries (Fig. 1)

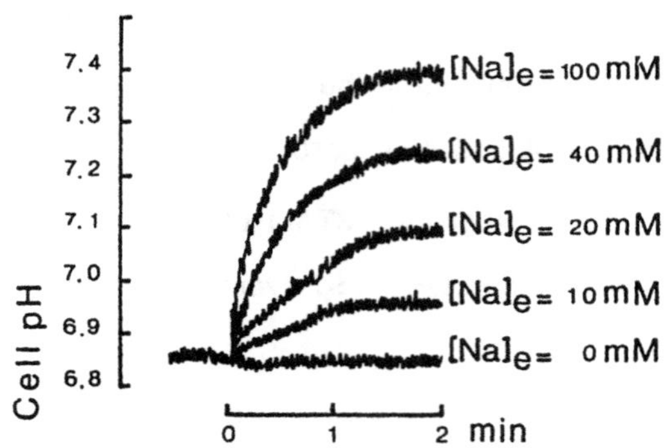

Fig.1. Sodium-dependency of pH recovery of acidified and Na-depleted rat MTAL cells at 25°C.

The initial rate of Na-dependent pHi recovery rose as the external sodium concentration ([Na]e) was augmented to reach a plateau at [Na]e = 35-40 mM, which indicated that Na^+:H^+ exchange was a saturable process. Lineweaver-Burk plot of the data led to a calculated value for the NaKm of 11 mM. The Na^+-H^+ exchange mechanism has been identified as a plasma membrane Na^+:H^+ antiport by its sensitivity to amiloride (amiloride Ki of 2.8 $\times 10^{-5}$M) and electroneutrality (unaltered by valinomycin at different external potassium concentrations). The basolateral or luminal localization of the Na^+:H^+ antiporter cannot be assessed in a suspension like ours in which the cell basolateral and luminal membranes are both in contact with the external medium. In this regard, Na^+-H^+ antiport activity has been localized to the luminal and not basolateral side of the isolated and perfused mouse MTAL (Kikeri 1990).

The Na^+-H^+ antiporter is not, however, the sole mechanism of proton transport in rat MTAL cells. The evolution of the cell pH in function of time at 37°C is shown in Fig. 2 before and after a brisk intracellular acidification caused by rapid entry within the cell of undissociated acetic acid following addition of potassium acetate to the external medium.

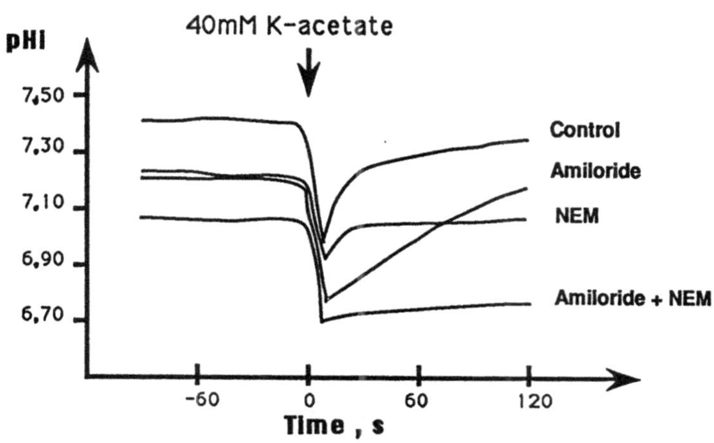

Fig. 2. Evolution of cell pH of rat MTAL cells at 37°C in the presence of 2×10^{-3}M amiloride and/or 10^{-4} M NEM.

Control cells rapidly recovered from acidification within one or two minutes ; in the presence of 2 x 10^{-3} M amiloride, cell pH recovery was slower (inhibition of the Na^{+}:H^{+} antiporter) but nevertheless occurred in a linear fashion, which clearly demonstrated the presence of a Na-independent mechanism of proton extrusion. In the presence of both amiloride and 10^{-4} M NEM, an inhibitor of non-mitochondrial proton pumps, cell pH recovery was abolished. Finally, NEM alone lowered the resting cell pH and inhibited cell pH recovery. The activity of a plasma membrane H^{+} -ATPase has been confirmed by various experimental results concerning the amiloride-insensitive cell pH recovery :

- inhibition by maneuvers known to alter the intracellular production of ATP (potassium cyanide, iodoacetic acid, absence of energetic substrates) ;

- inhibition by 10^{-4} M omeprazole, a known inhibitor of the H^{+}-K^{+}-ATPase of the gastric mucosa ;

- insensitivity to vanadate and oligomycin, which indicates that the plasma membrane H^{+}-ATPase of MTAL cells is of the vacuolar type.

These results are in agreement with data obtained in other laboratories that have documented the presence of an important non-mitochondrial H^{+}-ATPase activity in the MTAL (Brown et al. 1988, Sabatini et al. 1990). Particularly, Brown et al. (1988) have localized by immunocytochemistry to the luminal membrane the plasma membrane H^{+}-ATPase of the rat MTAL.

Thus the rat MTAL secretes protons through the luminal cell membrane by two mechanisms : a Na^{+}-H^{+}antipoter and a H^{+}-ATPase, each being responsible of ~ 50 % of proton extrusion under our experimental conditions (Froissart et al. 1990).

Bicarbonate transport : electroneutral K^{+}: HCO_3^{-} symport

We have determined the membrane mechanism of bicarbonate transport in the same preparation by studying HCO_3^{-} efflux from the cells, the direction of transport that normally occurs in vivo. This was achieved by abruptly diluting 200 µl of a medium containing bicarbonate-loaded cells into 3 ml of bicarbonate-free medium. Because of the dilution, medium PCO_2 abruptly fell from 40 to 2.5 mmHg and external bicarbonate concentration from 25 to 1.6 mM. A similar protocol has been sucessfully used by others to study bicarbonate transport in renal cells (Zeidel et al. 1986). The evolution of cell pH during 60 s under three different experimental conditions is shown in Fig. 3.

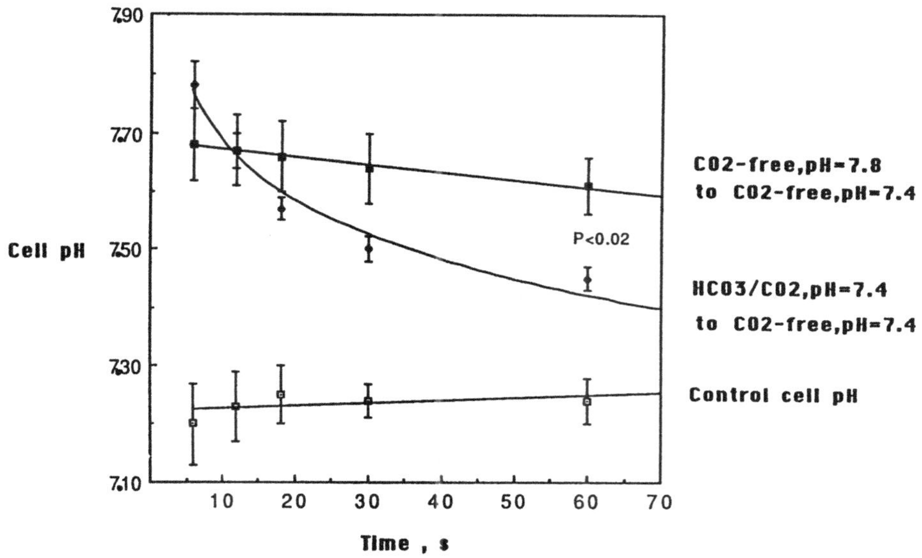

Fig. 3. Evolution of cell pH of rat MTAL cells abruptly diluted in various experimental media.

When cells were preincubated and then diluted in the same medium containing 25 mM bicarbonate, pHe 7.4, the control cell pH was 7.20-7.25 which corresponded to a mean intracellular bicarbonate concentration of 17 mM. When these same cells were diluted into the nominally CO_2-free medium (final PCO_2 = 2.5 mmHg), the cell pH response could be divided into two phases :

- a brisk alkalinization first occurred due to very rapid CO_2 exit from the cells ; because the first cell pH measurement was delayed to the 6th second, this initial alkalinization could not be observed directly ;

- then a secondary acidification followed specifically due to bicarbonate efflux ; the first mean cell pH value measured at the 6th second was about 7.80, which corresponded to a mean intracellular bicarbonate concentration of about 6 mM ;

Finally, when cells were alkalinized to the same extent by preincubation at pHe 8.0 in the absence of exogenous HCO_3^-/CO_2, and then diluted into the same CO_2-free medium but at pHe = 7.40, the cells did not acidify as rapidly as they did when they contained bicarbonate.

Using this protocol under various experimental conditions, we have shown (Leviel et al. 1990) that bicarbonate efflux was :

- independent of external or internal chloride (chloride replaced with gluconate) ;

- electroneutral because unaltered by abruptly removing chloride from or adding 2 mM barium to the medium which depolarized the cell membrane by 50% and 15%, respectively ;

- unaltered by the complete absence of sodium both inside and outside the cell ;

- dependent on intracellular potassium content and affected by the transmembrane potassium gradient,

- and, finally, only slightly inhibited by 10^{-4}M and 5 x 10^{-4}M DIDS.

These results provided evidence that the main bicarbonate transport mechanism in rat MTAL cells is an electroneutral K^+-HCO_3^- symport, which is described for the first time in this cell type. The apparent discrepancy with other works that have concluded to the presence of a Na^+:(HCO_3^-)n symport in rat and mouse TAL (Krapf 1988, Kikeri et al. 1990) will be discussed in our original paper that is in preparation.

Effects of AVP on H^+/HCO_3^- transport in rat MTAL cells

AVP has been shown both in vivo in our laboratory (Bichara et al. 1987) and in vitro (Good 1990) to inhibit the transepithelial bicarbonate transport by the rat MTAL. We have therefore begun the study of the effects of AVP on pHi in our preparation in order to find out what H^+/HCO_3^- transport mechanisms are affected by AVP.

AVP could affect the H^+/HCO_3^- transport either directly or indirectly. Indeed, the known AVP-induced stimulation of the Na^+:K^+:$2Cl^-$ symport could secondarily alter luminal proton secretion and basolateral bicarbonate transport through changes in intracellular sodium and potassium concentrations and/or in the membrane potential. The changes in pHi of cells incubated in a bicarbonate-free medium induced by 10^{-8} M AVP are shown in Fig. 4.

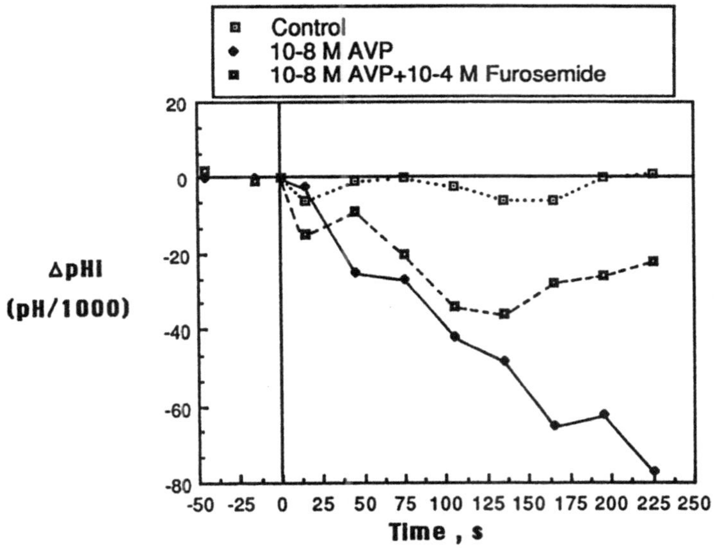

Fig. 4. AVP-induced intracellular acidification in rat MTAL cells incubated at 37°C in a HCO_3-free medium.

10^{-8} M AVP rapidly acidified the cells ; in the presence of 10^{-4} M furosemide to inhibit the AVP-sensitive Na^+:K^+:$2Cl^-$ symport, the AVP-induced cell acidification was of less magnitude but was nevertheless significant. 10^{-10}M AVP also significantly acidified MTAL cells.

Since 2 x 10^{-5} mM forskolin, an activator of adenylate cyclase, quantitatively reproduced the effects of 10^{-8} M AVP on cell pH, it is clear that AVP inhibits luminal proton secretion both directly and indirectly mainly through cyclic AMP-dependent mechanisms. This result is in agreement with that of Good (1990) who observed that 2.8 x 10^{-10}M AVP inhibited bicarbonate absorption by the isolated perfused rat MTAL both in the presence and absence of luminal 10^{-4}M furosemide. Whether AVP acts on Na^+:H^+ antiport and/or H^+-ATPase is under current investigation.

When 10^{-8} M AVP was added to a MTAL cells-containing bicarbonate-medium, no change in pHi was observed, which could suggest that AVP also inhibits basolateral K^+-HCO_3^- symport.

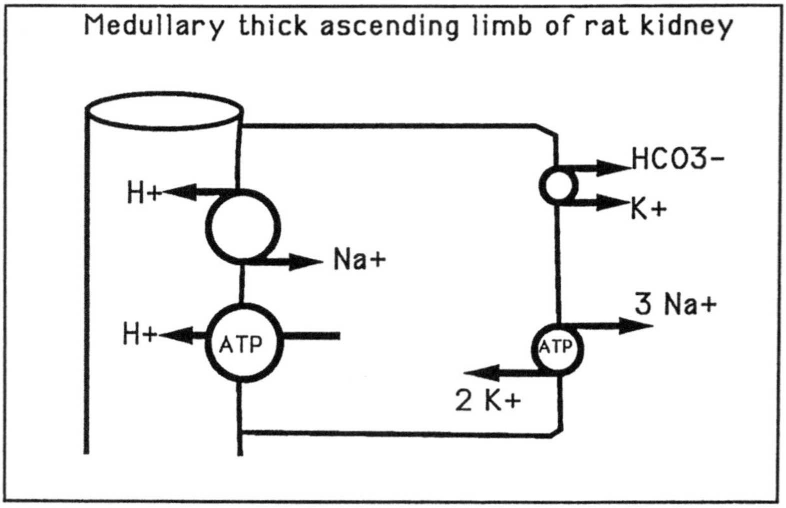

Fig. 5. Model of H^+ and HCO_3^- transport mechanisms of rat MTAL cells.

In conclusion, we proposed the following cellular model to account for bicarbonate absorption by the rat MTAL (Fig.5). H^+/HCO_3^- transport by this nephron segment is under hormonal control, particulary by peptide hormones such as AVP, glucagon, and PTH (Paillard and Bichara 1989, Good 1990).

REFERENCES

Bichara M, Mercier O, Houillier P., Paillard M, Leviel F (1987) Effects of antidiuretic hormone on urinary acidification and on tubular handling of bicarbonate in the rat. J Clin Invest 80:621-630.

Bichara M, Paillard M, Corman B, de Rouffignac C, Leviel F (1984) Volume expansion modulates $NaHCO_3$ and NaCl transport in the proximal tubule and Henle's loop. Am J Physiol 247:F140-F150.

Brown D, Hirsch S, Gluck S (1988) Localization of a proton-pumping ATPase in rat kidney. J Clin Invest 82:2114-2126.

Froissart M, Marty E, Leviel F, Bichara M, Poggioli J, Paillard M (1990) Plasma membrane H^+-ATPase and Na^+-H^+ antiporter in medullary thick ascending limb (MTAL) of rat kidney. Kidney Int 37:537.

Good DW (1985) Sodium-dependent bicarbonate absorption by cortical thick ascending limb of rat kidney. Am J Physiol 248:F821-F829.

Good DW (1990) Inhibition of bicarbonate absorption by peptide hormones and cyclic adenosine monophosphate in rat medullary thick ascending limb. J Clin Invest 85:1006-1013.

Good DW, Knepper MA, Burg MB (1984) Ammonia and bicarbonate transport by thick ascending limb of rat kidney. Am J Physiol 247:F35-F44.

Kikeri D, Azar , Sun A, Zeidel ML, Hebert SC (1990) Na^+-H^+ antiporter and Na^+-$(HCO_3^-)_n$ symporter regulate intracellular pH in mouse medullary thick limbs of Henle. Am J Physiol 258:F445-F456.

Krapf R (1988) Basolateral $H/OH/HCO_3$ transport in the rat cortical thick ascending limb. Evidence for an electrogenic Na/HCO_3 cotransporter in parallel with a Na/H antiporter. J Clin Invest 82:234-241.

Leviel F, Marty E, Bichara M, Poggioli J, Paillard M (1990) Electroneutral K^+ -HCO_3^- cotransport in medullary thick ascending limb (MTAL) of rat kidney. Kidney Int 37:541.

Paillard M, Bichara M (1989) Peptide hormone effects on urinary acidification and acid-base balance : PTH, ADH, and glucagon. Am J Physiol 256:F973-F985.

Sabatini S, Laski ME, Kurtzman NA (1990) NEM-sensitive ATPase activity in rat nephron : effect of metabolic acidosis and alkalosis. Am J Physiol 258:F297-F304.

Trinh-Trang-Tan MM, Bouby N, Coutaud C, Bankir L (1986) Quick isolation of rat medullary thick ascending limbs. Enzymatic and metabolic characterization. Pfluegers Arch 407:228-234.

Zeidel ML, Silva P, Seifter JL (1986) Intracellular pH regulation in rabbit renal medullary collecting duct cells. Role of chloride-bicarbonate exchange. J Clin Invest 77:1582-1688.

SECTION VI

CALCIUM-BINDING PROTEINS
CHARACTERISTICS AND STRUCTURE

STRUCTURE AND SELECTIVITY OF Ca-BINDING SITES IN PROTEINS: THE 5-FOLD SITE IN AN ICOSAHEDRAL VIRUS

George Eisenman and Osvaldo Alvarez
Dept. of Physiology
UCLA Medical School
Los Angeles, CA 90024-1751. U.S.A.
and
Dept. of Biology
Faculty of Sciences
Univ. of Chile, Santiago, Chile

INTRODUCTION AND SUMMARY

Ca binding sites in proteins utilizing a binding loop (EF hand) flanked by two helices have been well characterized both structurally (Kretsinger, 1980; Szebenyi and Moffat, 1986; Herzberg and James, 1985) and functionally (Vogel and Forsen, 1987). In such sites the ion is coordinated by a roughly octahedral shell of oxygen atoms from side chain carboxylates, the peptide backbone, and water molecules. In the classical EF hand three side-chain carboxylates participate in Ca liganding; but in the pseudo-EF hand of calbindin only one carboxylate participates because a proline insertion and loop rotation cause two of the carboxylates to rotate away from the Ca so that carbonyl oxygens from the peptide backbone replace the carboxyl oxygens.

Theoretical calculations of the free energies underlying binding and selectivity in such sites are made difficult by the unknown ionization state of the carboxylic ligands, as well as by H-bond mediated cooperativity between pairs of EF-hands. It is therefore of great interest that a simpler Ca binding site composed of a ring of five backbone carbonyl oxygens has been found in the protein capsid of an icosahedral virus (Jones and Liljas, 1984). The ligands of this Ca binding site are of only one chemical type; and the site is also of interest as the only known example in a protein of a cation binding site using solely backbone carbonyl ligands, resembling the macrocycles in this regard (cf. Eisenman and Dani, 1987; Jullien and Lehn, 1988; Perutz, 1989). Since this binding ring constitutes the narrowest part in an hourglass shaped channel-like structure it is an almost literal embodiment of the "selectivity filter" postulated for cell membrane channels (Hille, 1984); and the entire viral structure is sufficiently analogous to a membrane channel that it is appropriate to call it a "**channelog**".

The symmetry of this molecule and the absence of ionizable groups directly involved as ligands make it a relatively promising structure on which to begin to deduce functional behavior theoretically using a computational chemical approach. The results of such computations are presented here with particular focus on the energetics of equilibrium ion binding and on the kinetics of loading and unloading. The following questions of general interest to Ca binding and permeation are addressed and some will be answered. What are the consequences of having truly "frozen" versus rearrangeable structures? In particular, what are the consequences of structural rearrangements when an ion is replaced by another of differing size, or as an ion moves into and out of a binding site? Under what conditions will strictly isomorphous vs. locally non-isomorphous, or even cooperative, replacement occur? And what are the energetic consequences of the differing structural rearrangements (i.e., energy costs, kinetic benefits, likelihood of occurrence)? Also, what are the effects of varying the fixed dipolar charge on the carbonyl ligands as well as of the ionization state of rings of glutamic and lysine residues some 6 angstroms away from the site?

First, the results for ion binding and permeation of a relatively simple calculation for "frozen" crystallographic coordinates (Eisenman, et al, 1988) using Warshel and Russell's (1984) Protein-dipole Langevin-dipole (PDLD) algorithm are summarized.

NATO ASI Series, Vol. H 48
Calcium Transport and
Intracellular Calcium Homeostasis
Edited by D. Pansu and F. Bronner

Then certain limitations on these calculations chiefly due to the restriction to frozen coordinates and the use of the Langevin dipole approximation to represent water molecules are removed by computationally much more laborious Molecular Dynamics and Free Energy Perturbation (MD/FEP) calculations using Warshel and Creighton's (1989) MOLARIS program in the presence of 134 explicit flexible simple point charge (SPC) water molecules having -0.8 electronic charges at the oxygen center and +0.4 electronic charges at the H loci and with bond angles and stretches determined by the conventional protein force fields (cf. Aqvist and Warshel, 1990).

STRUCTURE OF THE Ca BINDING SITE

For background the organization of 2STV, the smallest icosahedral virus (diameter 170 angstroms) is described. In this virus 60 identical protein subunits, each consisting of 195 residues, self-assemble into a spherical, membrane-like shell, the capsid, encapsulating a single RNA molecule (Jones and Liljas, 1984, Montelius et al, 1988). A cross section through the center of this structure is shown in Fig. 1, taken from Montelius et al (Montelius et al, 1988). The solid lines in the upper half show the alpha carbon chains in the normal state; while the dotted lines show these in an expanded state produced by EDTA removal of Ca from the virus. The 3-fold and 5-fold symmetry axes are indicated by triangles and pentagons; and the locations of the bound cations are shown by dots or open circles. The dots are for the normal state, the open circles for a Ca-depleted expanded state in which only the 5-fold site is occupied by an ion. The Ca binding sites are involved in holding the capsid together and probably serve as sensors to tell the virus when it has entered the submicromolar Ca environment of a cell interior so it can release its nucleic acid (Durham et al, 1977; Hull, 1978; Rossman et al, 1983).

Three different types of divalent ion binding sites have been found in 2STV: "5-fold", "3-fold", and "general" (Jones and Liljas, 1984). The 3-fold and general types of cation-binding sites contain charged carboxylate side chains and resemble other known Ca binding sites (Kretsinger, 1980; Herzberg and James, 1985; Vogel and Forsen, 1987) in this respect. There are twenty 3-fold sites per capsid in 2STV, lying on the icosahedral 3-fold axis with the cation liganded by the carboxylate oxygens of the 3-fold related Asp55 residues. There are sixty general sites per capsid lying near, but not on, the icosahedral 3-fold axis, which consists of two main chain carbonyls (from Ser61 and Gln64) and one carboxylate (OD1 from Asp194) of the same subunit, together with a second carboxylate from a 3-fold related subunit (OE1 of Glu25). The locations of these sites can be seen in Fig. 1. Unfortunately, almost nothing is known about their ionic selectivity beyond the known affinity of the 3-fold and general sites for micromolar Ca concentrations (and the even stronger affinity for Ca of the 5-fold sites) as well as the apparent preference for Ca relative to other cationic species in the crystallization liquor (Jones and Liljas, 1984, Montelius et al, 1988).

The most interesting site from the point of view of the present paper, and the only one whose structure-selectivity relationships will be examined here, is the 5-fold site lying in the channelog on the 5-fold axis between 5 monomers which make up the 12 pentamers that form the capsid. There are 12 such sites per capsid; and their molecular structure can be seen in the stereoscopic views in Fig 20 of Jones and Liljas (1984) and in color plate 1 of Eisenman et al (1988). The ion binding backbone carbonyl ligands that form a binding site come from a sharp bend at the narrow end of the wedge-shaped beta barrel where a beta strand reverses its direction. Similar structures are found in the human Rhinovirus, the Mengo virus, and the Polio virus (Rossman et al, 1983).

The aperture is formed by a ring of 5 backbone carbonyls contributed by THR_{138} from each of the pentamerically related monomers. This site is an example of the ability of proteins to bind cations solely by interactions with the peptide backbone, as in small peptides like the valinomycin and gramicidin analogues (Eisenman and Dani, 1987) or cyclo(-l-Pro-Gly-)$_3$ (Sussman and Weinstein, 1989). Indeed, this site is the only presently known example of such a protein binding site composed solely of backbone carbonyl ligands.

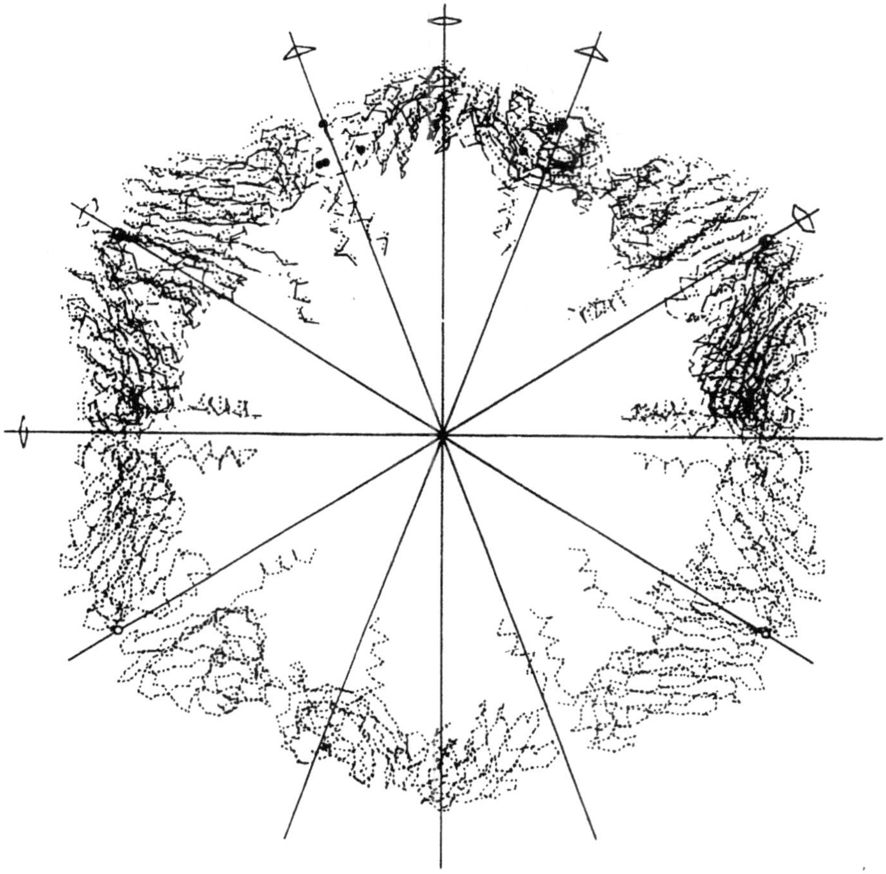

Fig.1. A cross-section through the center of the protein shell of an expanded virus particle showing the alpha-carbon skeleton of 16 subunits (dotted lines). The corresponding alpha carbon chains in the native virion are shown as continuous lines. Open circles are Ca ions that remain in the 5-fold sites in the expanded virion; filled circles are Ca ions in all three types of sites in the native virion. (Reproduced from Montelius, et al (1988), with permission).

This kind of architecture is encountered in all the known icosahedral viruses; and a pentameric ring of backbone carbonyl oxygens also lines the analogous structure in the human *rhinovirus* (Rossman et al, 1985), although the hole in the ring is twice as wide as in 2STV (being 4 angstroms rather than 2 angstroms in internal diameter). Similar carbonyl rings are found in the Mengo and Polio Viruses.

Also lining the channelog further toward the interior of the capsid are two adjacent rings of polar residues. These consist of a wide ring of 5 Glu_{140} gamma carboxylate side chains about 5.5 angstroms internal to the selectivity filter and a wider and slightly more distant ring of 5 Lys_{143} gamma amino side chains (Jones and Liljas, 1984). The ionization state of these residues will be ignored in most of the PDLD calculations since they are likely to neutralize each other; but they will be considered in the MD/FEP section. Numerous positively charged residues line the channel further to its interior in juxtiposition to the RNA. Because of these complexities we will emphasize the loading and unloading behavior toward the outside medium rather than to the interior.

It is clear from the Xray data alone that a ring of backbone carbonyl oxygens can provide a binding site for Ca (and presumably other cations) and that such ligands can replace a major portion of the primary hydration shell. In addition, ion depletion studies (Montelius et al, 1988) indicate that this site is highly selective for Ca, with a dissociation constant in the micromolar to sub-micromolar range.

ENERGY CALCULATIONS

To understand both equilibrium affinity of the site (i.e., binding) and the kinetics of loading and unloading, it is necessary to characterize the free energy profile for moving an ion through the structure. The procedure for computing such a free energy profile is conceptually quite simple in a structure whose atomic coordinates are known. For the Satellite Tobacco Necrosis Virus, Jones and Liljas' (1984) coordinates for the monomer are available in the Brookhaven Data Base (Bernstein et al, 1977); and the pentamer was generated using Robinson and Crofts' (1988) PDV program by sequential 72° rotations around the 5-fold axis.

In both the approximate Protein Dipole Langevin Dipole (PDLD) calculations for the frozen coordinate case presented in the next section and the more definitive Molecular Dynamics and Free Energy Perturbation (MD/FEP) calculations presented subsequently one computes the energy of the system as a probing sphere containing the ion is moved through the protein from a great distance on one side to a great distance on the other (see Fig. 2). The sphere eclipses a variable number of water and protein atoms whose energies are explicitly calculated as described below. The change of free energy of the system as a function of the position of the probe gives the desired free energy profile with reference to the aqueous solution. Typical profiles for Ca, Na, and Cl^- are shown in Fig. 3.

PDLD CALCULATIONS

Fig. 2 illustrates how the profile is calculated. This figure schematizes a section along the 5-fold axis of a typical viral channel (actually a tracing of a longitudinal section of 4SBV (Silva et al, 1987). At the right is shown a sphere of 12 angstroms radius which is moved as a probe along the channelog axis from a distant position at the right through the protein to a distant position at the left. The energy of all the atoms included in the sphere at each position is computed microscopically using Warshel's (Warshel and Russell, 1984; Warshel and Creighton, 1989) PDLD algorithm (the energy due to the charge of the ion is calculated for each position as the difference in the energy of the system when there is an ionic charge at the center of the probing sphere and when the charge is zero). The computations are summarized elsewhere (Eisenman et al, 1988) and were performed on a Definicon DSI/780+ 32-bit coprocessor in an IBM/AT environment. The calculations evaluate the energetics involved in the movement of ions of differing size and charge through a "frozen" channelog structure which retains strictly its original crystallographic coordinates and in which water molecules are modelled by a grid of Langevin dipoles. Note that if one is interested only in binding selectivity one does not need to evaluate the total profile but only the difference of total energy when the probe is at the selectivity filter (which is the binding energy) and at a great distance from it (which is the hydration energy). For the latter value we can use experimental values for the hydration energies (see Eisenman et al, 1988, Eisenman and Villarroel, 1990).

Typical energy profiles for Ca, Na, and Cl^- moving along the channel axis are presented in Fig. 3. The curve labelled "Total" is the total energy of the system plotted as a function of the distance of the probing ion from the selectivity filter. The water in the system had been modelled using the Langevin-dipole approximation; and computations were carried out in the "frozen" coordinate approximation, where energies are calculated using measured crystallographic coordinates without allowing any structural rearrangement to take place as an ion is moved through the protein. The ion has also been constrained to the 5-fold axis. These approximations will be removed in the MD/FEP calculations.

The "Total" energy profiles for the cations Ca and Na show an energy minimum (i.e., a binding site) at the selectivity filter separated from the fully hydrated reference levels by two relatively symmetrical energy maxima (i.e., barriers) to entering and leaving the site from or to the adjoining solutions. In channel parlance (Lauger, 1973), this corresponds to the 2-barrier 1-site situation. If the permeant species is an anion, the profile, typified by that for Cl^-, is seen to have a sharp energy maximum (i.e., a barrier) at the filter instead of the energy minimum found for a cation. Thus, the carbonyl selectivity filter in this channelog excludes Cl; and the binding site is expected to be cation selective. In addition, the depth of the energy well is much greater for Ca than for Na, indicating that the site is expected to bind Ca more strongly.

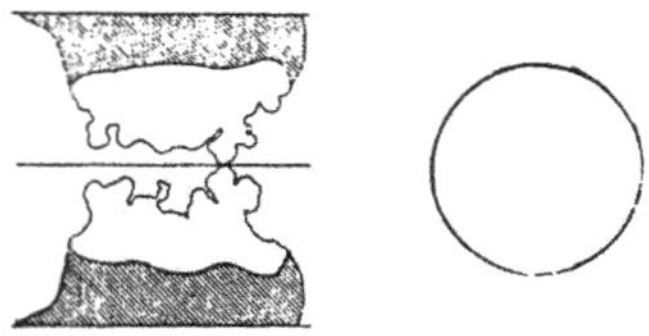

Fig 2. Diagram of procedure for calculating energy profile for permeation. A section along the 5-fold axis of a typical viral channelog (Southern Bean Mozaic Virus, after Silva, et al, 1987) is shown at the left. At the right is shown a sphere of 12 angstroms radius containing in its center an ion of given size and charge and filled with a grid of Langevin dipoles. The sphere is moved as a "probe" along the 5-fold axis from a position far to the right, through the protein, to a position far to the left; and the total energy of the system is computed at each point.

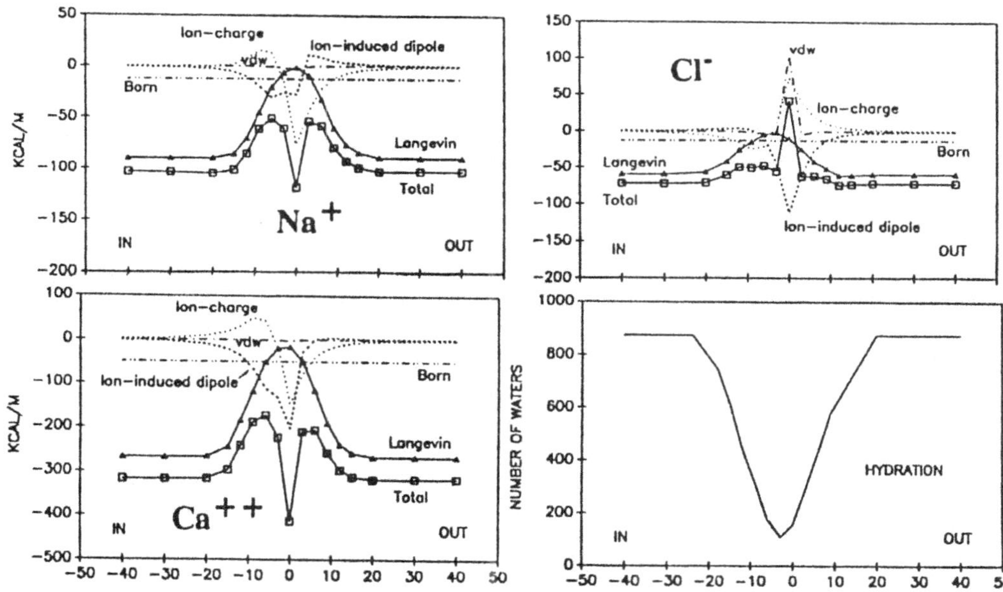

Fig 3. Energy profiles and extent of hydration as a function of distance from the Ca binding site (selectivity filter) of the 5-fold channelog in the Satellite Tobacco Necrosis Virus , 2STV (after Eisenman, et al, 1988). The total energy at any position is the sum of the following individual terms: a coulombic "Ion-charge" energy; an "Ion-induced dipole" energy; the Van der Waals ("vdw") energy; the "Langevin" dipole energy; and the macroscopic "Born" energy external to the 12 angstroms sphere. Each of these energies is computed microscopically by the PDLD program for all atoms within this sphere as the difference between the energies of the system when the ion is uncharged and charged. The total energy also includes a macroscopic Born energy external to the sphere.

The components of the energy profile

The several energy terms that contribute to the total energy profile are: (1) the *van der Waals energy* between the ion and the immediately adjacent ligands of the selectivity filter. (2) the *ion-charge energy* between the ion and the formal charges assigned to all the atoms of the protein. (3) the *ion-induced dipole energy* between the ion and the polarizable dipoles of the protein (this represents differences in dipole-induced dipole energy in the absence of the ion and in its presence). (4) the *Langevin energy* representing the energy of interaction of water molecules with the ion and with the protein. And (5) the *Born energy* for a continuous medium external to the 12 angstrom probing sphere.

The *Langevin energy* in Fig. 3 has its maximal (negative) value at distances more remote than 20 angstroms from the selectivity filter. As the probing ion moves into the channel, and as the channel narrows down toward the selectivity filter, the Langevin energy becomes smaller and smaller because water is being sterically excluded from the vicinity of the ion. This can be seen in the lower right plot of "hydration", which shows the number of Langevin model "waters" available to hydrate the ion. It is apparent that fewer and fewer water molecules are available to hydrate the ion as the selectivity filter is approached, which is why the Langevin dipole energy for the ions decreases as they enter the narrower regions of the channel. (The water exclusion in the narrowest regions is exaggerated since Langevin dipoles are not permitted within 2.6 angstroms of the protein surface.

The *Ion-charge energy* (thin dotted line) represents the coulombic interactions between the ion and all the formal charges of the protein. For Ca (and Na) this energy is seen to have its most negative (attractive) value at the selectivity filter, falling off monotonically toward either solution, somewhat more sharply toward the inside and becoming repulsive at about -8 angstroms. This is because of the predominantly attractive interactions between cations and the dipoles of the backbone carbonyl ligands making up the selectivity filter. The opposite situation can be seen to occur for an anion.

The *Ion-induced dipole energy* for Ca (heavy dotted line), adds a favorable interaction which is maximal at the filter and decrements to either side. Internally it contributes sufficient favorable energy (reflecting favorable interactions with the larger number of protein atoms near the channel axis internally) to offset the repulsive Ion-charge term at -8 angstroms.

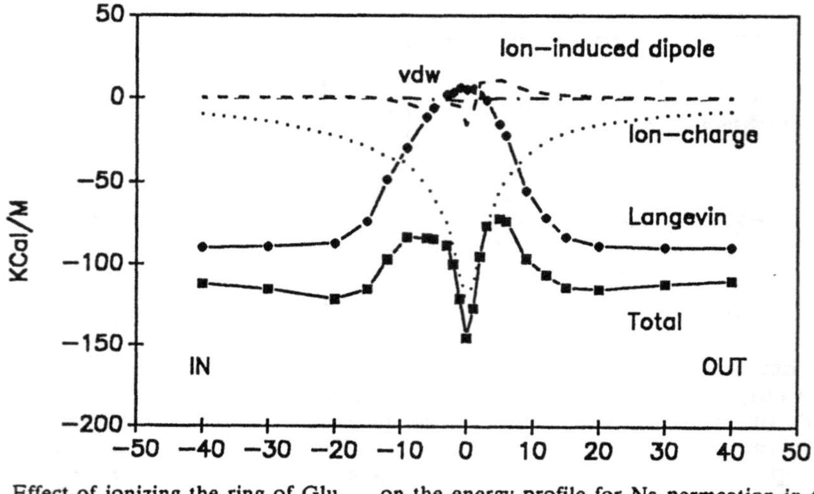

Fig. 4. Effect of ionizing the ring of Glu_{140} on the energy profile for Na permeation in the 5-fold channelog of the Satellite Tobacco Necrosis Virus. Plotted in the same manner as Fig. 3, but with an excess negative charge of -1 distributed uniformly over the carboxylate oxygens of all 5 Glu_{140} residues.

The *Van der Waals energy* (labelled "vdw") is weakly attractive and essentially negligible for both Ca and Na, but is repulsive for Cl^-. Part of the exclusion of Cl^- from this channel is therefore due to a steric effect. The parameters used in this calculation were chosen to represent the Pauling crystal radii but probably underestimate the ion size.

The *Born energy* is the macroscopic energy outside the 12 angstroms radius calculated by the Born equation. It is appxroximately constant.

Modulatory effects of more distant groups

One can examine theoretically the effect of ionizing the ring of side chain carboxylate oxygens from Glu_{140} noted above to be located some 5.5 angstroms internal to the selectivity filter. For the calculations in Fig. 3 this ring was assumed to be unionized (i.e., the dissociable H^+ is assumed still to be bound). The interactions of an ion on the channel axis with this ring can be seen in the ion-charge and ion-induced dipole energies at -5.5 angstroms from the selectivity filter (the interactions produce the second dip in the ion-induced dipole energy seen for Na and Ca). The ion-charge and ion-induced dipole effects roughly cancel each other, with the ion-charge term being weakly repulsive and the ion-induced dipole term being weakly attractive.

However, if dissociation of a single proton from this ring of oxygens is assumed to occur, this can be modelled by distributing an excess negative charge of -1 over all the carboxylate oxygens. Fig. 4 shows the profile for Na^+ in this case. It can be seen that the Ion-charge energy, which was repulsive toward the inside of the capsid in Fig. 3, becomes strongly attractive, and remarkably symmetrical, in Fig. 4. Paralleling this change, the Ion-induced dipole energy becomes almost negligible throughout (which, intriguingly, is the kind of behavior expected in a system where ion-charge interactions have been optimized (cf. Aqvist and Warshel, 1989). As a consequence of ionizing the Glu_{140} ring, the affinity of the site for Na is increased by about 20 KCal/M and the heights of both barriers relative to the aqueous solution are reduced by more than 10 KCal/M. The permeability of the channel, which is inversely related to the height of the largest barrier relative to the aqueous solution (Hille, 1984), is correspondingly increased. The rate determining barrier has also changed since the internal barrier which was the higher one in Fig. 3 is now the lower one.

Summary and Discussion of PDLD calculations

The energy profile for cations is seen to consist of an energy minimum (i.e., a binding site) at the selectivity filter separated from the fully hydrated reference levels by two relatively symmetrical energy maxima (i.e., barriers) to entering and leaving the site from or to the adjoining solutions. In channel parlance, this corresponds to a 2-barrier 1-site situation (Hille, 1984). The particular shape of the pròfile arises from a competition between binding and hydration. The loss of hydration energy is compensated by energy gained from increasing interactions with the protein as the selectivity filter is approached. Thus, the binding energy reaches its maximum at the selectivity filter, where the hydration energy reaches its minimum. These results indicate that a ring of backbone carbonyl oxygens is indeed expected to bind cations more favorably than anions. In addition, the minimum is deeper for Ca than for Na even for the relatively low value of partial charge of 0.15 for the carbonyl oxygen and carbon used in these calculations because polarizability of the ligand oxygens was not suppressed. This calculation is in agreement with experimental findings (Jones and Liljas, 1984; Montelius et al, 1988). However, since the relative preference for divalent over monovalent cations is a function of the value of the partial charge on the carbonyl oxygen (Eisenman, et al, 1988), this result is merely suggestive, though it will be borne out by the MD/FEP calculations of the next section.

Despite the simplicity of the calculated energy profile, and the much more reasonable values for the energy barriers and wells found using the Langevin dipole approximation for water molecules than were found in previous calculations on a related viral structure by Silva et al (1987) which did not include water molecules, the PDLD calculations are still not realistic; for it can be seen that the energy barriers for loading and unloading an ion are too large (about 50 KCal/M for Na and over 100 KCal/M for

Ca) for this structure to function as a channel, or even to load and unload ions on a time scale of hours as observed experimentally (Montelius et al, 1988). There are two likely reasons for this. First, restricting the protein atoms to their original frozen coordinates does not allow for the structure to optimize itself as the ion is moved through it. This restraint can be removed in a molecular dynamics calculation, as is done in the next section. Second, and more importantly, the Langevin Dipole approximation for the hydration energy is too imprecise to give the correct spatial dependence for this term. Since the energy profile is exquisitely sensitive to the balance between loss of ion-water energy and its replacement by ion-protein energy, a relatively small error in the Langevin Dipole estimate of hydration energy can lead to an unrealistic energy profile. One solution to this problem, and the one adopted here is to use explicit water molecules and carry out the computation by molecular dynamics simulation, as is done in the next section.

MOLECULAR DYNAMICS/FREE ENERGY PERTURBATION

Procedure

This section presents preliminary results of ion-water and ion-protein interactions for 2STV calculated using the well known well known Molecular Dynamics and Free Energy Perturbation (MD/FEP) procedure (Zwansig, 1954; see for example, Warshel, Sussman and King, 1986) using the MOLARIS program (Warshel and Creighton, 1989), following Aqvist and Warshell (1989). The MOLARIS code was provided by Arieh Warshel and run on an Ardent Titan minisupercomputer.

The program calculates the energy of charging the ion inside the channel by changing the charge in small steps from fully charged to zero while allowing the neighboring atoms to continuously rearrange their positions so as to optimize their energy. The system consists of the ion, the atoms of the protein, and a sphere of explicit water molecules centered at the position of the ion and filling all the space not occupied by the ion or protein atoms. As the ion is charged adiabatically in succesive small steps, the protein atoms and the water molecules within the sphere are allowed to move until their energy is optimized. The energies and the coordinates are stored for each step of charging and the energies are then mapped for the charging and discharging processes to yield the free energy.

The starting structure for MD/FEP calculations is a cylinder extracted from the crystallographic coordinates of Jones and Liljas (1984) containing all complete residues which had any atom within a 20 Angstrom distance of the 5-fold axis. This structure was "annealed" at 5 degrees Kelvin to remove possible bad contacts using 1000 steps of molecular dynamics and a step size of 0.001 picoseconds with no constraints on the ion, waters, or protein atoms within 30 angstroms of the ion. The coordinates generated by this procedure were the starting point for all subsequent runs.

Two kinds of calculations are presented here. In the first, the *energy profile* is calculated for an ion moving along its optimum free energy path in a rearrangeable channel. In these calculations the ion is constrained harmonically to a series of positions along the channel's axis but allowed to move in a plane normal to the axis. In this way it finds its free energy minimum which need not lie on the channel's axis. In the second, the *selectivity* is calculated for ions of differing size and charge constrained to lie in the plane of the carbonyl ligands. For each position of the ion in the first case, and for each species of ion in the second, the energy of the system was calculated for 11 values of the partial ion charge varying from fully charged to zero in ten steps, each decrementing the charge by 10 %. The reversibility of this calculation was checked by repeating the procedure so as to recharge the ion from zero to fully charged.

A typical MD/FEP calculation for a given ion position involves twelve molecular dynamics simulations, each of 1500 steps at 298° K with a step size of 0.02 picoseconds. The ion is constrained in a plane at its desired position and all protein atoms and water molecules within 10 angstroms of it are allowed to move. All protein atoms more distant are quadratically constrained to their annealed positions. Within the ten angstrom radius of the probing sphere a weak harmonic constraint of 0.5 $KCal/A^2$ is included. This constraint is conservative for the present purpose of assessing the barriers for entering and leaving the site because it will cause their height to be overestimated. **Run 1** is an initialization run of 1500 steps with the full charge and appropriate Lennard-Jones parameters for the ion in question. **Run 2** is a further 1500 steps for the fully charged ion which produces the first set of energies to be used in the free energy mapping and the coordinates to be used for the next run in which the charge is reduced to 90% of the total. **Run 3** is performed at 80% of total charge, **Run 4** at 70%, etc, until **Run 12** which is carried out for zero charge.

In each run, 1500 different sets of coordinates for the ion, water molecules and protein atoms are obtained. The first 50 structures were discarded and a weighted average energy of the remaining 1450 structures is computed. The energy of charging the ion is calculated from the integration of the energy differences between the successive steps. 33 hours of computer time on an Ardent Titan minisupercomputer were required to obtain the energy for each ion position. The total energy includes the cost of rearranging the water and the protein atoms and all the coulomb interactions of the ions in the system. It leaves out the energy of the interactions of the ion with the induced dipoles in the structure.

Lennard-Jones 6-12 Parameters for ions in water

Before carrying out calculations for the protein, it was first necessary to determine a set of Lennard-Jones 6-12 parameters for ions in water which can accurately reproduce simultaneously the experimentally known free energies of hydration as well as the radial distribution of water molecules. This was done using the MD/FEP procedure on a surface constrained all-atoms solvent model (Warshel and King, 1985) for a sphere of 10 Angstroms radius containing 134 flexible simple point charge (SPC) water molecules. This model incorporates angular and radial constraints in order to compensate for the artificial surface created as a result of using a finite number of water molecules. If one assumes the Lennard-Jones parameters for carbonyl oxygens to be the same as for the water oxygens, as we have done here, it is then possible to use these parameters to carry out the same procedure for any location in the protein.

For the Lennard-Jones energy in Kcal/M:

$$U = A_1A_2/r^{12} - B_1B_2/r^6$$

the following A_2 and B parameters were determined, assuming the same values of 774 ($A^{12}[Kcal/M]^{1/2}$) and 24 ($A^6[Kcal/M]^{1/2}$) for the oxygen in the water molecule and in a carbonyl group (1 Kcal = 4.18 KJ):

	A	B
Li^+	17.3	1.2
Na^+	150	5
K^+	730	14
Rb^+	1206	20
Cs^+	2250	30
Mg^{++}	12.5	0.15
Ca^{++}	182	5.7
Sr^{++}	340	8.6
Ba^{++}	783	14.9
"La^{+++}"	182	5.7

The species labelled "La" is a generic lanthanide having the Lennard-Jones parameters of Ca but a charge of +3.

The use of the same 6-12 parameters for a water and carbonyl oxygen is an approximation which needs to be refined further. Though these parameters quite accurately reproduce the hydration free energies, they probably understate the ionic radii for the divalent and trivalent cations. The values presented are preliminary but are sufficiently accurate to illustrate the *trends* with varying size and charge that are relevant for the present paper.

Kinetics

This section is directed to answering the question raised under **PDLD calculations** (see Fig. 2) as to whether the barriers for entering and leaving the binding site are so large that it could not exchange ions fast enough to function as a selectivity filter in a channel. Because the simulations are more reliable for singly charged than for doubly charged species, we performed them for Na.

Fig. 5 presents the energy profile calculated by the MD/FEP procedure for this species. The upper curve plots the free energy profile without the Born contribution. The lower curve includes a crude estimate of the Born energy outside the probing sphere to give the total free energy. An interesting detail is that the site has been "split" into two energy minima on either side of the ring's equator. Notice in either plot that the barrier for an ion entering or leaving the site from the right (external solution) side has now been reduced substantially (to about 20 KCal/M instead of 50 KCal/M). Such a site would exchange ions with the external solution with a time constant in the order of seconds. Extending the radius of the probing sphere to 16 Angstroms further reduces this barrier to about 14 Kcal/M. It therefore seems likely that the ring of carbonyl oxygens can actually load and release monovalent ions with sufficient rapidity that such a structure could actually function as the selectivity filter in an ion-selective channel.

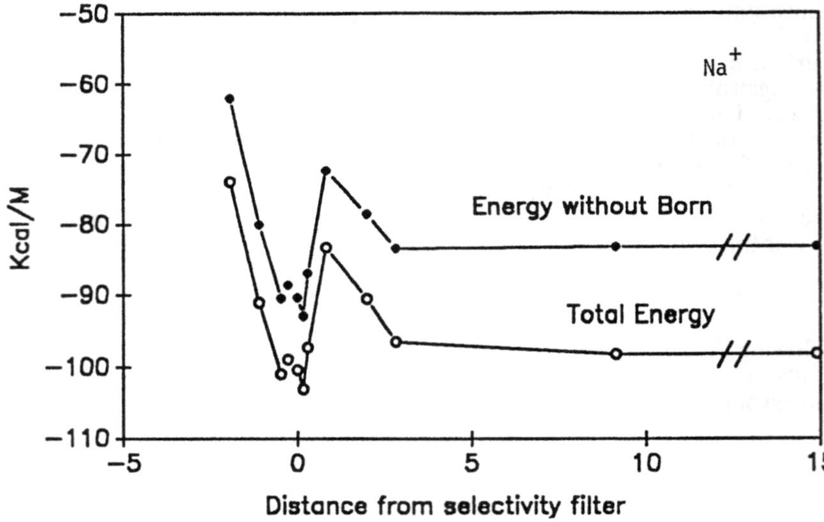

Fig. 5. Optimized Free Energy Profile for Na through the 5-fold channelog in the Satelite Tobacco Necrosis Virus calculated by the MD/FEP procedure. The upper curve does not include the Born energy. The lower curve calculates the total free energy including an approximation for the Born energy which varies slightly for different positions in the Protein. The extreme right hand data point is for the hydration of the ion in pure water.

The positions were probed initially in the sequence 0, +1.0, +2.0, +3.0, 0, -1, -2, using coordinates of the preceding position as the starting point for the next one. Intermediate distances were filled in subsequently and it was verified that the energies were independent of the direction from which they were approached. Several points in the profile were also checked by starting from different initial positions to verify that the profile was independent of the sequence in which the simulations were performed. We will not discuss the barrier toward the inside of the virus here because the height of this is strongly influenced by the (unknown) degree of ionization of the rings of Glu_{140} and

Lys_{143} residues 5.5 Angstroms away toward the interior; although it seems likely that even if both rings were fully ionized, H-bonds between them would largely neutralize their local charges (see below).

Equilibrium Ion Binding and Selectivity for Ca, Mg, Sr, Ba

Previous PDLD calculations for the expected selectivity with the protein atoms "frozen" at their crystallographic location indicated that the selectivity of the 5-fold binding site in 2STV should be in the lyotropic Ba > Sr > Ca > Mg sequence (Eisenman et al, 1988); but we suspect that the van der Waals parameters used in these calculations underestimated the sizes of these species. Fig. 6 presents the results of MD/FEP calculations using the present Lennard-Jones parameters. This figure plots the free energy of exchanging a Ca ion in the plane of the selectivity filter for the other group IIa cations, Mg, Sr and Ba, all relative to their free energies of hydration in the aqueous reference state. These calculations were all performed by the MD/FEP procedure starting with the annealed structure for Ca and carrying out the full discharging cycle for each species constrained to the plane of the annealed Ca ion. An alternative strategy (which would have avoided passing through uncharged states) would have been to use the MD/FEP procedure to "mutate" the Ca ion to Mg, Sr, or Ba by mapping the energy changes as the size of the ion was slowly changed at constant (full) charge.

The results are shown in Fig. 6, where the most striking finding is the existence of an optimum of selectivity for an intermediate sized ion. The pattern in Fig. 6 can be seen to consist of a monotonic curve, convex upward, which corresponds to an "Eisenman Coulomb Topology" (see Eisenman and Horn, 1983). With the present parameters the optimum occurs for Sr, with a selectivity sequence of Sr > Ca > Ba> Mg, which is one of the Eisenman Coulomb sequences (Eisenman, 1965). It is likely that, with further refinement of the 6-12 parameters, the optimum will shift toward Ca since the present set of parameters probably understates the sizes of the ions or the carbonyl oxygen.

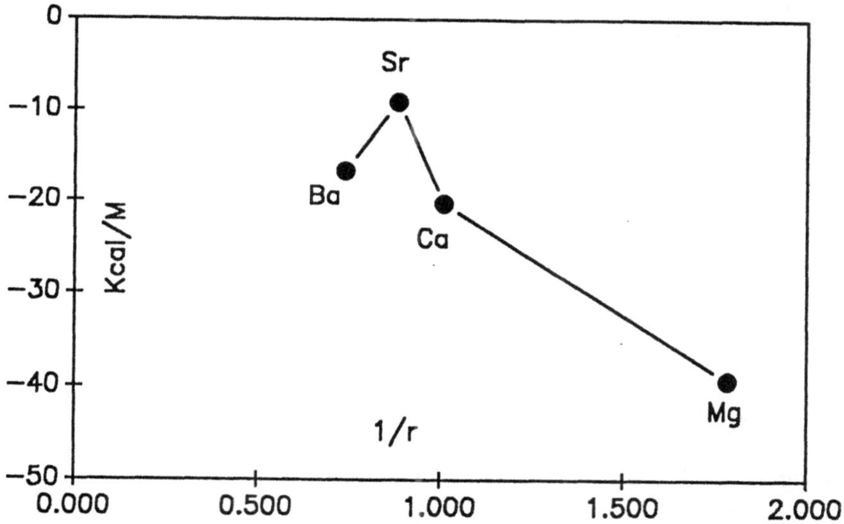

Fig. 6. Binding Selectivity for the 5-fold Ca binding site of the Satellite Tobacco Necrosis Virus. The differences in Free Energy between in water and the site are plotted for the indicated ions as a function of the reciprocal of their Pauling crystal radii. Increasing preference by the site is plotted upwards. Note the optimum for Sr, with the binding sequence being Sr > Ba > Ca > Mg.

Some General Comments on Selectivity in Proteins

Some comments need to be made about what has been learned using MD/FEP methods about the selectivity in a rearrangeable filter within a protein structure as compared to previous calculations on rigid filters (Eisenman et al, 1988, Eisenman and

Villarroel, 1990). Somewhat surprisingly, because the van der Waals repulsive energy was the parameter which determined selectivity in rigid filters, whereas in rearrangeable (small) structures it was the "field strength", the computed selectivity in rigid filters was found to exhibit the same pattern of so-called "Eisenman sequences" as that developed in previous "field strength" modelling of binding sites with rearrangeable structure (Eisenman, 1961, 1965). Indeed, narrow rigid filters showed a formal equivalence to sites having high field strength; while wide filters showed a formal equivalence to sites having low field strength.

By contrast, the present molecular dynamics simulations, in agreement with the experience of computational chemists, indicate that van der Waals repulsions are easily relaxed through small structural rearrangements which cost relatively little energy; so that size differences between different ions are easily accomodated. However, the structural adjustments propagate to regions more distant than the nearest neighbors; and, consequently, selectivity is no longer a locally determined property. Indeed, by analogy to the distribution of energy in a system composed of springs of various strengths, the 12th power van der Waals repulsive interaction, which is a very "hard spring" indeed, immediately distributes itself into the "softer springs" of lower power interactions (e.g., coulomb, torsional, etc.). Thus, the species dependent contribution due to van der Waals repulsions in a rigid filter is no longer confined to interactions with the nearest neighbor atoms but becomes distributed over more distant regions of the protein. Thus, in a protein one can see in a general way how both "field strength" and steric effects can contribute to the actual selectivity.

Effects of ionizing the Glu_{140} and Lys_{143} rings.

The energies in Fig. 6 indicate that the most preferred divalent cation, Sr, is still 9 Kcal/M disfavored in the site relative to solution; and the Ca ion is disfavored by about 20 Kcal/M. Thus, this result is not consistent with the observed high affinity of Ca (Montelius et al, 1988). However, this calculation was done ignoring the effects of the ionizable Glu_{140} and Lys_{143} rings some 5.5 angstroms internal to the site. Because the Glu_{140} ring is closer to the site than the Lys_{143} ring, full ionization of these rings might be approximated by assigning a small excess negative charge to the Glu_{140} ring. This was done in a preliminary way by assigning a single excess charge to one of the Glu_{140} residues. The results of an MD/FEP calculation for Ca done under this condition are tabulated below:

	Relative Affinity	
	Charge on Glu_{140}	
	+0	-1
Ca^{2+}	+52	-42
"La^{3+}"	+78	-62

Comparing the "Relative Affinity" (that is the free energy of the ion in the site minus that in water) when the excess charge on Glu_{140} is 0 vs -1, it is clear that a negative charge on this ring causes the affinity to shift from a value disfavoring Ca to one strongly favoring Ca in the site. K, the binding constant in L/M, is defined by RT ln K = -Free Energy so that the more negative the free energy, the stronger the binding. It will be interesting to see if a similar finding will result from allowing both Glu_{140} and Lys_{143} rings to be fully ionized.

These calculations were done with a newer version of MOLARIS on a "weaker" protein with somewhat different H-bonding interactions and with the harmonic constraint within 10 angstroms of the ion reduced by factor of 50. The resulting protein energies (which contribute to the "Relative Affinity" above) are therefore not directly comparable with those in figure 6. Only the trends with assigning the charge to the Glu_{140} residue are to be noted:

Lanthanide vs Ca Selectivity

Because recent studies by Burroughs and Horrocks (personal communication) using laser activated Eu luminescence on the 5-fold site of 2STV indicate that the Lanthanides can be used as luminescent probes for the selectivity of this site, and because their preliminary results imply a nanomolar binding constant for Eu and a much stronger binding for Lanthanides than for Ca, it is of interest to examine the behavior calculated above for a "generic lanthanide" ion, namely an ion with the Lennard-Jones parameters of a Ca but with a charge of +3.

From the above tabulation, it is clear that, neglecting any effects of ionization of the Lys_{143} and Glu_{140} rings (i.e., for Glu_{140} charge = 0) the lanthanide would be strongly disfavored by the site relative to the solution. Ca would also be disfavored, but less so, with the result that under this condition the site would favor Ca over La. This of course is in disagreement with what Burroughs and Horrocks find. However, it can be seen that for an excess charge of -1 on one of the Glu_{140} residues, the Lanthanide is strongly favored by the site over the solution, and now even more so than Ca. Thus, under the influence of the field due to one excess charge on Glu_{140} the site now prefers the lanthanide over Ca by 20 Kcal/M.

This result suggests that some ionization of the Glu_{140} and Lys_{143} rings is needed to produce expectations in accord with the recent findings of Burroughs and Horrocks of a strong affinity of this site for Lanthanides and a preference for Lanthanides over Ca.

Effect of field strength of the C=O group

One of the variables that controls selectivity in classical theory (Eisenman, 1961, 1962) is the field strength of the ion-binding ligands. It is possible to examine this by assigning a different value of partial charge to the 5 carbonyl ligands of the selectivity filter. We compare below, using the newer version of MOLARIS) the effect of normal partial charge (0.283) vs a high value (0.6) on the energies of interaction with Ca:

C=O charge	
normal	high
-271	-314

It is apparent that doubling the partial charge of the 5 carbonyl ligands can increase the Ca affinity by 43 Kcal/M.

DISCUSSION OF SELECTIVITY

Selectivity is always a balance between energy of hydration and energy of interaction with the filter, including any conformational energy that occurs in structures that rearrange. We previously found (Eisenman et al, 1988) in a rigid selectivity filter, where ion-exchange is strictly isomorphous, that selectivity is determined solely by local interactions, most particlularly by van der Waals repulsions. If the size of the filter exceeds that of the largest ion, selectivity is expected to be inversely related to the hydration free energy (i.e., lyotropic). Only when the size of an ion exceeds the size of the filter do inversions of selectivity sequence occur in which smaller ions become favored. In this extreme situation it is the van der Waals repulsion energy that causes the selectivity reversal among ions of the same valence type, and not the partial charge (*i.e.* "field strength") of the ligands, as in the earliest treatments of selectivity (Eisenman, 1961, 1962, 1965). In a rigid filter, field strength is relevant only for selectivity between ions of differing valence and/or sign. On the other hand, when structures can rearrange in the presence of ions of differing size and charge, van der Waals energy differences among differing species become negligible; and the major contributions to selectivity come from two factors. One is the "field strength" of the (optimized) binding ligands, variations in which lead to species difference in mostly local electrostatic energies of the ions with the ligands as well as more distantly. The other is the structural energy change of the protein itself, which is less local and can be quite widespread, or even allosteric under appropriate circumstances.

It should be noted that a completely frozen selectivity filter is probably not attainable in any protein system since the MD simulations of this paper indicate that, even in the highly organized viral selectivity filter, structural rearrangements take place at very little energy cost. These rearrangements relieve any large van der Waals repulsions. The species dependent contribution due to size remains, but it is no longer confined to van der Waals repulsions with the immediate nearest neighbor atoms. Instead, it becomes distributed over more distant atoms. Therefore, previous selectivity treatments, which dealt only with the rearrangements of small, reorganizable ensembles of atoms (Eisenman, 1961, 1965) or the modelling of larger ensembles in terms of rigid structures (Mullins, 1956,1960; Eisenman et al, 1988; Eisenman and Villarroel, 1990), must now be superceded by calculations that take into account more distant effects. Fortunately, such calculations have become feasible with the advent of minisupercomuters and powerful computational chemistry algorithms such as AMBER (Wiener and Kollman, 1981), CHARMM (Brooks et al, 1983), DELPHI (see Gilson and Honig, 1987), MM2 (Burkert and Allinger, 1982), DREIDING (Mayo, Olafson and Goddard, 1990), POLARIS (Warshel and Russell, 1984), or MOLARIS (Warshel and Creighton, 1989).

Acknowledgement. This work was supported by USPHS Grant GM 24749, FONDECYT grant 1112-1989, and DTI grant B-2805. We thank Arieh Warshel, Johan Aqvist, Francisco Bezanilla, and Terry Lydon for their valuable help with theory and methodology.

References

Aqvist J, Warshel, A, 1989. Energetics of ion permeation through membrane channels. The solvation of Na^+ by gramicidin A. *Biophys. J.* **58**:171-182

Bernstein, F.C., Koetzle, T.F., Williams, G.J.B., Meyer, E.F., Brice, M.D., Rodgers, J.R., Kennard, O., Shimanouchi, T., Tasumi, M. 1977. The protein data bank: A computer-based archival file for macromolecular structures. *J. Mol. Biol.* **112**:535-542

Brooks, B.R., Brucoleri, R.E., Olafson, B.D., States, D.J., Swaminathan, S., Karplus, M. 1983. CHARMM: A program for macromolecular energy minimization and dynamics calculations. *J. Comp. Chem.* **4**:187-217

Burkert, U., Allinger, N. 1984. *Molecular Mechanics,* ACS Monograph 177. ACS. Publ. Washington, D.C.

Durham, A.C.H., Hendry, D.A., Von Wechmar, M.B. 1977. Does calcium ion binding control plant virus disassembly? *Virology* **77**:524-533

Eisenman, G. 1961. On the elementary atomic origin of equilibrium ionic specificity. *In*: Symposium on Membrane Transport and Metabolism. A. Kleinzeller, A. Kotyk, editors. pp. 163-179. Academic Press, New York

Eisenman, G. 1962. Cation selective glass electrodes and their mode of operation. *Biophys.J.* **2, Pt 2**:259-323

Eisenman, G. 1965. Some elementary factors involved in specific ion permeation.

Proceedings of the XXIIIrd International Congress of Physiological Sciences, Tokyo, **87**: 489-506

Eisenman, G., Horn, R. 1983. Ionic selectivity revisited: The role of kinetic and equilibrium processes in ion permeation through channels. *J. Membrame Biol.* **76**:197-225

Eisenman, G., Dani, J.A. 1987. An introduction to molecular architecture and permeability of ionic channels. *Ann. Rev. Biophys. Biophys. Chem.* **16**:205-226

Eisenman G, Oberhauser A, Bezanilla F 1988. Ion selectivity and molecular structure of binding sites and channels in icosahedral viruses. *In*: Transport Through Membranes: Carriers, Channels and Pumps. A. Pullman, J. Jortner, B. Pullman, editors. pp. 27-50. Kluwer Academic Publ. Dordrecht, Boston, London

Eisenman, G., Villarroel, A. 1990. Ion selectivity of pentameric protein channels: Backbone carbonyl ligands as cation binding ligands and side chain hydroxyls as "ambidextrous" ligands for cations and anions in viral capsids. *In*: Monovalent Cations in Biological Systems. C.A. Pasternak, editor. pp 1-29. CRC Press. Boca Raton, FL.

Gilson, M.K., Honig, B.H. 1987. Calculation of electrostatic potentials in an enzyme active site. *Nature (Lond.)*, **330**:84-86

Herzberg, O., James, M.N.G. 1985. Common structural framework of the two Ca^{2+}/Mg^{2+} binding loops of troponin C and other Ca^{2+} binding proteins. *Biochem.* **24**:5289-5302

Hille, B. 1984. *Ionic Channels of Excitable Membranes.* pp. 426. Sinauer Assoc. Inc., Sunderland, Mass.

Hull, R. 1978. The stabilization of the particles of turnip rosette virus. III. Divalent cations. *Virology* **89**:418

Jones, T.A., Liljas, L. 1984. Structure of satellite tobacco necrosis virus after crystallographic refinement at 2.5A resolution. *J. Mol. Biol.* **177**:735-767

Jullien, L., Lehn, J.M. 1988. The "Chundle" approach to molecular channels: synthesis of a macrocycle-based molecular bundle. *Tetrahedron Lett.* **29**:3803-3806.

Kretzinger, R.H. 1980. Structure and evolution of calcium-modulated proteins. *CRC Crit. Rev. Biochem.* **8**:119-174

Lauger, P. 1973. Ion transport through pores: a rate-theory analysis. *Biochim Biophys Acta* **311**:423-441

Mayo, S.L., Olafson, B.D., Goddard, W.A., 1990. DREIDING: A generic force field for molecular simulations. *J. Phys. Chem.* In press.

Montelius, I., Liljas, L., Unge, T. 1988. Structure of EDTA-treated satellite tobacco necrosis virus at pH 6.5. *J. Mol. Biol.* **201**:353-363

Mullins, L.J. 1956. The structure of nerve cell membrane. *In:* Molecular Structure and Functional Activity of Nerve Cells. R.G. Grennel, L.J. Mullins, editors. pp. 123-166. American Institute of Biological Sciences, Washington

Mullins, L.J. 1960. An analysis of pore size in excitable membranes. *J. Gen. Physiol.* **43** (Suppl.1):105-117

Perutz, M.F. 1989. Mechanisms of cooperativity and allosteric regulation in proteins. *Quart. Reviews of Biophys.* **22**:139-236

Robinson, H., Crofts, A. 1988. Exploring protein structures on an IBM-PC. *Biophys. J.* **53**:404a

Rossmann, M.G., Abad-Zapatero, C., Murthy, M.R.N., Liljas, L., Jones, T.A., Stranberg, B. 1983. Structural comparisons of some small spherical viruses. *J. Mol. Biol.* **165**:711-736

Rossmann, M.G., Arnold, E., Erickson, J.W., Frankenberger, E.A., Griffith, J.P., Hecht, H.-J., Johnson, J.E., Kamer, G., Luo, M., Mosser, A.G., Rueckert, R.R., Sherry, B., Vriend, G. 1985. The structure of a human common cold virus (Rhinovirus 14) and its functional relations to other picornaviruses. *Nature (Lond.)* **317**:145-153

Russell, S.T., Warshel, A. 1985. Calculations of electrostatic energies in proteins. The energetics of ionized groups in bovine pancreatic trypsin inhibitor. *J. Mol. Biol.* **185**:389-404

Silva, A.M., Cachau, R.E., Goldstein, D.J. 1098. Ion channels in southern bean mosaic virus capsid. *Biophys. J.* **52**:595-602

Sussman, F., Weinstein, H. 1989. On the ion selectivity in Ca-binding proteins: The cyclo(-l-Pro-Gly-)$_3$ peptide as a model. *Proc. Natl. Acad. Sci. USA*. **86**:7880-7884

Szbenyi, D.M.E., Moffat, K. 1986. The refined structure of vitamin D-dependent Calcium-binding protein from bovine intestine. *J. Biol. Chem.* **261**:8761-8777

Vogel, H.J., Forsen, S. 1987. NMR studies of calcium-binding proteins. *In:* Biological Magnetic Resonance, **Vol. 7**. L.J. Berliner , J. Reuben, editors. Plenum Press, New York and London

Warshel, A. 1982. Dynamics of reactions in polar solvents: Semiclassical trajectory studies of electro-transfer and proton-transfer reactions. *J. Phys. Chem.* **86**:2218-2224

Warshel, A., Levitt, M. 1976. Theoretical studies of enzymic reactions: dielectric, electrostatic and steric stabilization of the carbonium ion in the reaction of lysozyme. *J. Mol. Biol.* **103**:227-249

Warshel, A., Russell, S.T. 1984. Calculations of electrostatic interactions in biological systems and in solutions. *Quart. Rev. Biophys*. **17**:283-422

Warshel, A., King, G. 1985. Polarization constraints in molecular dynamics simulation of aqueous solutions: the surface constraint all atom solvent (SCAAS) model. *Chem. Phys. Lett.* **21**:124-129

Warshel, A., Sussman, F., King, G. 1985. Free energy charges in solvated proteins: microscopic calculations using a reversible charging process. *Biochemistry* **25**:8368-8372

Warshel, A., Creighton, S. 1989. Microscopic free energy calculations in solvated macromolecules as a primary structure-function correlator and the MOLARIS program. *In*: Computer Simulation of Biomolecular Systems. W.F. van Gunsteren, P.K. Weiner, editors. pp. 120-137. Escom, Leiden

Weiner, P.K. and Kollman, P. A. 1981. Amber: assisted model building with energy refinement. A general program for modelling molecules and their interactions. *J. Comput. Chem.* **2**, 287-320

Zwansig, R.W. 1954. High-temperature equation of state by a perturbation method, I. Nonpolar gases. *J. Chem. Phys.* **22**:1420-1426

MOLECULAR AND FUNCTIONAL ASPECTS OF SOME CYTOSOLIC CALCIUM-BINDING PROTEINS

Claus W. Heizmann
Department of Pediatrics
Division of Clinical Chemistry
University of Zürich
Steinwiesstr. 75
CH-8032 Zürich

The calcium signal is transmitted into an intracellular response via the families of calcium-binding proteins which are thought to be involved in the regulation of many cellular activities.

Calcium-binding proteins may be divided into the following groups:
(1) The calcium-binding EF-hand proteins such as calmodulin, troponin-C, parvalbumin, S-100 proteins, calbindins, oncomodulin and many newly described members (Persechini et al 1989; Heizmann and Hunziker 1990a,b) (some are listed in TABLE 1). Parvalbumin and oncomodulin will be discussed in more detail below.

(2) Proteins which contain sequences with obvious resemblance to the EF-hand. One example is dystrophin, the protein product of the Duchenne/Becker muscular dystrophy gene. The cysteine-rich domain contains two EF-hand loop structures but otherwise is different from the EF-hand consensus sequence (Monaco 1989). Other examples of proteins with putative EF-hand domains are human lysozyme (Kuroki et al 1989) and bovine glial fibrillary acid protein (Yang et al 1988).

NATO ASI Series, Vol. H 48
Calcium Transport and
Intracellular Calcium Homeostasis
Edited by D. Pansu and F. Bronner

(3) A distinct family consisting of Ca^{2+}-binding proteins which bind certain phospholipids in a calcium-dependent manner (Crompton et al 1988; Haigler et al 1989).
Several names have been proposed for members of this family, including lipocortins, annexins, calpactins, chromobindins, calcimedins and others. Although no biological function is known for any of these proteins, they are attracting considerable interest because of their potential involvement in calcium-mediated stimulus-response coupling. Proteins of this family are postulated to be involved in phospholipase regulation, membrane trafficking and cytoskeletal organization. Furthermore, the relevance of the phosphorylation of two of these proteins (calpactin I and II) in the context of cell growth and transformation is presently under investigation.
(4) Other proteins which are found to bind Ca^{2+} under various conditions but whose primary sequences have not been elucidated. Their relationship (if any) to each other, to the EF-hand- or to the calcium-dependent and phospholipid binding proteins is unclear.
One example is an integral membrane calcium-binding protein (IMCAL) whose concentration is regulated by vitamin D and which is possibly involved in the mediation or regulation of calcium translocation through the plasma membrane. Interestingly, IMCAL was found to be significantly decreased in plasma membranes of various tissues of spontaneously hypertensive rats (Kowarski et al 1986). Membranes bind less calcium in these rats than in control animals. This difference may result from a decreased content of IMCAL and could underlie the pathogenesis of the hypertension.

In this review I will summarize some recent data both about EF-hand proteins in general, and more specifically, about molecular and functional aspects of parvalbumin in the CNS and endocrine tissue and oncomodulin in some tumor cells.
Calcium-binding proteins whose functions are known, such as

calmodulin or troponin-C, are far outnumbered by those whose roles are not well understood.
However, these proteins have attracted considerable interest recently because of their altered concentrations in many diseased states of the central nervous system (e.g. Parkinson's and Alzheimer's diseases, epilepsy), hypertension, acute and chronic inflammatory lesions, cystic fibrosis, and in several tumor cells.

TABLE 1 SOME NOVEL CALCIUM-BINDING PROTEINS OF THE EF-HAND TYPE

CALRETININ
UVOMORULIN
α-ACTININ (non-muscle cells)
CA^{2+}-BINDING PROTEINS from invertebrates
CALPAIN
LSP_1
22kd PROTEIN (sorcin/V19)
MYELOPEROXIDASE
THIOREDOXIN REDUCTASE
p24 (calcyphosine)
pMP41
β and δ -CRYSTALLINS
SPEC-PROTEINS
LSP_1
CA^{2+}-BINDING PROTEIN from *Streptomyces erythreus*
CALTRACTIN from *Chlamydomonas*
CDC31 GENE PRODUCT from *Saccharomyces cerevisiae*
CA^{2+}-BINDING PROTEINS (TCBP'-S) from *Tetrahymena*
GALACTOSE-BINDING PROTEIN (GBP) of bacterial transport and chemotaxis
PHOTOSYSTEM II Ca^{2+}-BINDING PROTEINS
CA^{2+}-BINDING PROTEIN in *Schistosoma mansoni*
FLAGELLAR CA^{2+}-BINDING PROTEIN (FCaBP)
S-100 PROTEIN FAMILY: S-100a; S-100b; S-100L; calbindin 9K; MRP-8(p8); MRP-14(CFAg;p14); 18A2; p9Ka; p11; pEL98; 42A; 42C;

(For reviews see Kretsinger 1980; Kretsinger et al 1988; Kligman and Hilt 1988; Persechini et al 1989; Strynadka and James 1989; Heizmann and Hunziker 1990 a,b; Heizmann C.W. ed. 1990/91)

DISTRIBUTION / LOCALIZATION AND SOME FUNCTIONAL IMPLICATIONS

A. Brain

Parvalbumin, calbindin D-28K, calcineurin and S-100 proteins have been found in high concentrations in the central and peripheral nervous system of many species including man (for review see Heizmann and Braun 1989). Many processes e.g. release of neural neurotransmitters, electric activities, memory storage, fast axonal flow, depend on calcium. However, the expression and developmental appearance of these calcium-binding proteins in the central nervous system is quite different. Several of these proteins have been found in distinct subpopulations of neurons.

TABLE 2 IMMUNOCYTOCHEMICAL LOCALIZATION OF CA^{2+}-BINDING PROTEINS IN MAMMALIAN CEREBELLUM

	rat			cat		
Brain area	PV	Calbindin	CaN	PV	Calbindin	CaN
Cerebellum						
-Purkinje cells	+++	+++	++	+++	++	+
-Basket/stellate cells	+++	-	+	+++	-	
-Granule cells	-	-	-	-	-	+
-Golgi cells		-				
-Cerebellar nuclei	+f	+f				

PV,parvalbumin; calbindin, calbindin D-28K; CaN, calcineurin; f,fibrous neutropil; +,++,+++, density of stained structures; -, negative (for details and further localizations see Heizmann and Braun, 1989)

Parvalbumin, for example, is present in a subpopulation of mammalian neurons containing the inhibitory neurotransmitter -aminobutyric acid (GABA). Some exceptions to the association with GABA have been found, however, in the retina of the cat and in some nuclei of birds. Generally, parvalbumin is associated with neurons that have a high firing rate and a high oxidative metabolism. The presence of parvalbumin in fast-spiking cells in the rat hippocampus (CA_2 region) has been demonstrated by injection of Lucifer yellow *in vitro* in combination with postembedding parvalbumin immunohistochemistry. It is suggested that parvalbumin may be involved in the buffering/ transport of calcium in a subset of neurons with specialized electrophysiological properties.The likely functions of neural Ca^{2+}-binding proteins was also raised at the First European Symposium on Calcium Binding Proteins in Normal and Transformed Cells (1989) (Pochet et al eds 1989; Rogers 1989). It was suggested that parvalbumin may be a slow but high-affinity "calcium sink" taking up calcium in the cytosol during intense neuronal activity whereas calbindin or calretinin could bind calcium more quickly acting as a "diffusion catalyst" to redistribute it in the cell.There are also indications that parvalbumin and other cytosolic Ca^{2+}-binding proteins may be distributed inhomogeneously in the cell.These results are in agreement with several measurements of intracellular calcium $[Ca^{2+}]i$, using the Fura 2 technique.Many experiments prove $[Ca^{2+}]i$ dynamics to be highly compartmentalized (Tank et al 1988). The complex spatio-temporal control of $[Ca^{2+}]i$ and of Ca^{2+}-dependent processes especially in Purkinje cells may be regulated by members of the EF-hand protein family.

Ca^{2+}-binding proteins·have also been detected in the nuclei of some cells. Parvalbumin-immunoreactivity, mostly present in the cytosol, has also been found in the nuclei of some neurons e.g. in the dorsal lateral geniculate nucleus of the cat brain suggesting a participation of parvalbumin in Ca^{2+}-dependent nuclear events.

B. Endocrine glands

Parvalbumin and S-100 proteins have been detected in the testis and some other endocrine glands. The localization and the expression of these proteins in the rat testis was studied during development. High parvalbumin immunoreactivity in Leydig cells was found when testosterone production was highest. When Leydig cell activity was low, parvalbumin levels also remained low. This correlation suggests that parvalbumin may be involved in processes associated with the Ca^{2+}-dependent production of testosterone in Leydig cells.
When rats were hypophysectomized at mid-puberty (postnatal day 30-35) hormone levels in the serum dropped below detection and the Leydig cells lost their parvalbumin. Upon administration of FSH in combination with LH, testosterone levels rose and parvalbumin staining reappeared. Leydig cells of rats before puberty (aged 23-24 days) were devoid of parvalbumin immunoreactivity before and after hypophysectomy. Replacement therapy of the ablated animals with FSH and LH restored hormone and parvalbumin levels. This close correlation of parvalbumin expression in Leydig cells with serum hormone levels supports the view that parvalbumin may be associated with the Ca^{2+}-dependent production of testosterone in Leydig cells (Heizmann and Kägi, 1989).

Parvalbumin-immunoreactivity was also found in the adrenal gland of the rat (unpublished results), a tissue which produces three major types of hormones, namely mineralocorticoids, glucocorticoids and androgens. The medulla represents 10% of the gland, whilst the remaining 90% is made up of cortical tissue - the zonae glomerulosa, fasciculata and reticularis.

TABLE 3 LOCALISATION OF PARVALBUMIN IN THE ADRENAL GLAND OF THE RAT

	CORTICAL ZONES			MEDULLA
Tissue layer:	glomerulosa	fasciculata	reticularis	
hormones produced	⊢aldosterone⊣	⊢androgens	⊣	⊢epinephrine⊣ norepinephrin
		⊢glucocortico-steroids	⊣	
PV-immuno-reactivity		⊢ large patches of stained cells in between areas lacking PV-IR; many stained cells at the border to medulla	⊣	⊢ single cells ⊣

Within these layers, parvalbumin differs in distribution from calbindin D-28K, S-100 proteins and calmodulin (not shown). Only parvalbumin is expressed in layers where androgens are produced.

C. Tumor cells

The Ca^{2+}-binding protein oncomodulin was found in a wide variety of tumors of the rat but not in normal tissues except in extraembryonic human and rat placental cells (MacManus et al 1989).

The primary sequences of rat oncomodulin and rat parvalbumin are very similar and alignment of the sequences revealed 55 homologous amino acid positions. Oncomodulin from rat has been cloned and the structure of the rat oncomodulin gene has been elucidated (Banville and Boie 1989; Furter et al 1989).

Analysis of the promotor sequence of the oncomodulin gene revealed that the gene is under the control of a solo long

terminal repeat (LTR) element related to intracisternal-A particles (IAP), a family of endogenous retroviral elements. However, so far neither IAPs nor IAP LTRs have been characterized in the human genome. Therefore, it is questionable if oncomodulin gene expression in man is controlled by mechanisms similar to those proposed for the rat oncomodulin gene. An analysis of the oncomodulin promotors from various species will answer the question whether the proposed germ line insertion was an early event during evolution or if it was acquired after the divergence of the species.
Preliminary data indicate that oncomodulin mRNA levels in human tumors are very low compared to rat and it has been suggested that this is probably due to the lack of the strong LTR promotor in man.
We have obtained the same results at the protein level (Huber et al, submitted). A polyclonal antibody was prepared against a synthetic peptide (taken from the rat oncomodulin sequence (AA.99-108) coupled to hemocyanin. The antibody obtained was specific for rat oncomodulin and did not cross-react with other Ca^{2+}-binding proteins.

Using this antibody we specifically detected a prominent immunoreaction at the site of oncomodulin (Mr=12K) in extracts of a chemically transformed rat fibroblast cell line (T14c) but never in extracts of over 25 human tumor cell lines of various origins. This confirmed the very low expression of oncomodulin in human tumor cell lines, and oncomodulin may therefore not be suitable as a tumor marker in human tissues.
Since oncomodulin has been detected in visceral and parietal extraembryonic endoderm and placenta (MacManus et al 1989) it might be more appropriate to study its role in prenatal development. First functional studies in this direction have been initiated (Chalifour et al 1989) by microinjection of a metallothionein-oncomodulin DNA into fertilized mouse embryos. Oncomodulin expression was found to be correlated with fetal

mortality and an interruption of either cellular differentiation or organogenesis before day 9 in development was observed.

Acknowledgements

I would like to thank Dr. A. Rowlerson and Mrs M. Killen for correcting and Mrs R. Kuster for typing the manuscript.
This study was supported by the Swiss National Science Foundation (31-9409.88), and the EMDO-, and Hartmann Müller Stiftungen.

REFERENCES

Banville D, Boie Y (1989) Retroviral long terminal repeat is the promotor of the gene encoding the tumor-associated calcium-binding protein oncomodulin in the rat. J Mol Biol 207:481-490

Chalifour LE, Gomes ML, Mes-Masson A (1989) Microinjection of metallothionein-oncomodulin DNA into fertilized mouse embryos is correlated with fetal lethality. Oncogene 4:1241-1246

Crompton MR, Moss SE, Crumpton MJ (1988) Diversity in the lipocortin/calpactin family. Cell 55:1-3

Furter C, Heizmann CW, Berchtold MW (1989) Isolation and analysis of a rat genomic clone containing a long terminal repeat with high similarity to the oncomodulin mRNA leader sequence. J Biol Chem 264:18276-18279

Haigler HT, Fitch JM, Jones JM, Schlaepfer DD (1989) Two lipocortin-like proteins, endonexin II and anchorin CII, may be alternate splices of the same gene. TIBS 14:48-50

Heizmann CW, Braun K (1989) Calcium-binding proteins. Molecular and functional aspects. In: Anghileri LJ (ed) The Role of Calcium in Biological Systems. CRC Press, Vol 5, Boca Raton, Florida, in press.

Heizmann CW, Kägi U (1989) Structure and function of parvalbumin. In: Hidaka H (ed) Calcium Protein Signalling.

Plenum Press, New York and London, p 215-222

Heizmann CW, Hunziker W (1990a) Intracellular calcium-binding molecules. In: Bronner F (ed) Intracellular Calcium Regulation. Alan R. Liss, Inc., p 211-247

Heizmann CW, Hunziker W (1990b) More and more Ca^{2+}-binding proteins are looking for a job. TIBS, in press

Heizmann CW (ed) (1990/91) Novel Ca^{2+}-binding Proteins in Health and Pathology (tentative title). Springer, Berlin Heidelberg New York, in press

Kligman D, Hilt DC (1988) The S-100 protein family. TIBS: 437-443

Kowarski S, Cowen LA, Schachter D (1986) Decreased content of integral membrane calcium-binding protein (IMCAL) in tissues of the spontaneously hypertensive rat. Proc Natl Acad Sci 83:1097-1100

Kretsinger RH (1980) Structure and evolution of calcium modulated proteins. CRC Crit Rev Biochem 8:119-174

Kretsinger RH, Moncrief ND, Goodman M, Czelusniak J (1988) In: Morad M, Naylor WG, Kazda S, Schramm M (eds) The Calcium Channel, Structure, Function and Implication. Springer, Berlin Heidelberg New York, p 16-34

Kuroki R, Taniyama Y, Seko C, Nakamura H, Kikuchi M, Ikehara M (1989) Design and creation of a Ca^{2+}-binding site in human lysozyme to enhance structural stability. Proc Natl Acad Sci 86:6903-6907

MacManus JP, Brewer LM, Banville D (1989) Oncomodulin in normal and transformed cells. In: Pochet R, Lawson DEM, Heizmann CW (eds) Calcium Binding Proteins in Normal and Transformed Cells. Plenum Press, New York and London, in press

Monaco AP (1989) Dystrophin, the protein product of the Duchenne/Becker muscular dystrophy gene. TIBS 14:412-415

Persechini A, Moncrief ND, Kretsinger RH (1989) The EF-hand family of calcium-modulated proteins. TINS 12:462-467

Pochet R, Lawson DEM, Heizmann CW (eds) (1989) Calcium Binding Protein in Normal and Transformed Cells. Plenum Press, New York and London

Rogers J (1989) Calcium-binding proteins. The search for a function. Nature 339:661-662

Strynadka NCJ, James MNG (1989) Crystal structures of the helix-loop-helix calcium-binding proteins. Ann Rev Biochem 58:951-998

Tank DW, Sugimori M, Connor JA, Llinas RR (1988) Spacially resolved calcium dynamics of mammalian Purkinje cells in cerebellar slices. Science 242:773-777

Yang ZW, Kong FC, Babitch JA (1988) Characterization and location of divalent cation binding sites in bovine glial fibrillary protein. Biochemistry 27:7045-7050

THE MAJOR PROTEIN OF FROG OTOCONIA IS A HOMOLOG OF PHOSPHOLIPASE A2

Kenneth G. Pote and Robert H. Kretsinger
Department of Biology
University of Virginia
Charlottesville, VA, 22901, U.S.A.

We have purified the major protein of the otoconia of *Xenopus laevis*; its molecular weight is about 22,000. The corresponding otoconial protein from *Rattus norvegicus* has a molecular weight of 90,000. Polyclonal antibodies raised against both proteins do not cross react. Both antibodies are highly specific for tissues of the inner ear of the respective animal. A cDNA library has been made from *X. laevis* inner ear tissue and positive clones have been identified using DNA probes designed after the partial amino acid sequence.

Partial amino acid sequence has been determined for the *X. laevis* otoconial protein using Edman microsequencing and mass spectrometry. The N-terminal 38 residues are 50% identical to the N-termini of phospholipases A2. The nonidentical residues are easily accommodated in the crystal structure of the western diamondback rattlesnake (*Crotalus atrox*) venom phospholipase A2. We propose that the N-terminal portion of the saccular otoconial protein from *X. laevis* is homologous to phospholipase A2 and has a similar tertiary structure.

INTRODUCTION: The sensory detectors within the inner ears of vertebrates have a variety of morphologies. Their functions include detection of linear and angular acceleration as well as low frequency vibration. Apical projections of the hair cells within the sensory epithelia of the peripheral portion of the vestibular system are displaced during acceleration, thereby initiating a nervous impulse. In the maculae these apical processes, the stereocilia, are imbedded in an extracellular matrix, the otoconial membrane. This matrix contains small crystals of calcium carbonate, otoliths (ear stones) or otoconia (ear dust). Three polymorphs of $CaCO_3$ mineralize these structures. Vaterite is present in otoconia of chondrostean fishes, aragonite in fish, calcite in mammals and birds, and both aragonite and calcite in reptiles and amphibians. For all three the morphology observed in the inner ear differs somewhat from that of the $CaCO_3$ crystals grown in vitro. Most, if not all, biominerals contain a small amount of protein. It is assumed that these proteins somehow regulate the initiation of crystal growth and/or the subsequent rates of growth of various crystal faces that in turn determine the morphology and size of the final crystal.

We have begun to characterize the proteins of the aragonitic otoconia of the African clawed frog (*Xenopus laevis*) and of the calcitic otoconia of the Norway rat (*Rattus norvegicus*) with two goals in mind. Antibodies and cDNA probes to the proteins can be used to follow the developmental expression within the inner ear. Knowing the structures of the

NATO ASI Series, Vol. H 48
Calcium Transport and
Intracellular Calcium Homeostasis
Edited by D. Pansu and F. Bronner

proteins and having their encoding clones available for expression in bacteria will permit us to study their effects on calcium carbonate crystal formation in vitro.

MATERIALS and METHODS: Otoconia were dissected from young adult *X. laevis* and *R. norvegicus* as described by Pote and Ross (1986, 1990). The saccule of *X. laevis* contains aragonitic otoconia; it is easily dissected free of the utricle, which contains calcitic otoconia. Yields are cited in terms of 100 animals, whose dissections required 20 to 30 hours.

	wet mass of otoconia	total protein	purified major protein
X. laevis 22 kDa	~800 mg	~300 µg	~250 µg
R. norvegicus 90 kDa	~2 mg	~100 µg	~70 µg

The pooled otoconia were washed three times in 0.1% sodium dodecyl sulfate (SDS) in 100 mM Na acetate, pH 7.4. To free the protein contained in the mineral the otoconia were demineralized in 100 mM ethylene diamine tetraacetic acid (EDTA) or by dropwise addition of 12 M HCl at 0° C. This solution was dialyzed against 10 mM MgCl followed by about eight changes of deionized H_2O. The dialysate was lyophilized and the protein was measured by Bradford microassay using the protocol supplied with the reagent purchased from Biorad. Purity was determined on a 10% SDS-polyacrylamide (PAGE) gel for the *R. norvegicus* protein and a 15% gel for the protein from *X. laevis*. The lyophilized proteins were dissolved in 0.1% trifluoroacetic acid (TFA) and purified by high pressure liquid chromatography (HPLC) using a C-8 reverse phase column. The separation conditions were a 0-60% gradient of 0.1% TFA in water to 0.085% TFA in acetonitrile, over 40 minutes at a flow of 200 µl per minute. Over 90% of the *X. laevis* otoconial protein elutes in one peak.

The center fraction of the otoconial protein peak was collected and lyophilized. After reduction with dithiothreitol the *X. laevis* otoconial protein was hydrolyzed by trypsin added to the protein at a ratio of 1:100 by mass in ammonium bicarbonate buffer for four hours at 37° C. The tryptic fragments were purified using the same HPLC conditions and the sequences were determined by use of a tandem quadrapole mass spectrometer using the methods of Hunt et al, 1986. In addition, a HPLC purified intact sample of the protein was sequenced through 38 cycles using Edman chemistry with an Applied Biosystems automated microsequenator. This successfully identified 37 of the N-terminal amino acids.

Antibodies were raised to both the 90 kDa and the 22 kDa proteins by injecting ten week old guinea pigs with 1.0 µg of SDS-PAGE purified protein suspended in Freund's complete adjuvant. The animals were given two subcutaneous injections over the shoulder blades and two intramuscular injections in the thigh for each immunization. Booster injections in Freund's incomplete adjuvant were given three weeks later and at subsequent three week intervals. Blood was withdrawn by cardiac puncture one week following injection of the antigen. After each blood draw the sera were tested for immunoreactivity using western blots. These were performed on the proteins following separation using SDS-PAGE and electrophoretic transfer to nitrocellulose in 5% methanol in 25 mM Tris-HCl, and 129 mM glycine at pH 7.5 for five hours at 150 mA constant current.

A model of the N-terminal fragment of the *X. laevis* otoconial protein was built using the coordinates for the carbon backbone of the western diamond back rattlesnake (*Crotalus atrox*) venom phospholipase A2. The coordinates of Renetseder et al (1985) were used and the amino acid side chains were altered to that of the otoconial protein using the University of California, San Diego, Molecular Modeling System software. The resulting structure was put through 160 cycles of minimization using the X-plor program (Brünger, 1987). The resulting structure was then compared to the otoconial protein before minimization and compared to the *C. atrox* phospholipase A2.

RESULTS: Over 0.2 mg of the major protein of *X. laevis* (22 kDa) and nearly 0.1 mg of *R. norvegicus* (90 kDa) otoconial protein can be purified in several days (Fig. 1).

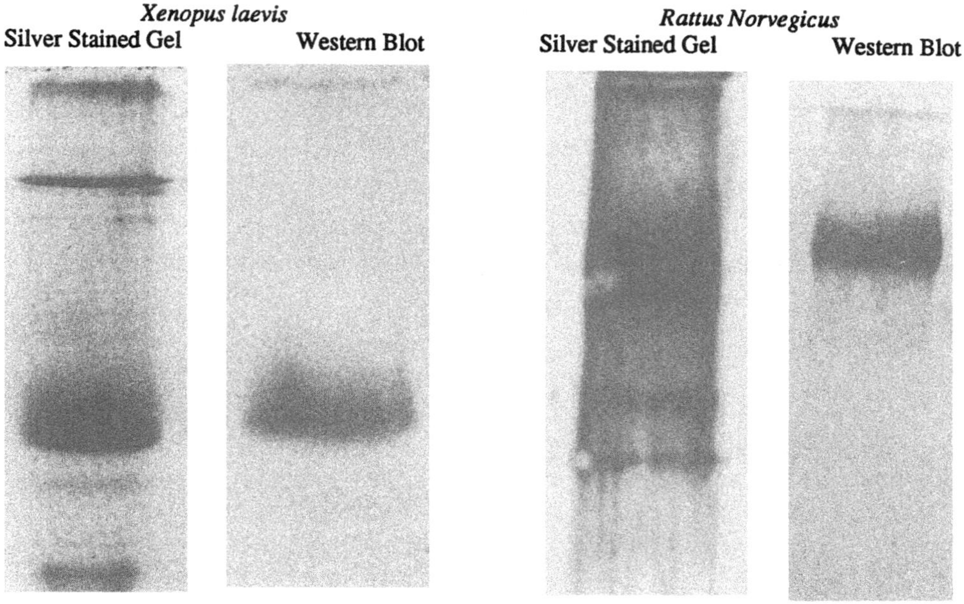

Figure 1. SDS-PAGE gel electrophoresis of the proteins extracted from the otoconia of *Xenopus laevis* and *Rattus norvegicus*. The 22 kDa protein accounts for about 80% of the total protein. When this mixture of proteins is purified by high pressure liquid chromatography the 22 kDa peak is well resolved from the minor components and is judged to be over 95% pure. Polyclonal antibodies were raised against the 22 kDa *X. laevis* protein and the 90 kDa *R. norvegicus* protein. Their specificities are shown on westerns. Neither antisera shows interspecies or interprotein cross-reactivity (not shown).

As shown in figure 1 both antisera are specific for their respective inducing proteins. Neither cross reacts with minor proteins in the same otoconia or with otoconial proteins from the other species (data not shown).

Tryptic Peptide 1: (near C-terminus)	Phe	Tyr	Val	Glu	Glu	Gln	Lys			
Tryptic Peptide 2: (near C-terminus)	Glu	*Leu*	Glu	Asp	Tyr	Asn	*Leu*	Tyr	Phe	Arg
Tryptic Peptide 3:	Cys	Glu	Cys	Asp	Glu	Lys				
Tryptic Peptide 4: (N-terminus)	Thr	(Thr)	(Ala)	(Gln)	Phe	Asp	Glu	Met	*Leu*	Lys
N-Terminal Portion:	Thr	Pro	Ala	Gln	Phe	Asp	Glu	Met	Ile	Lys
(continuation)	Val	Thr	Thr	Ile	Ile	Tyr	Gly	Leu	Ala	?
(continuation)	Phe	Ser	Asp	Tyr	Gly	Cys	His	Cys	Gly	Leu
(continuation)	Asn	Asn	Gln	Gly	Met	Thr	Val	Asp		

Figure 2. Partial Sequence of *Xenopus laevis* Otoconial Protein. *Leu* indicates either leucine or isoleucine, which have identical masses and therefore cannot be distinguished by mass spectrometric analysis. The question mark indicates a residue missed during Edman sequencing. This may be a site of glycosylation (see discussion).

The partial amino acid sequence derived from the Edman degradation and mass spectrometric analysis is shown in figure 2. The first three peptides were HPLC purified from a tryptic digest of the HPLC purified, intact, otoconial protein from *X. laevis*. The N-terminal portion was sequenced using automated Edman microsequencing of the protein purified by reverse phase HPLC. This sequence is shown in figure 2. The alignment of the 38 amino acid N-terminal fragment with several phospholipases A2 is shown in figure 3.

The crystal structures of both *Bos taurus* pancreatic and *C. atrox* venom phospholipases A2 have been determined and refined with high resolution data (*B. taurus*, Dijkstra, et al, 1981; *C. atrox*, Brunie et al, 1985). We built a model of the tertiary structure of the first 38 residues of the 22 kDa protein in which its side chains replaced those of the *C. atrox* phospholipase A2 structure. This structure was refined using X-plor (Brünger et al, 1987). Unacceptable van der Waals contacts could be relieved with almost no movement of the main chain and with only small movements of the side chains (Fig. 4)

```
                                  1                   2                   3
                1 2 3 4 5 6 7 8 9 0 1 2 3 4 5 6 7 8 9 0 1 2 3 4 5 6 7 8 9 0 1 2 3 4 5 6 7 8 9
Xenopus OP:     T P A Q F D E M I K V T T I I Y G - L A ? F S D Y G C H C G L N N Q G M T V D
Crotalus PA2:   S L V Q F E T L I M K I A G R S G - L L W Y S A Y G C Y C G W G G H G L P Q D
Bos PA2:        A L W Q F N G M I K C K I P S S E P L L D F N N Y G C Y C G L G G S G T P V D
Naja PA2:       N L Y Q F K N M I K C T V P S R S - W W D F A D Y G C Y C G R G G S G T P V D
                      * *     * * *   *         *   *     * * * * * *   * * *       *     * *
                                                I I I I I I   I I   I       I
```

Figure 3. Alignments of the N-terminal portion of *Xenopus laevis* otoconial protein (OP) with the N-termini of several phospholipases A2. *Crotalus PA2= Crotalus atrox, Bos PA2=Bos taurus* pancreatic enzyme, *Naja PA2=Naja naja naja* (Davidson and Dennis, 1990) venom derived phospholipases. The numbering system honors the convention adopted for the phospholipases, many of which have an additional Pro-18 relative to *X. laevis* 22 kDa otoconial protein. The stars indicate residues common between the otoconial protein and one of the phospholipases. An I identifies the residues that are invarient in the calcium binding domain of the phospholipases A2.

DISCUSSION: Our studies of the otoconia have two goals: tracing the development of the tissue(s) that produce otoconia and understanding the function of the otoconial proteins in biomineralization. Our initial results using polyclonal antibodies indicate that only the inner ear epithelial tissues produce the major otoconial proteins in both *R. norvegicus* and *X. laevis*. The availability of these antibodies as well as oligonucleotide probes should permit us to map the temporal expression of both the mRNA's and the otoconial proteins.

The discovery that the first 38 residues of *X. laevis* 22 kDa otoconial protein are about 50% identical to the N-termini of phospholipases A2 has several implications. Given the precedents of many other homolog families, this degree of similarity strongly implies homology. This inference is supported because this alignment occurs at the N-termini of both proteins. It is frequently observed that the tertiary structures of homologous proteins are nearly superimposable over the region(s) of sequence similarity. Hence, it should be rewarding to examine the structures and functional characteristics of the phospholipases A2 to focus our future investigations of the otoconial proteins.

The extracellular phospholipases examined from mammalian pancreas and snake venom are about 122 amino acids long and form one large homolog family. A standardized numbering system has been adopted to accommodate insertions; we honor this in describing the phospholipases A2 as well as 22 kDa. There are two groups of phospholipases A2; both have seven disulfide bonds, six of which are common:

Sources	11-77	27-126	29-45	44-105	50-133	51-98	61-91	84-96
1. Pancreas, *Elapidae* & *Hydrophidae*	+	+	+	+		+	+	+
2. *Crotalus* & *Viperidae*		+	+	+	+	+	+	+

The partial amino acid sequence of 22 kDa (Fig. 3) is 38 residues long; it is numbered 1 through 39 because a gap in inserted following position 17 to accommodate a Pro occuring in many group 1 enzymes. The *X. laevis* 22 kDa otoconial protein is tentatively assigned to group 2 because it lacks Cys at position 11.

The Edman degradation was initially performed on unreduced 22 kDa. Blanks were obtained at positions 27 and 29. These were shown to be cysteine residues by mass spectrometric sequencing of the reduced and vinyl pyridinated protein. This may indicate Cys's in disulfide linkage, inferred to be 27-126 and 29-45. This observation is consistent with extensive CNBr digestion of the unreduced protein producing low yields of peptides. Phospholipases A2 are very stable because of the seven disulfide crosslinks; our preliminary observations of the 22 kDa otoconial protein also indicate great stability.

The molecular weight of 22 kDa was inferred from SDS-PAGE mobility. If correct it would contain nearly 200 residues. Glycosylation may, however, account for the increase in

Figure 4. Proposed tertiary structure of the N-terminal portion of the 22 kDa otoconial protein of *Xenopus laevis*. The backbone coordinates are those of *Crotalus atrox* phospholipase A2. The figure on the left shows the theoretical backbone of the refined otoconial protein superimposed on the phospholipase A2 (double line trace). The figure on the right shows the N-terminal 39 residues of the otoconial protein as they were placed by refinement.

mass seen on an acrylamide gel. Both group 1 and 2 phospholipases A2 are about 122 residues long. Cys-126 is C-terminal for group 1 and seven from the C-terminus for group 2. If 22 kDa does form a 27-126 disulfide bond, then its entire first ~120 residues may be homologous to phospholipase A2.

The last 80 (~200 - ~120) residues, if present, may not be homologous to phospholipase A2. The two short sequences (Fig. 2) determined by mass spectroscopy show no similarity to phospholipase A2. They probably are near the C-terminus of 22 kDa. We are continuing the mass spectrometric sequencing and have HPLC purified an additional six peptides generated by tryptic digestion following extensive reduction with dithiothreitol.

Pancreatic phospholipase A2 is synthesized as a proenzyme and the *B. taurus* crystal structure has been determined (Dijkstra et al, 1981). We have no evidence for a pro-form of 22 kDa but should search for one. Both phospholipase A2 and the *X. laevis* 22 kDa otoconial protein are extracellular proteins.

Periodic acid Schiff staining of the otoconial membranes show the otoconial proteins of *R. norvegicus* might be glycosylated (Belanger, 1960). In contrast none of the studied phospholipases A2 are glycosylated. The *X. laevis* otoconial protein may be glycosylated at residue 21 because it consistently gives a blank cycle in the Edman sequenator. This is true following extensive reduction with dithiothreitol, indicating the missed residue is likely not a cysteine. We have analysed the protein for carbohydrate content. It contains galactose, mannose, fucose, N-acetyl-galactosamine, N-acetyl-glucosamine, and sialic acid. This is important because the sugar groups on the otoconial protein might interact with the $CaCO_3$ within otoconia. We plan to study the *in vitro* interactions of the otoconial proteins expressed in bacteria. Since bacteria lack glycosyl transferases, the 22 kDa protein expressed in bacteria might not duplicate its *in vivo* function. We will also express the protein using the Baculovirus expression system, which will glycosylate the protein.

Some members of class 2 function as dimers; some as monomers. The dimer interface as seen in the crystal structure of *C. atrox* phospholipase A2 has contacts involving hydrophobic sidechains. The tryptophan at position 31 in *C. atrox* extends into a hydrophobic pocket in the other monomer to hold the monomers together. Position 31 is a leucine in the 22 kDa *X. laevis* aragonitic otoconial protein. This may not enhance the formation of the dimers. We anticipate that 22 kDa is monomeric; however, this is certainly not proven.

Most phospholipases A2 bind a Ca^{2+} ion with millimolar affinity; it is essential for enzymatic acitvity. In the crystal structure from *B. taurus* (Dijkstra et al, 1981) calcium is coordinated to two water molecules, the carbonyl oxygen atoms of Tyr-28, Gly-30, and Gly-32, as well as both oxygen atoms of the carboxylate group of Asp-49. These four residues are invariant in phospholipases A2. Five glycines, including positions 30, 32 and 33, have dihedral angles disallowed for other amino acids. In the 22 kDa otoconial protein positions 32 and 33 are both Asn. In our modelling and refinement studies we find that we can substitute Asn-32 and Asn-33 and maintain the general course of the main chain; however, the carbonyl at position 32 no longer points at the putative calcium binding site. There is some controversy about the residues involved in catalysis and the conformational changes involved in the binding of lipid bilayers. Even so, the Ca^{2+} ion is not involved in catalysis or in the binding of substrate or product. Radioactive ^{45}Ca is not bound by a band of *X. laevis* 22 kDa otoconial protein transferred to nitrocellulose (Pote and Ross, 1990). Tentativley we infer that the binding of a Ca^{2+} ion per se is not essential to the biomineralization function of this protein.

Phospholipase A2 is especially interesting in that its velocity of reaction is much higher when an intact lipid bilayer, rather than a single diacylglyceride, is the substrate. This implies a major change of conformation for either the enzyme or the bilayer and probably both. The inferred or determined structure of isolated *X. laevis* 22 kDa otoconial protein may be significantly different from that found in the $CaCO_3$ crystal lattice of otoconia.

Obviously we have a long way to go to understand the molecular mechanism and control of biomineralization. Even so, we feel that determining the tertiary structure of the major otoconial protein from *X. laevis* will be a major step toward this goal.

ACKNOWLEDGEMENTS: The partial sequence presented was obtained through collaboration with Donald Hunt and Charles Hauer, Chemistry Department, University of Virginia.

REFERENCES:

Achari, A., Scott, D., Barlow, P., Vidal, J.C., Otwinowski, Z., Brunie, S. and Sigler, P.B. (1987) "Facing up to Membranes: Structure/ Function Relationships in Phospholipases" Cold Spring Harb. Symp. Quant. Biol. 52:441-452.

Belanger, L.F. (1960) "Development, Structure, and Composition of the Otolithic Organs of the Rat" In Calcification in Biological Systems. Amer. Assoc. Adv. Sci., Washington, D.C. R.F. Sognnaes (Ed.) pp. 151-162.

Brünger, A.T. (1987) "Crystallography Refinement by Simulated Annealing" Lecture Notes for the International School of Crystallographic Computing, Adelaide, (Isaacs, N., Ed.)

Brunie, S., Bolin, J., Gewirth, D. and Sigler, P.B. (1985) "The Refined Crystal Structure of Dimeric Phospholipase A2 at 2.5 A. Access to a Shielded Catalytic Center" J. Biol. Chem. 260:9742-9749.

Davidson, F.F. and Dennis, E.A. (1990) "Amino Acid Sequence and Circular Dichroism of Indian Cobra (Naja naja naja) Venom Acidic Phospholipase A2" Biochem. Biophys. Acta 1037:7-15.

Dijkstra, B.W., Kalk, K.H. Hol, W.G.J. and Drenth, J. (1981) "Structure of Bovine Pancreatic Phospholipase A2 at 1.7 A Resolution" J. Mol. Biol. 147:97-123.

Hunt, D.F., Yates, J.R. III, Shabanowitz, J., Winston, S., and Hauer, C.R. (1986) "Protein Sequencing by Tandem Mass Spectrometry" PNAS (USA) 83:6233-6237.

Pote, K.G. and Ross, M.D. (1986) "Ultrastructural Morphology and Protein Content of the Internal Organic Material of Rat Otoconia" J. Ultrastructure Molec. Structure Res. 95:61-70.

Pote, K.G. and Ross, M.D. (1990) "Each Otoconial Polymorph has Proteins Unique to that Polymorph" Manuscript in Preparation.

Renetseder, R., Brunie, S., Dijkstra, B.W., Drenth, J. and Sigler, P.B. (1985) "A Comparison of the Crystal Structures of Phospholipase A2 from Bovine Pancreas and *Crotalus atrox* Venom" J. Biol. Chem. 260:11627-11634.

PHOSPHORYLASE KINASE FROM BOVINE STOMACH SMOOTH MUSCLE: A Ca^{2+}-DEPENDENT PROTEIN KINASE ASSOCIATED WITH AN ACTIN-LIKE MOLECULE

V.G. Zevgolis, T.G. Sotiroudis[1] and A.E. Evangelopoulos
Institute of Biological Research
The National Hellenic Research Foundation
48 Vassileos Constantinou Avenue
Athens 116 35
Greece

It is widely recognized that Ca^{2+} is a major second messenger mediating the action of neurotransmitters, hormones, growth factors and mitogens (Rasmussen, 1986). Although the actions of Ca^{2+} do not appear to be mediated by a universal biochemical mechanism there is experimental evidence that certain of the physiological effects of Ca^{2+} are mediated or modulated by a protein phosphorylation cascade via stimulation of a number of Ca^{2+}-dependent protein kinases. Biochemical studies of the signalling systems that utilize Ca^{2+} as a second messenger in several tissues have revealed the presence of at least six protein kinases that can be regulated by Ca^{2+}. Ca^{2+}/calmodulin dependent protein kinase II and protein kinase C appear to be general or multifuntional protein kinases, while myosin light chain kinase and phosphorylase kinase are considered to be dedicated protein kinases. Two additional Ca^{2+}-regulated protein kinases, Ca^{2+}/calmodulin-dependent protein kinase I and III, may also have a very limited substrate specificity but have not been extensively studied (Schulman, 1988).

In the last few years there has been a surge of interest in smooth muscle biochemistry and especially in the regulation of contractile process by protein phosphorylation (Kamm and Stull, 1989). Glycogen, which is the most conspicuous source of endogenous carbohydrate metabolism observed in the smooth muscle cells of different organs has been considered as one of the main available energy stores, which takes part in maintaining the vascular tone under stimulated conditions in the absence of glucose (Sotiroudis et al, 1986). In addition, it has been suggested that metabolism in smooth muscle was functionally compartmentalized and that

[1]Presenting author

NATO ASI Series, Vol. H 48
Calcium Transport and
Intracellular Calcium Homeostasis
Edited by D. Pansu and F. Bronner

this compartmentation reflected a particular enzymic localization, with a glycolytic enzyme cascade associated with the plasma membrane and glycogenolysis linked to contractile filaments or cytosolic elements (Paul, 1989). Thus, it is obvious that for a better understanding of the coordination of metabolic and contractile activity in smooth muscle, the purification of the enzymes involved in the corresponding regulatory mechanisms is needed.

Phosphorylase kinase is one of the regulatory enzymes involved in the cascade of reactions associated with glycogenolysis and it has been studied in a number of tissues and species (Carlson et al, 1979; Chan and Graves, 1984; Pickett-Gies and Walsh, 1986). However, only the rabbit skeletal muscle enzyme has been thoroughly characterized in terms of its physicochemical, enzymatic and regulatory properties. It is a large oligomeric enzyme (1.3 MDa) with the subunit structure $(\alpha\beta\gamma\delta)_4$. It is regulated in a complex way by phosphorylation and Ca^{2+} and it sits at the crossroad of glycogenolysis linking glycogen breakdown to both nervous and endocrine stimulation (Carlson et al, 1979; Chan and Graves, 1984; Pickett-Gies and Walsh, 1986). δ-subunit is identical to calmodulin and confers Ca^{2+} sensitivity to the enzyme (Cohen et al, 1978). γ-subunit has catalytic activity (Skuster et al, 1980) and is similar to other protein kinases (Reimann et al, 1984). The two large subunits, α and β, are homologous proteins (Zander et al, 1988; Kilimann et al, 1988), they cary all phosphorylation sites and at least one of their function is regulatory. The primary structures of γ- and δ-subunits have been determined and cDNAs encoding α-, β- and γ- subunits have been cloned (Zander et al, 1988; Kilimann et al, 1988; daCruz Silva and Cohen, 1987; Bender and Emerson, 1987).

In our effort to better understand the hormonal control of glycogen metabolism in smooth muscle we have purified and characterized phosphorylase kinase from chicken gizzard smooth muscle (Nikolaropoulos and Sotiroudis, 1985). It was interesting as well as puzzling that gizzard phosphorylase kinase showed a completely different subunit pattern to that of rabbit skeletal muscle enzyme upon SDS-PAGE (one main protein band of 61 kDa), although the Mr of both kinases were similar. In addition, the gizzard kinase was activated neither by cAMP-dependent protein kinase nor by autophosphorylation. Because of the complexity of the regulation of glycogenolysis by phosphorylase kinase it appeared of interest to investigate if the peculiar behaviour of gizzard phosphorylase kinase is a unique feature of primitive vertebrate tissues or it is a property shared also by mammalian smooth muscle cells. The aim of the present report is to describe recent data on the purification and characterization of bovine stomach smooth muscle phosphorylase kinase and to discuss the puzzling association of this kinase with an actin-related molecule.

Purification and characterization of phosphorylase kinase from bovine stomach smooth muscle

Phosphorylase kinase was purified 110-fold from bovine stomach smooth muscle by a procedure involving three steps: DEAE-cellulose chromatography at pH 7.5, ammonium sulfate fractionation and glycerol density ultracentrifugation. The initial fractionation gives a single peak of phosphorylase kinase activity at about 0.2 M NaCl. It is essential to include protease inhibitors in the homogenization buffer, otherwise phosphorylase kinase is transformed to a new kinase form, eluted from the anion exchange column at 0.1 M NaCl, suggesting proteolytic degradation. Ammonium sulfate fractionation step takes advantage of the precipitation of the kinase at low salt concentrations (25% saturation), while ultracentrifugation is an efficient purification step because of the high molecular mass of smooth muscle kinase (1 MDa, as revealed by gel filtration analysis). The purified enzyme migrates on SDS-PAGE as a single protein band of 43 kDa and shows a close similarity with bovine aortic actin as revealed by amino acid analysis and sequencing of a tryptic decapeptide fragment (His-Gln-Gly-Val-Met-Val-Gly-Met-Gly-Gln) (Vandekerchove and Weber, 1979). Moreover, the kinase preparation containing the 43 KDa protein band exhibits two other actin characteristics: (i) it shows a polydispersed profile upon glycerol density ultracentrifugation and non-denaturing PAGE and (ii) it is able to polymerize and hydrolyze ATP in presence of KCl and $MgCl_2$. In contrast, the 43 KDa protein is not immunoprecipitated by a polyclonal anti-actin antibody, it does not inhibit DNase I either before or after treatment with 1.5 M guanidine-HCl (Cooper and Pollard, 1982), while it shows a high sensitivity to proteolytic degradation. At this point it must be emphasized that although the proteolysis of our 43 KDa protein by trypsin is a very rapid process, producing a polydispersed mixture of low Mr peptides within a few minutes, F actin is totally resistant to trypsinolysis even after 19 h incubation

(Kay et al, 1982), while G actin produces a main final product of 33 KDa (Movnet and Ue, 1984). Thus, the 43 kDa protein associated with bovine stomach phosphorylase kinase preparations differs widely from native actin in several important features, suggesting that in our case we deal with an actin-related protein.

In our effort to separate phosphorylase kinase activity from the 43 kDa protein we used a variety of chromatographic media, but both catalytic activity and protein either eluted superimposably (phosphorylase b-Sepharose, Blue-Sepharose) or they could not be eluted (phenyl-Sepharose, hydroxylapatite, protamine-Sepharose). Furthermore, protamine which is known to induce supramolecular actin-protamine structures (Grazi et al, 1982) totally precipitated the 43 kDa protein together with the kinase activity at a molar ratio of protamine to actin-like molecule of 0.5:1. In addition, in a similar experiment, our actin-like preparation was partially aggregated in presence of Mg^{2+} (6 mM) and Ca^{2+} (1 mM). After centrifugation (37000 xg, 60 min) about 10% of both protein and phoshorylase kinase activity were recovered in the pellet.

Bovine stomach phosphorylase kinase shows a pH 6.8/8.2 activity ratio of 0.23, it has an absolute requirement for Ca^{2+}, as demonstrated by almost complete inhibition in presence of EGTA, and it is activated 1.8-fold by Ca^{2+}/calmodulin and 3,0-fold by short-time trypsinolysis, at pH 6.8. It can utilize ATP as phosphoryl donor and rabbit phosphorylase b as substrate; by contrast, it cannot use GTP and the proteins histone II-AS, casein or phosvitin. The protein kinase activity which is isolated as homogenous 43 kDa protein is not inhibited by concentrations of antibodies against rabbit skeletal muscle phosphorylase kinase which completely inhibit an equivalent amount of purified rabbit skeletal muscle phosphorylase kinase activity. Incubation of smooth muscle kinase preparation with ATP/Mg^{2+} in presence or absence of the catalytic subunit of cAMP-dependent protein kinase results in a marginal stimulation of phosphorylase kinase activity (1.5-fold), without time-dependent correlation between activation and ^{32}P-incorporation into the 43 kDa protein band. These results suggest that smooth muscle phosphorylase kinase is rather insensitive to activation by protein phosphorylation.

Bovine stomach smooth muscle phosphorylase kinase: A new isoform of the enzyme?

Our inability to separate phosphorylase kinase activity from the 43 kDa actin-related protein strongly suggests a functional association. On the other hand, the biochemical and immunological characterization of smooth muscle kinase described above indicates a structural difference of phosphorylase kinase from bovine stomach and rabbit skeletal muscle. Thus, two main questions about smooth muscle phosphorylase kinase need to be addressed: (i) Does the enzyme from smooth muscle represent a new phosphorylase kinase entity and (ii) what would be the functional significance of the association of bovine stomach phosphorylase kinase with an actin-like protein?

There are several reports in the literature suggesting the existence of different phosphorylase kinase isoforms: (a) A protein kinase possessing phosphorylase b to a transforming activity was isolated from the trailing edge of a DEAE-cellulose chromatography fraction, referred to as Peak II, during purification of cAMP-dependent protein kinases. This kinase differed in several properties from both phosphorylase kinase and cAMP-dependent kinase (Reimann et al, 1971; Graves et al, 1975). (b) A Ca^{2+}-dependent protein kinase was detected in skeletal muscle membranes of I-strain and wild type mice, which converted phosphorylase b to a, and exhibited a high pH 6.8/8.6 activity ratio and a low affinity for antibodies against rabbit muscle phosphorylase kinase (Varsanyi et al, 1978). (c) A phosphorylase b to a converting activity has been shown associated with isolated homogenous calsequestrin which can be differentiated from skeletal muscle phosphorylase kinase (Varsanyi and Heilmeyer, 1979). (d) A trace residual phosphorylase kinase activity in mixed skeletal muscle of phosphorylase kinase deficient ICR/IAn mice had quite different properties from normal enzyme: It had a much higher pH 6.8/8.2 activity ratio, its activity was less dependent on Ca^{2+}, it was not activated by cAMP-dependent protein kinase or by trypsin and its elution behaviour on Sepharose 4B was remarkably polydispersed (Cohen et al, 1981). (e) Recently, Bender and Emerson (1987) demonstrated that skeletal muscle phosphorylase kinase catalytic subunit mRNAs are expressed in heart tissue but not in liver. This result suggests that the liver γ-subunit and skeletal muscle γ-subunit are encoded by separate genes. In light of the above observations and our experimental results it is reasonable to assume that bovine stomach phosphorylase kinase associated with an actin-like protein is a new isoform different from that associated with skeletal muscle glycogen complex.

A similar association of phosphorylase kinase activity with actin has been suggested by Fischer et al (1978), who showed a considerable homology between the γ-subunit of dogfish phosphorylase kinase and dogfish actin with respect to several structural and functional properties of this component of the contractile apparatus. The postulation by Paul (1989) that glycogenolysis in smooth muscle is linked to contractile filaments permits us to suggest that another form of phosphorylase kinase exists in this muscle type having a suitable structure which could allow the enzyme to position itself properly within the network of actin filaments. In this case phosphorylase kinase either utilizes a Ca^{2+}-binding actin-like molecule as a regulatory subunit and/or it serves as an actin-binding protein to regulate the arrangement and turnover of actin (Pollard and Cooper, 1986). An analogous hypothesis was also made recently by Putnam-Evans et al (1989) who indicated that a novel monomeric Ca^{2+}-dependent protein kinase is associated with actin filaments in plant cells. Of course one cannot exclude the possibility that the 43 kDa smooth muscle actin-like molecule itself possesses phosphorylase kinase activity. In this context, an actin like molecule, may (under certain conditions) present Ca^{2+}-dependent protein kinase catalytic activity. This appears reasonable, since it has nucleotide binding site and binding sites for divalent cations, although it has not the sequence motifs which are characteristic of other protein kinases (Pollard and Cooper, 1986). Nevertheless, it is known that there is at least one example of a protein kinase (isocitrate dehydrogenase kinase/phosphatase), which does not exhibit the typical characteristics of other protein kinases (LaPorte et al, 1989).

In summary, our data suggest that phosphorylase kinase from bovine stomach smooth muscle is associated with an actin-like molecule and it is possibly a different entity from the normal rabbit skeletal muscle enzyme. Further progress in our understanding of phosphorylase kinase function in smooth muscles will require the preparation of the corresponding catalytic subunit cDNA(s).

Acknowledgments

V.G.Zevgolis was supported by EMBO short-term fellowship 6120. We are grateful to Professor L.M.G. Heilmeyer Jr.(Ruhr-Universitat Bochum) for accepting V.G.Zevgolis to his Institute to perform the amino acid analysis and sequence experiments and for valuable discussions. We are indebted to Dr. H.E. Meyer for the assistance with the sequence experiments and to Dr. M.Varsanyi for helping in immunological experiments and for providing the antibodies to phosphorylase kinase. Material described herein is part of the Ph.D. Thesis of V.G.Zevgolis.

REFERENCES

Bender PK, Emerson CP Jr (1987) Skeletal muscle phosphorylase kinase catalytic subunit mRNAs are expressed in heart tissue but not in liver. J Biol Chem, 262:8799-8805

Carlson GM, Bechtel PJ, Graves DJ (1979) Chemical and regulatory properties of phosphorylase kinase and cyclic AMP-dependent protein kinase. Adv Enzymol 50:41-115

Chan KFJ, Graves DJ (1984) Molecular properties of phosphorylase kinase. In: Cheung Y (ed) Calcium and cell function vol 5 Academic Press, New York, p1

Cohen P, Burshell A, Foulkes JG, Cohen PTW, Vanaman TC, Nairn AC (1978) Identification of the Ca^{2+}-depoendent modulator protein as the fourth subunit of rabbit skeletal muscle phosphorylase kinase. FEBS Lett 92:287-293

Cohen PTW, Le Marchand Brustel Y, Cohen P (1981) Regulation of glycogen phosphorylase and glycogen synthase by adrenalin in soleus muscle of phosphorylase kinase-deficient mice. Eur J Biochem 115:619-625

Cooper JA, Pollard TD (1982) Methods to measure actin polymerization. Methods Enzymol 85:182-210

da Cruz EF, Cohen PTW (1987) Isolation and sequence analysis of a cDNA clone encoding the entire catalytic subunit of phosphorylase kinase. FEBS Lett, 220:36-42

Fischer eH, Alaba JO, Brautigan DL, Kerrick WGL, Malencik DA, Moeschler HJ, Picton C, Pocinwong S (1978) Evolutionary aspects of the structure and regulation of phosphorylase kinase. In: Li CH (ed) Versatility of proteins, Academic Press, New York, p133

Graves DJ, Carlson GM, Skuster JR, Parrish RF, Carty TJ, Tessmer GW (1975) Pyridoxal phosphate-dependent conformational states of glycogen phosphorylase as probed by interconverting enzymes. J Biol Chem 250: 2254-2258

Grazi E, Magri E, Pasquali-Ronchetti I (1982) Multiple supramolecular structures formed by interaction of actin with protamine. Biochem J 205:31-37

Kamm KE, Stull JT (1989) Regulation of smooth muscle contractile elements by second messengers. Ann Rev Physiol 51:299-313

Kay J, Siemankowski LM, Siemankowski RF, Greweling JA, Goll DE (1982) Degradation of myofibrillar proteins by trypsin-like serine proteinases. Biochem J 201:279-285

Kilimann MW, Zander NF, Kuhn CC, Crabb JW, Meyer HE, Heilmeyer LMG Jr (1988) The α and β subunits of phosphorylase kinase are homologous: cDNA cloning and primary structure of the β subunit. Proc Natl Acad Sci USA, 85:9381-9385

LaPorte DC, Stueland CS, Ikeda TP (1989) Icocitrate dehydrogenase kinase/phosphatase. Biochimie 71:1051-1057

Mornet D, Ue K (1984) Proteolysis and structure of skeletal muscle actin. Proc Natl Acad Sci USA 81:3680-3684

Nikolaropoulos S, Sotiroudis TG (1985) Phosphorylase kinase from chicken gizzard. Partial purification and characterization. Eur J Biochem 151: 467-473

Paul RJ (1989) Smooth muscle energetics. Ann Rev Physiol 51:331-349

Pickett-Gies CA, Walsh DA (1986) Phosphorylase kinase. In: Boyer P, Krebs EG (eds) The Enzymes vol 17. Academic Press, New York, p395

Pollard TD, Cooper JA (1986) Actin and actin-binding proteins. A critical evaluation of mechanisms and functions. Ann Rev Biochem 55: 987-1035

Putnam-Evans C, Harmon AC, Palevitz BA, Fechheimer M, Cormier MJ (1989) Calcium-dependent protein kinase is localized with F-actin in plant cells. Cell Motil Cytosk 12:12-22

Rasmussen H (1986) The calcium messenger system. N Eng J Med 314:1094-1101 and 1164-1170

Reimann EM, Titani K, Ericsson LH, Wade RD, Fischer EH, Walsh KA (1984) Homology of the γ subunit of phosphorylase b kinase with cAMP-dependent protein kinase. Biochemistry 23:4185-4192

Reimann EH, Walsh DA, Krebs EG (1971) Purification and properties of rabbit skeletal muscle adenosine 3',5'-monophosphate-dependent protein kinases. J Biol Chem 246:1986-1995

Schulman H (1988) The multifunctional Ca^{2+}/calmodulin-dependent protein kinae. In: Greengard P, Robinson GA (eds) Advances in second messenger and phosphoprotein research, vol 22. Raven Press, New York, p39

Skuster JR, Chan KFJ, Graves DJ (1980) Isolation and properties of the catalytically active γ subunit of phosphorylase b kinase. J Biol Chem 255:2203-2210

Sotiroudis TG, Nikolaropoulos S, Evangelopoulos AE (1986) Glycogen metabolism in smooth muscle. In: Heilmeyer LMG (ed) Signal transduction and protein phosphorylation, Plenum Press, New York, p243

Vandekerchove J, Weber K (1979) The complete amino acid sequence of actins from bovine aorta, bovine heart, bovine fast skeletal muscle and rabbit slow skeletal muscle. Differentiation 14:123-133

Varsanyi M, Groschel-Stewart U, Heilmeyer LMG Jr (1978) Characterization of a Ca^{2+}-dependent protein kinase in skeletal muscle membranes of I-strain and wild-type mice. Eur J Biochem 87:331-340

Varsanyi M, Heilmeyer LMG Jr (1979) The protein kinase properties of calsequestrin. FEBS Lett 103:85-88

Zander NF, Meyer H, Hoffmann-Posorske E, Crabb JW, Heilmeyer LMG Jr, Kilimann MW (1988) cDNA cloning and complete primary structure of skeletal muscle phosphorylase kinase (α subunit). Proc Natl Acad Sci USA 85:2929-2933

SECTION VII

THE CALBINDINS - GENE STRUCTURE AND FUNCTION

STRUCTURE AND FUNCTIONAL ANALYSIS OF THE CHICK CALBINDIN GENE

E. Muir, M. Harding, P. Wilson & D.E.M. Lawson
AFRC Institute of Animal Physiology and Genetics Research, Babraham, Cambridge, CB2 4AT, U.K.

Calcium-binding protein (calbindin D-28K) has been known since its discovery to be dependent upon the presence of 1,25-dihydroxyvitamin D (1,25-$(OH)_2D$) for its synthesis. As with other steroid hormones 1,25-$(OH)_2D$ is accumulated in its target tissues due to the presence of a receptor protein. The complex of receptor and steroid is found in the mucosal cell nuclei and it is shortly after the arrival of the steroid in the nucleus that calbindin mRNA is produced. An involvement of calbindin in vitamin D-stimulated calcium absorption was also one of the first recognised physiological properties of this protein. Unfortunately, over the past 15 years since the relationship between 1,25-$(OH)_2D$, calbindin and calcium absorption was made, only small pieces of information have been added to our understanding of the mechanisms involved in calcium absorption and in the regulation of this process. (for review see Davie & Lawson 1979)

There are, perhaps, three main reasons for the difficulties which have been encountered. One, it is not possible to maintain fully differentiated intestinal epithelial cells in culture. Two, no cultured cell line is known which contains calbindin in experimentally useful amounts. Three, 24 h after a single injection of a physiological dose of 1,25-$(OH)_2D$ to rachitic animals, calcium absorption has declined to near basal levels after the increase in activity in response to the hormone whereas at this time intestinal calbindin has reached its maximum value which is maintained over the subsequent 24 h. That is, rachitic animals dosed with 1,25-$(OH)_2D$ 24 h up to 48 h previously have significant amounts of calbindin in the intestine but active calcium transport cannot be detected (Spencer et al, 1978). An explanation for this observation has not been provided. Possibly there is a small, membrane bound, rapidly turning over pool of calbindin which has not been identified with certainty (Feher & Wasserman, 1979). Alternatively, there is an intestinal constituent involved in calcium absorption which either has a short half-life or requires the continued presence of 1,25-$(OH)_2D$.

The techniques of molecular biology provide a means of answering these questions, in particular whether 1,25-$(OH)_2D$ directly stimulates calbindin gene transcription and the nature of the molecular role of calbindin in cells.

NATO ASI Series, Vol. H 48
Calcium Transport and
Intracellular Calcium Homeostasis
Edited by D. Pansu and F. Bronner

For the longer term it should be possible to produce transgenic mice constitutively expressing calbindin in both wild-type and mutant forms.

CALBINDIN GENE STRUCTURE AND EVOLUTION

In a series of studies we have isolated the chick and human calbindin genes and sequenced substantial lengths (Wilson et al., 1988; Parmentier et al., 1989). The calbindin genes are at least 19 kb containing 10 introns, most of which do not fall at homologous positions, neither with respect to the sixfold repeating structure of calbindin, nor with respect to previously sequenced genes for calmodulin and other calcium-binding proteins. The genes of the calmodulin superfamily can be divided into 3 families by amino-acid sequence homology and by intron distribution in their genes. One family includes calmodulin, parvalbumin and myosin alkali light chain. A second family consists of the calpain gene where introns fall between the domains. The third family consists of calbindin and calretinin; this latter protein is found exclusively in brain and has 50% homology with calbindin. In this last family introns are found in all three phases of the reading frame. It appears that these three gene families evolved by separate gene duplication events from a common 2-domain ancestor to give three different 4-domain ancestors. There was a further gene duplication event to give the 6-domain ancestor of the calbindin family. Because of the positions of the introns in the reading frames of all the genes in this family the conclusion must be that the introns have been inserted independently into the three gene families since the 4-domain ancestors were created. Comparison of amino acid homologies of the domains within calbindin shows the highly conserved nature of this protein with an evolutionary rate slower than insulin and similar to cytochrome C suggestive of an important role for calbindin in cells (Parmentier et al., 1987).

The calbindin gene sequence provided a possible explanation of the origin and proportions of the three forms of its mRNA which differ only in size: 2.1, 2.8 and 3.1 kb and all being 1,25-$(OH)_2$D-dependent. Faithful and efficient 3′ end processing of RNA transcription requires not only a cleavage/polyadenylation signal (consensus: AATAAA) but additional sequences about 50 bases further downstream consisting of G+T-rich and T-rich elements (Giles & Proudfoot, 1987). Analysis of the 3′ flanking region of the gene identifies the three polyadenylation sequences used to form the three calbindin mRNAs and that only the signal used to form the 2.1 kb mRNA is followed by a poly$(T)_{18}$ sequence located 40 b from the cleavage signal.

Neither of the other signals are followed by sequences with good homology to the consensus transcription termination signals. Interestingly, there are other polyadenylation signals in the 3′ non-translated region of the 2.1 kb mRNA which are not used presumably because they are not followed by transcription termination signals.

CALBINDIN GENE PROMOTER

There is little homology shown by promoters of the same gene from different species and this applies to the chick and human calbindin genes. Both the chick and the human gene contain the same RNA polymerase II binding sites -ATAAATA- and the surrounding regions are GC rich with at least one potential Sp1 binding site. The fragment from -1 to -760b of the chick promoter contains no palindromic sequences which is the common feature of steroid response elements in eukaryotic promoters. However, at -260b and -466b in the human gene there are sequences of perfect and imperfect palindromes some of which are overlapping. The promoter activity in the 5′-flanking region of the chick calbindin gene was assessed by sub-cloning different lengths into the multiple cloning site of the expression vectors pBLCAT2 and pBLCAT3 (Luckow & Schutz, 1987). The inclusion of an active promoter in the latter plasmid drives the gene for chloramphenicol transacetylase which can be readily measured. pBLCAT2 contains the 'TATA' box from thymidine kinase gene and can be used to test for DNA fragments with inhibitory or stimulatory activity.

Fragments of the calbindin promoter from +52b to -760b were inserted in both directions into pBLCAT3 and the resulting recombinant transfected into a number of cell types. Cells were treated so as to give both transient and stable transfectants but most observations were made with the former. CAT activity was detected only if the inserted calbindin fragment contained the TATA box and was in the correct direction. If the fragment contained neither the TATA box nor was inserted in the 5′-3′ direction, CAT activity was undetectable. Promoter activity was found with the sequence from + 52 b to - 81 b inserted in the 5′-3′ direction; for MDCK cells the activity was 80 m units/mg protein/h. The activity was reduced to 20 munits/mg protein/h for the fragment +52 b to -760 b.

In another series of experiments chick calbindin gene promoter fragments of varying length were inserted either into pBLCAT3 or, following removal of a fragment containing the TATA box, into pBLCAT2. These constructs were assessed for 1,25-$(OH)_2D$ stimulation of the CAT reporter gene using a transient expression assay with a culture of primary chick kidney cells. To correct for variation of transcription efficiency the cells were co-transfected with a plasmid expressing the lacZ gene. CAT activity was recorded with all fragments correctly inserted but an effect of $1,25-(OH)_2D$ could not be detected with any of the them. Similar experiments were carried out using MDCK, LLCPK and GH_3 cells, all of which have the $1,25-(OH)_2D$ receptor, and Ltk^- cells but without any effect of $1,25-(OH)_2D$ being observed.

The levels of expression observed with 2.3 kb and 4.0 kb promoter fragments were 20-30% lower than those observed with the 810 b fragment. This may be an effect of the length of the DNA on the conformation of the fusion gene rather than an indication of a repressor sequence. These results indicate either that the vitamin D response element is not contained in the 4 kb region 5′ to the transcription start point or that appropriate calbindin expression requires additional transcription factors. The latter explanation would imply that control of calbindin expression can only be investigated _in vivo_. To address this possibility expression of constructs containing the calbindin promoter in transgenic mice is being investigated.

TRANSGENIC MICE CONTAINING THE CALBINDIN-CAT FUSION GENE

Ova were injected with the 4.2 kb fragment consisting of the 2.3 kb calbindin promoter and the 2.0 kb CAT gene with appropriate introns and transcription termination signals. 22 eggs were implanted into nine foster mothers. These ova resulted in 34 mice of which 16 were transgenic, 14 survived and 12 transmitted the gene in a Mendelian fashion to their off-spring. Southern analysis showed these 12 founder mice to contain from 1 to 200 copies of the fusion gene which was present as a 4.2 kb fragment. Since our 4.2 kb calbindin-CAT gene produces an active enzyme in transfected cells and is present in the mice in an unaltered form, we expect the mice to produce CAT mRNA in those tissues where all necessary transcription factors are present. RNA was extracted from various tissues of the offspring of founder mice, blotted onto nitrocellulose and probed with a specific CAT gene fragment. At present RNA transcripts reacting with the probe have been found only in the spleen of the offspring of two founder mice. No positive reaction product was found in any tissue from another four founder mice. This analysis is not complete but should appropriate transcription not be found it would indicate that additional nucleotide sequences to the 2.3 kb used for these observations are required for correct activation of the calbindin promoter.

CALBINDIN EXPRESSION IN EUKARYOTIC CELLS

We have made several attempts to produce calbindin containing cells by transfecting cultured cell lines with eukaryotic expression vectors into which our full-length cDNA (2.1 kb) was inserted. We have used the SV40 promoter to drive calbindin transcription but transcripts were not detected by specific calbindin probes. Recently we have observed calbindin mRNA expression on GH_3 cells and 3T3 cells transiently transfected with a vector in which calbindin was ligated between the c-fos promoter and a β-globin gene fragment containing signals for termination of transcription and polyadenylation. Analysis of the extracted RNA gave 3 bands hybridising to specific calbindin probes of length 1.6, 3.5 and 4.5 kb. The two larger bands reacted with a specific probe for B-globin showing that correct termination of the calbindin gene is not well regulated.

Subsequently 3T3 cells were stably transfected and a number of clones produced from single cells. These clones were analysed for calbindin mRNA expression and one clone used for subsequent experiments. All cells in this clone stained immunocytochemically for calbindin and Western analysis of the cytoplasm of this clone showed only one band co-migrating with authentic calbindin and reacting with anti-calbindin.

The promoter of the fos gene can be transiently stimulated by exposure to serum proteins due to the presence of a 22b sequence located about 300b upstream from the fos TATA box. GH_3 and 3T3 cells transiently transfected with the fos-calbindin expression vector were treated with 10-15% serum after a period with 0.5% serum in the culture medium. Substantial amounts of calbindin mRNA were detected in cells after only 15 min exposure to the serum. Interestingly calbindin mRNA levels declined thereafter. In preliminary experiments the $t_{1/2}$ of calbindin mRNA in both these cells was between 2h and 4h which is similar to the value reported for egg shell gland calbindin mRNA by Dr Y Nys elsewhere in this volume.

It is our intention to transfect a range of cultured cell-lines including those derived from the intestine and kidney with the calbindin expression vector and record the effect of the protein's presence on the cells ability to handle calcium. In particular we will measure intracellular calcium levels and calcium fluxes. The plasmid will also be used to study the factors affecting the $t_{1/2}$ of calbindin mRNA which appears to be surprisingly short compared to the $t_{1/2}$ of calbindin.

Acknolwedgements. We are very grateful to Dr A Surani, S. Barton and M. Norris of this Institute for preparing the transgenic mice and their assistance with the analysis.

Davie M, Lawson DEM (1979) Aspects of the metabolism and function of vitamin D. Vitamins & Hormones 37, 1

Feher JJ, Wasserman (1979) Studies on the subcellular localisation of the membrane-bound fraction of intestinal calcium binding protein. Biochim Biophys Acta 585:599-618

Gil A, Proudfoot, NJ (1987) Position dependent sequence elements downstream of AAUAAA are required for efficient rabbit β-globin mRNA 3' end formation. Cell 49:399-406

Luckow B, Schutz G (1987) CAT constructions with multiple unique restriction sites for the functional analysis of eukaryotic promoters and regulatory elements. Nucleic Acids Res 15:5490

Parmentier M, Lawson DEM, Vassart G (1987) Human 27-kDa calbindin cDNA sequence. Evolutionary and functional implications. Eur. J. Biochem. 170, 207-215

Parmentier M, De Vijlder JJM, Muir E, Kessel GV, Lawson DEM, Vassart G (1989) The human calbindin gene: structural organisation of the 5' and 3' regions, chromosomal assignment and restriction length poly-morphism. Genomics, 4, 309-319

Spencer R, Charman M, Wilson P, Lawson DEM (1976) Vitamin D-stimulated intestinal calcium absorption may not involve calcium-binding protein directly. Nature 263, 161

Wilson PW, Rogers J, Harding M, Pohl V, Pattyn G, Lawson DEM (1988) Structure of chick chromosomal genes for calbindin and calretinin. J. Mol. Biol. 200, 615-625

REGULATION OF CALBINDIN-D_{28K} GENE EXPRESSION

S. Christakos, A. M. Iacopino, H. Li, S. Lee and R. Gill
Department of Biochemistry and Molecular Biology
UMDNJ-New Jersey Medical School
185 South Orange Avenue
Newark, New Jersey 07103

I. Introduction

Calbindin belongs to the calmodulin family of intracellular calcium modulated proteins that reversibly bind calcium in the micromolar range (Van Eldik et al, 1982). Since the discovery of avian intestinal calbindin-D_{28k} (Wasserman and Taylor, 1966) similar calcium binding proteins have been reported in other species and in many other tissues including kidney pancreas and brain (Christakos et al, 1989), suggesting multiple effects of the vitamin D endocrine system. This chapter will focus on the regulation of mammalian calbindin-D_{28k}. The tissue specificity and regulation by 1,25 dihydroxyvitamin D_3 (1,25$(OH)_2D_3$) of calbindin will be addressed as well as receptor regulation as a possible mechanism for modulating calbindin's response to hormone. In addition, the regulation of calbindin by factors other than 1,25$(OH)_2D_3$ will be considered.

II. Regulation of rat renal calbindin-D_{28k} gene expression

The mammalian 28,000 M_r vitamin D dependent calcium binding protein (calbindin-D_{28k}; CaBP) is present in highest concentrations in kidney and brain. In kidney, calbindin-D_{28k}, which is present in the distal tubule, is regulated by 1,25$(OH)_2D_3$. The time dependence of renal calbindin-D_{28k} gene expression in the vitamin D deficient rat following a single injection of 1,25$(OH)_2D_3$ indicates that the peak of rat renal calbindin gene transcription (2h) precedes the peak of mRNA accumulation (12h) (Varghese et al, 1989). These data reflect the involvement of both transcriptional and post transcriptional mechanisms in the regulation of renal calbindin-D_{28k} by 1,25$(OH)_2D_3$. Although administration of 1,25$(OH)_2D_3$ to vitamin D deficient rats results in an increase in renal calbindin-

NATO ASI Series, Vol. H 48
Calcium Transport and
Intracellular Calcium Homeostasis
Edited by D. Pansu and F. Bronner

D_{28k} gene expression, no change in renal vitamin D receptor mRNA has been observed, suggesting that the induction of the calbindin gene by $1,25(OH)_2D_3$ in the kidney of vitamin D deficient rats is not modified by a corresponding alteration in new receptor synthesis. Similar results are observed for calbindin-D_{9k} and $1,25(OH)_2D_3$ receptor mRNA in vitamin D deficient rat intestine. It should be noted however, that administration of $1,25(OH)_2D_3$ to vitamin D replete rats results in an induction of both renal and intestinal calbindin and $1,25(OH)_2D_3$ receptor mRNAs, suggesting the presence of an inhibitor of $1,25(OH)_2D_3$ mediated receptor upregulation in the vitamin D deficient rat. Although glucocorticoid administration results in an inhibition of intestinal calbindin-D_{9k} gene expression (which may be related to the glucocorticoid mediated inhibition of intestinal calcium absorption) no change in renal calbindin-D_{28k} gene expression is observed with glucocorticoid treatment suggesting differential regulation of calbindin gene expression.

III. Regulation of calbindin-D_{9k} and calbindin-D_{28k} in mouse kidney

In mouse, both calbindin D_{9k} and calbindin-D_{28k} are present in the distal tubule of the kidney. Although they are regulated similarly developmentally (a sharp increase in calbindin mRNA between birth and one week of age and a peak at three weeks of age), the time course of response to $1,25(OH)_2D_3$ administration is markedly different. The time dependence of renal calbindin-D_{28k} gene expression following $1,25(OH)_2D_3$ administration to vitamin D deficient mice is similar to that observed for calbindin-D_{9k} in rat and mouse intestine and for calbindin-D_{28k} in rat kidney (a maximum at 12 hours and a decrease at 24 h after $1,25(OH)_2D_3$ treatment). Unlike a calbindin-D_{28k} and other steroid regulated gene products, a delayed response of calbindin-D_{9k} mRNA to hormone adminstration is observed (a progressive increase in calbindin-D_{9k} gene expression from 6-24 hours). These findings suggest the possibility of different transcription factors regulating the expression of the two different vitamin dependent calcium binding proteins in mouse kidney. Future studies involved in elucidating these control mechanisms should provide new insights concerning $1,25(OH)_2D_3$ regulated gene expression. Both proteins in mouse kidney do not respond to glucocorticoids. It is possible that both proteins, although localized in the same cells, have different functions. One CaBP might be important in the control of intracellular calcium levels whereas the other CaBP might be involved in allocation of calcium within the cell, such as organellar shuttling of calcium. The real physiological significance of the coexistence of two vitamin D regulated calcium binding proteins in the same cells remains to be determined.

IV. Regulation of calbindin-D_{28k} gene expression in brain

In brain calbindin-D_{28k} is not altered by $1,25(OH)_2D_3$ administration. However, in the aging rat and human brain specific, significant decreases (50-88%) in calbindin mRNA and protein levels are observed in cerebellum, corpus striatum and in the nucleus basalis. Comparison of diseased human brain tissue with age and sex matched controls has indicated significant decreases (60-88%) in calbindin protein and mRNA in substantia nigra (Parkinson's), corpus striatum (Huntington's) and nucleus basalis (Alzheimer's) and in the hippocampus and nucleus raphe dorsalis (Parkinson's, Huntington's and Alzheimer's) but not in cerebellum, neocortex, amygdala or locus ceruleus (see Figure 1).

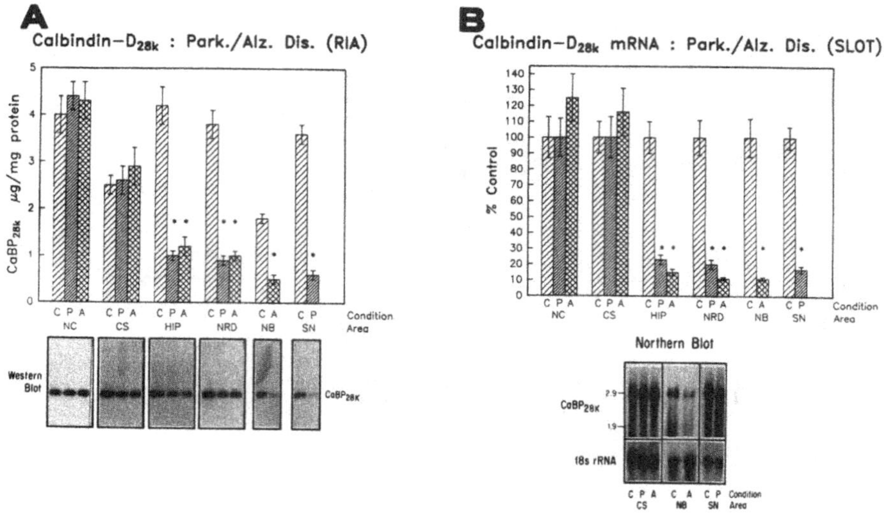

FIGURE 1: Neuronal calbindin and calbindin mRNA in Parkinson's

and Alzheimer's diseases. A. Graphic representation (mean ± SEM) of RIA results (above) and representative Western blot confirming the results (below). Parkinson's (P) and Alzheimer's (A) samples were compared to age and sex matched controls (C). Brain areas shown are neocortex (NC), corpus striatum (CS), hippocampus (HIP), nucleus raphe dorsalis (NRD), nucleus basalis (NB) and substantia nigra (SN). Asterisk represents $p < 0.01$. B. Graphic representation (mean ± SEM) of densitometric quantitation of slot blot hybridization analysis (above). Representative Northern blot (below).

In the aging cerebellum there is a 15-20% loss of immunoreactive Purkinje cells and a 73% reduction of calbindin mRNA (grain/cells, <u>in situ</u> hybridization; Figure 2) suggesting that the decrease in calbindin mRNA is not primarily due to neuronal loss but rather to a decrease in calbindin gene transcription in individual Purkinje cells. Since calbindin gene expression decreases specifically in brain areas known to be particularly affected in aging and in each of the neurodegenerative diseases, these findings suggest that decreased calbindin gene expression may lead to a failure to maintain intraneuronal calcium homeostasis which may contribute to calcium mediated irreversible cytotoxic events during the pathological processes.

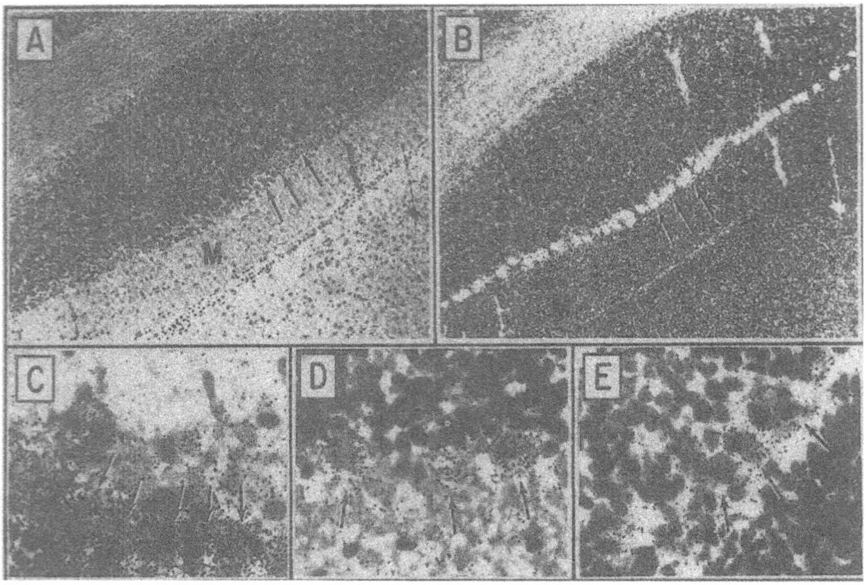

FIGURE 2: <u>In situ</u> hybridization demonstrating changes in Purkinje cell calbindin-D_{28k} mRNA during development and

aging. A: bright-field view of Purkinje cells (arrows) between molecular layer (M) and granular layer (G) (32 X); B: dark field view of A. C-E: higher magnification (100 X) of Purkinje cells (arrows) showing silver grains which represent positive hybridization of probe to calbindin-D_{28k} mRNA (C=2 weeks postnatal, D = 8 weeks, E = 120 weeks). Note decrease in silver grains/cell from C through E.

Besides changes with aging and disease, the possibility of hormonal regulation of neuronal calbindin was also investigated. The Type I corticosterone preferring receptors are localized almost exclusively in the hippocampus and like calbindin-D_{28k} are found almost entirely in area CA1 and dentate gyrus. In light of the receptor localization studies the possibility that corticosterone regulates brain calbindin-D_{28k} was investigated in rat hippocampus. We have found that adrenalectomy results in an 80% depletion of calbindin mRNA in hippocampus (Figure 3) and a marked depletion of calbindin immunoreactivity in the dentate gyrus. Administration of corticosterone for 7 days to adrenalectomized rats restores the levels of calbindin protein and mRNA in hippocampus to those observed in intact controls and induces calbindin gene expression 2.8 fold in hippocampus (Figure 3). No changes in calbindin-D_{28k} protein and mRNA in cerebellum, striatum or cerebral cortex we noted in adrenalectomized rats or in intact rats treated with corticosterone when compared to controls, indicating the specificity of the effect for calbindin-D_{28k} in hippocampus. These studies present the first evidence of a regulator of calbindin gene expression in the brain. The particular corticosterone paradigm that we have utilized has been used by Sapolsky et al 1985 for studies involving hippocampal cell loss during stress. Administration of corticosterone has been shown to increase calcium conductance into hippocampal neurons and to cause cell loss preferentially in areas CA_{2-4} (non calbindin containing areas) with areas CA_1 and dentate gyrus essentially spared (calbindin containing areas). Thus hippocampal calbindin containing neurons may be less vulnerable to calcium mediated toxic events. The possibility exists that increasing the calcium binding capacity of a neuronal cell may render it less vulnerable to damage.

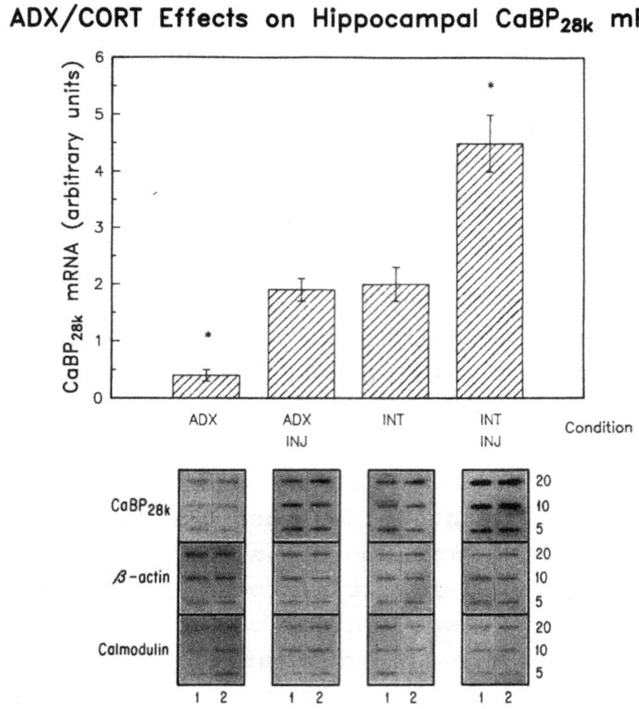

FIGURE 3: Effects of adrenalectomy and corticosterone treatment on rat hippocampal calbindin-D_{28k} mRNA levels. Note the specificity of the effect for calbindin-D_{28k} (no changes in actin or calmodulin mRNA under these conditions).

V. Analysis of the rat calbindin-D_{28k} promoter

We have recently isolated the chromosomal gene for calbindin by screening a rat genomic library in cosmid. The 5' untranslated region of the gene was identified by sequencing using a 33 mer oligoprobe (starting from the initiation site of rat calbindin-D_{28k}) as a primer. The TATA box is located at -30 from the Cap site and the promoter region is markedly G + C-rich. We have constructed recombinant plasmids in which the rat calbindin-D_{28k} promoter (from nt-1075 to nt +34) is fused to the reporter gene encoding chloramphenicol acetyl transferase (CAT). Ros 17/2.8 osteosarcoma cells were transfected with the plasmids. No induction by $1,25(OH)_2D_3$ could be demonstrated (Figure 4). Other CAT constructs and other cell lines are being tested in order to characterize response elements which are involved in modulating calbindin gene expression.

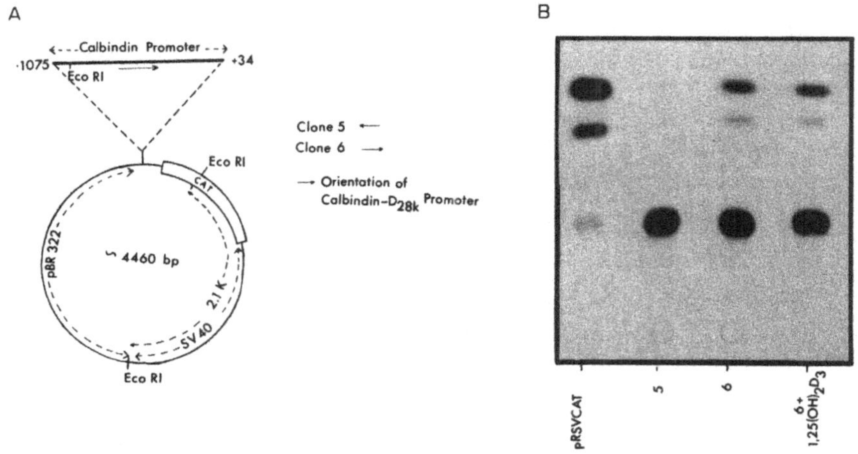

FIGURE 4: Chimeric gene constructs and CAT assay.

In summary, our results suggest tissue specific regulation of calbindin gene expression by 1,25$(OH)_2D_3$ and glucocorticoids which may be related to functional significance. *In vitro* studies using the chromosomal gene for calbindin, which are currently in progress, will be required in order to evaluate more accurately the mechanism whereby the calbindin gene is transcribed in only specific cells and the means whereby 1,25$(OH)_2D_3$ and other modulators can influence calbindin gene expression.

References

Christakos S, Gabrielides G, Rhoten WB (1989) Vitamin D dependent calcium binding proteins: chemistry, distribution, functional considerations and molecular biology. Endocrine Rev 10:3-26

Sapolsky RM, Pulsinelli WA (1985) Glucocorticoids potentiate ischemic injury to neurons:therapeutic implications. Science 229:1397-1399

Van Eldik LJ, Zendegui JG, Marshak DR, Watterson DM (1982) Calcium binding proteins and the molecular basis of calcium action. Int Rev Cytol 77:1-61

Varghese S, Deaven LL, Huang Y-C, Gill RK, Iacopino AM, Christakos, S (1989) Transcriptional regulation and chromosomal assignment of the mammalian calbindin-D_{28k} gene. Molecular Endocrinology 3:495-502

Wasserman RH, Taylor AN (1966) Vitamin D induced calcium binding protein in chick intestinal mucosa. Science 152:791-793

STRUCTURE AND REGULATION OF THE RAT DUODENAL CALBINDIN D 9k (CaBP-9k) GENE

M. Thomasset, N. Lomri, F. L'Horset, A. Bréhier, J.-M. Dupret, and C. Perret
INSERM U120, Alliée CNRS, Pharmacologie du Développement
44 Chemin de Ronde
78110 Le Vésinet, France

Introduction

The calbindin Ds are intracellular calcium-binding proteins (CaBP) which, like parvalbumin, calmodulin and troponin C, are involved in intracellular calcium homeostasis. The members of this superfamily of protein ligands thereby facilitate the second messenger role of calcium (Carafoli, 1987 ; Berridge, 1987). Each calcium binding site on these proteins has the EF-hand structure described by Kretsinger (1980).

The calbindin Ds were discovered by Wasserman's group (Wasserman and Taylor 1966 ; Kallfelz *et al.* 1967) and have calcium binding affinity constants of about 10^6 M^{-1}. The small, 9 000 dalton calbindin D (CaBP-9k) first found in the rat small intestine, has two calcium binding sites ; the larger, 30 000 dalton calbindin D (CaBP-28k), is most abundant in the mammalian cerebellum and kidney, and has 4 calcium-binding sites (Thomasset *et al.* 1982). While mammals contain both CaBP-9k and CaBP-28k, birds possess only CaBP-28k. The concentrations of CaBP-9k and CaBP-28k are as high as 2% of the total soluble proteins in these tissues. However, our quantitative studies on the distribution of these proteins and their mRNAs indicate that no tissue contains large concentrations of both molecules at the same time (Thomasset *et al.* 1982 ; Perret *et al.* 1985a,1985b ; Lomri *et al.* 1989). Both CaBP-9k and CaBP-28k are characteristically under the control

NATO ASI Series, Vol. H 48
Calcium Transport and
Intracellular Calcium Homeostasis
Edited by D. Pansu and F. Bronner

of the active vitamin D metabolite, 1,25-(OH)-$_2$D$_3$ (Thomasset *et al.* 1982 ; Perret *et al.* 1985b). The precise function of these two proteins remains a topic of considerable debate. Their high affinity for calcium and their high concentrations in certain tissues suggest that they may be calcium buffers, calcium carriers or modulators of enzyme activity (see Wasserman, this volume).

This article summarises the information provided by molecular and cellular biology on the structure and regulation of the rat CaBP-9k gene.

The first, essential steps in this work were the preparation of specific polyclonal antibodies to rat duodenal CaBP-9k (Thomasset *et al.* 1982) and of a cDNA probe specific for the CaBP-9k messenger RNA (Desplan *et al.* 1983), and the isolation of the CaBP-9k gene (Perret *et al.* 1988a).

Structure of the CaBP-9k, calcium-binding protein encoded by a gene distinct from that of CaBP-28k

The first addressed questions were : Was there a single gene, which selectively and tissue-specifically produced the two messenger RNAs by differential splicing ? Or were there two distinct genes, each directing the synthesis of one CaBP ?

The absence of immunological cross-reactivity between the two CaBPs and the lack of cross-hybridization between their mRNAs are in favour of two separate genes (Thomasset *et al.* 1982 ; Perret *et al.* 1985a ; Lomri *et al.* 1989). Polyclonal antibodies to rat CaBP-9k recognize (when diluted 1/450 000) only this particular protein in this species. They detect neither CaBP-9k from other species nor other calcium-binding proteins, such as CaBP-28k, S100, calmodulin, or parvalbumin (Thomasset *et al.* 1982). In contrast, polyclonal antibodies to CaBP-28k (diluted 1/120 000) cross react with CaBP-28k from several species, but not with either CaBP-9k or the other calcium-binding proteins listed above (Thomasset *et al.* 1982 ; Baudier *et al.* 1985 ; Intrator *et al.* 1985). However, these anti-CaBP-28k antibodies react with both CaBP-28k and another protein in the brain, which has a molecular weight 2 000 daltons greater than CaBP-28k. This was shown by immunoblotting after electrophoresis under denaturing conditions.

A single approximately 0.5 kb long transcript was found in the intestine, the cecum, the placenta, the vitelline membranes and the uterus by Northern hybridization with the cDNA probe to CaBP-9k mRNA (Perret *et al.* 1985b ; Warembourg *et al.* 1986; 1987). No longer messenger, which might code for CaBP-28k, was detected in either the kidney or the cerebellum. The CaBP-28k cDNA hybridized with a major 1.9 kb transcript and two minor ones of 2.8 kb and 3.2 kb in the kidney, the cerebrum, cerebellum, retina and hippocampus (Lomri *et al.* 1989). There was no hybridization with a messenger in either the liver or the duodenum (Lomri *et al.* 1989).

The sequence of rat CaBP-9k was deduced from that of cDNA (Desplan *et al.* 1983) and of the CaBP-9k gene (Perret *et al.* 1988a). It contains two calcium binding sites which satisfy the criteria defined by Kretsinger (Kretsinger, 1980). Site I is modified by the addition of two amino-acids. The structure of the protein is highly conserved ; the sequences of beef, pig and rat are very similar (Desplan *et al.* 1983 ; Perret *et al.* 1988a), despite the lack of immunological cross-reactivity. The sequences of CaBP-9k and S100 protein, another two-site calcium-binding protein, are 34% similar (Perret *et al.* 1988a; 1988b).

The sequence of rat CaBP-28k was deduced from that of the cDNA by Hunziker's group and our own (Lomri *et al.* 1989). This protein is highly conserved ; it is 98-99% homologous with human and bovine CaBP-28k and 95% homologous with chicken CaBP-28k (Lomri *et al.* 1989). CaBP-28k has 6 EF-hand structures, two of which are modified. This could explain why this molecule only binds 4 calcium atoms. It should be emphasized that there are no sequence homologies between CaBP-9k and CaBP-28k (Lomri *et al.* 1989 ; Perret *et al.* 1988a). In contrast, the sequences of CaBP-28k and calretinin (Rogers, 1987), another calcium-binding protein found in the retina, are 58% homologous (Lomri *et al.* 1989). Calretinin may be the second protein detected by immunoblots using anti-CaBP-28k antibodies. These observations suggest that results obtained only with polyclonal antibodies should be interpreted with caution and required confirmation by experiments using specific cDNA probes.

The recently established structures of the CaBP-9k and CaBP-28k genes confirm that each protein is encoded by a separate gene (Perret *et al.* 1988a ; Lomri *et al.* 1989). The rat CaBP-9k gene contains 3 exons interrupted by 2 introns. Each calcium site is encoded by a separate exon (Perret *et al.* 1988a). CaBP-9k gene belongs to the "EF hand" superfamily. Comparison of the intron/exon organization of the genes coding for these proteins (calmodulin, myosin alkali-light chain, myosin regulatory light chain, parvalbumin, calpains, CaBP-9k, calcyclin, MRP-8, MRP-14, CaBP-28k, calretinin) had led us to propose an hypothesis on the evolutionary origin of this protein family (Perret *et al.* 1988b ; Lomri *et al.* 1989). CaBP-9k gene is not arisen from CaBP-28k gene. CaBP-28k and calretinin genes have diverged very early from the other members (Perret *et al.* 1988b ; Lomri *et al.* 1989).

Regulation of CaBP-9k gene in the rat intestine

a. By 1,25-$(OH)_2$-D_3

1,25-$(OH)_2$-D_3 was shown to regulate the synthesis of CaBP-9k in rats fed a vitamin D-free diet for 5 weeks from weaning (Thomasset *et al.* 1982 ; Perret *et al.* 1985b). Under these conditions, the concentrations of CaBP-9k and its mRNA in the duodenum are greatly reduced in the duodenum, jejunum, ileum and caecum. A single injection of 1,25-$(OH)_2$-D_3 restores these concentrations (Thomasset *et al.* 1982 ; Perret *et al.* 1985b). The kinetics of CaBP-9k messenger accumulation in the duodenum in vitamin D-depleted rats given a single dose of 1,25-$(OH)_2$-D_3 shows that the level begins to increase 1 hour after injection and is maximal at 6 hours, while the CaBP-9k concentration increases later (Perret *et al.* 1985b).

Organotypic duodenal cultures, in which fetal rat duodenum is maintained in vitamin D-free, serum-free chemically defined medium for 10 days, were developed to determine the mechanism of regulation by 1,25-$(OH)_2$-D_3 (Bréhier and Thomasset, 1990). Under these experimental conditions, the cells differentiate into enterocytes. Addition of 10^{-8} M 1,25-$(OH)_2$-D_3 induces an increase in the level of CaBP-9k mRNA. The concentration doubles in one hour and quadruples in 6 hours, while the concentration of CaBP-9k itself increases later. Actinomycin D (1 uM) inhibits the steroid-induced increase in mRNA.

The synthesis of CaBP-9k mRNA, measured by "run-on" assay, increases within 15 minutes of calcitriol administration to vitamin D-deficient rats. This increase in synthesis is maximal 1 hour after the administration of hormone (Dupret *et al.* 1987).

The early increase in the synthesis and accumulation of CaBP-9k mRNA in response to 1,25-$(OH)_2$-D_3 and its inhibition by actinomycin D indicate that 1,25-$(OH)_2$-D_3 acts at the level of transcription. Thus calcitriol acts in the same way as other steroid hormones. Recent studies by the groups of Norman, Lawson and Christakos have shown that expression of the CaBP-28k gene is also positively regulated by calcitriol in the avian intestine and kidney and in the mammalian kidney. However, the expression of the CaBP-28k gene is insensitive to 1,25-$(OH)_2$-D_3 in the cerebellum of rat (Thomasset *et al.* 1982; Thomasset and Tenenhouse, 1988) and chick.

b. By calcium

However 1,25-$(OH)_2$-D_3 is probably not the unique effector implicated in the regulation of rat intestinal CaBP-9k. We have reported that 30-day-old 5th generation vitamin D-deficient rats maintained on a diet containing no detectable vitamin D and 0.4 % calcium had almost normal duodenal CaBP-9k levels (Thomasset and Tenenhouse, 1988). These rats had no detectable serum vitamin D metabolites, suggesting that effectors other than 1,25-$(OH)_2$-D_3 are responsible for this basal CaBP-9k level. It is possible that calcium modulates CaBP-9k synthesis and that the minimal effect of vitamin D deficiency in these experiments may be due to the relatively high dietary calcium level.

In order to test this hypothesis we have used fetal rat intestinal organ culture (Bréhier and Thomasset, 1990) and studied the effects of extracellular calcium concentration on the regulation of CaBP-9k gene expression.

Increasing the calcium concentration in the medium from 0 to 1.2 mM causes an increase in the production of CaBP-9k mRNA in both the presence and absence of 1,25-$(OH)_2$-D_3. The effect of calcium is blocked by actinomycin D. Thus, calcium also stimulates the expression of the CaBP-9k gene in the rat intestine (Bréhier and Thomasset, 1990).

Calcium has recently been shown to have a positive regulatory effect on the expression of the CaBP-28k gene in cultured renal cells (Clemens *et al.* 1989).

The CaBP-9k gene regulatory sequences

The rat CaBP-9k gene contains 3 exons and 2 introns (Perret *et al.* 1988a). Transcription is initiated at a single site. The promoter region contains the classical eucaryote gene promoter sequences : 1 TATA box and 4 CAAT box-type sequences. Several regulatory elements, such as "HRE", are also present (Perret *et al.* 1988a).

We are now in the process of using cellular transfection to identify the regulatory sequences actually implicated in the control of CaBP-9k gene transcription by 1,25-$(OH)_2$-D_3.

The sequence of the CaBP-28k gene has been determined by several groups, but, despite the work of a number of laboratories, the regulatory sequences remain unknown.

General conclusions

The calbindin Ds (CaBP-9k and CaBP-28k) are distinguished from the other calcium-binding proteins implicated in calcium homeostasis by their specific tissue distribution and their high concentrations in certain tissues. The synthesis of each CaBP is directed by its own gene. Hormonal regulation of the CaBP-9k and CaBP-28k genes varies with the tissue : calcitriol stimulates their expression in the intestine and kidney, in contrast oestradiol is the regulator in uterus. Extracellular calcium increases the expression of the CaBP genes in organ cultures of intestine and kidney. The calbindin Ds are therefore suitable models for studying the tissue-specific regulation of the same gene by several steroid hormones.

Acknowledgments

We are grateful to N. Gouhier, M. Eb and J. Emsellem for their expert technical assistance. We thank M. Courat and O. Parkes for their help in the preparation of the manuscript.

References

Baudier J, Glasser N, Strid L, Bréhier A, Thomasset M, Gerard D (1985) Purification, calcium-binding properties, and conformational studies on a 28-kDa cholecalcin-like protein from bovine brain. J Biol Chem 260:10662-10670

Berridge (1987) Inositol triphosphate and diacyglycerol : two interacting second messengers. Biochem 56:395-4331

Bréhier A, Thomasset M (1990) Stimulation of calbindin-D9K (CaBP9K) gene expression by calcium and 1,25$(OH)_2D_3$ in fetal rat duodenal organ culture. Endocrinology, in press

Carafoli E (1987) Intracellular calcium homeostasis. Ann Rev Biochem 56:395-433

Clemens TL, McGlabe SA, Garnett KP, Craviso GL, Hendy GN (1989) Extracellular calcium modulates vitamin D-dependent calbindin-D28K gene expression in chick kidney cells. Endocrinology 124:1582

Desplan C, Heidmann O, Lillie JW, Auffray C, Thomasset M (1983) Sequence of rat intestinal vitamin D-dependent calcium-binding protein derived from a cDNA clone. Evolutionary implications. J Biol Chem 258:13502-13505

Dupret JM, Brun P, Perret C, Lomri N, Thomasset M, Cuisinier-Gleizes P (1987) Transcriptional and post-transcriptional regulation of vitamin D-dependent calcium-binding protein gene expression in the rat duodenum by 1,25-dihydroxy-cholecalciferol. J Biol Chem 262:16553-16557

Intrator S, Elion J, Thomasset M, Bréhier A (1985) Purification, immunological and biochemical characterization of rat 28KDa cholecalcin (cholecalciferol-induced calcium-binding proteins). Biochem J 231:89-95

Kallfelz FA, Taylor AN, Wasserman RH (1967) Vitamin D-induced calcium factor in rat intestinal mucosa. Proc Soc Exp Biol Med 125:54-58

Kretsinger RH (1980) Evolution and function of calcium-binding proteins. Intern Rev Cytol 46:323-393

Lomri N, Perret C, Gouhier N, Thomasset M (1989) Cloning and analysis of calbindin-D28k cDNA and its expression in the central nervous system. Gene 80:87-98

Perret C, Desplan C, Bréhier A, Thomasset M (1985) Characterisation of rat 9-kDa cholecalcin (CaBP) messenger RNA using a complementary DNA. Absence of homology with 28-kDa cholecalcin mRNA. Eur J Biochem 148:61-66

Perret C, Desplan C, Thomasset M (1985) Cholecalcin (a 9-kDa cholecalciferol-induced calcium-binding protein) messenger RNA. Distribution and induction by calcitriol in the rat digestive tract. Eur J Biochem 150:211-217

Perret C, Lomri N, Gouhier N, Auffray C, Thomasset M (1988) The rat vitamin D-dependent calcium-binding protein (9-kDa CaBP) gene. Complete nucleotide sequence and structural organization. Eur J Biochem 172:43-51

Perret C, Lomri N, Thomasset M (1988) Evolution of the EF-hand calcium-binding protein family : evidence for exon shuffling and intron insertion. J Mol Evol 27:351-364

Rogers JH (1987) Calretinin : a gene for a novel calcium-binding protein expressed principally in neurons. J Cell Biol 105:1343-1353

Thomasset M, Parkes CO, Cuisinier-Gleizes P (1982) Rat calcium binding protein : distribution, development and vitamin D-dependence. Am J Physiol 243:E483-E488

Thomasset M, Tenenhouse A (1988) Vitamin D dependence of calbindin D 9K and calbindin D 28K synthesis in various rat tissues. In : Norman AW, Schaefer K, Grigoleit H-G, Herrath D-V (eds) Vitamin D. Molecular, Cellular and Clinical Endocrinology, Walter de Gruyter, Berlin New York p 516

Warembourg M, Perret C, Thomasset M (1986) Distribution of vitamin D-dependent calcium-binding protein messenger ribonucleic acid in rat placenta and duodenum. Endocrinology 119:176-184

Warembourg M, Perret C, Thomasset M (1987) Analysis and in situ detection of cholecalcin messenger RNA (9000 Mr CaBP) in the uterus of the pregnant rat. Cell Tissue Res 247:51-5719.

Wasserman RH, Taylor AN (1966) Vitamin D_3-induced calcium-binding protein in chick intestinal mucosa. Science 152: 791-793

COMPARATIVE REGULATION AT THE INTESTINAL AND UTERINE LEVEL OF CALBINDIN mRNA IN THE FOWL

Y. Nys*, R. Bouillon**, D.E.M. Lawson***

* INRA, 37380 Nouzilly, France

** AFRC, Babraham Cambridge, CB2 4AT, U.K.

*** LEGENDO, Leuven, Belgium

In the domestic hen, the egg shell gland secretes over a 14 h period large quantities of calcium (about 2g) and as a result the extracellular calcium pool turns over 12 minutes. Shell deposition is a discontinuous process and the diurnal periodicity, which is a consequence of the daily ovulatory cycle, imposes a severe challenge to the homeostatic regulation of calcium. Hens have adapted at all possible levels to meet the stresses imposed on the calcium regulatory mechanism including : having a specific appetite for calcium, increase the storage in the crop of food and calcium just before shell formation, having an increase in HCl secretion and in the level of soluble calcium in the intestinal contents (Mongin and Sauveur, 1979) so that the intestinal retention of calcium is doubled during the period of shell deposition (Hurwitz, 1989). Furthermore, during the 2–3 weeks preceeding egg laying, gonadal hormones induced the growth of the oviduct, the development of medullary bone and an enhancement of intestinal retention of calcium.

It is noteworthy that the calcium transport capacity of the uterus varies during shell deposition falling from its maximum values just after ovulation (Eastin and Spaziani, 1978). Transport of calcium across the shell gland wall involves both active and diffusional processes and, as with intestinal calcium absorption, is correlated with the concentration of calbindin D 28K (Hurwitz, 1989). CaBP–28K is present in both intestine and uterus but there is evidence that the calbindin is regulated in a tissue specific manner. It would appear likely that gonadal steroids and vitamin D are involved in the regulation of the calbindin synthesis since oviduct growth is under the control of gonadal hormones and since, in the intestine, the CaBP–28K level depends on vitamin D. Furthermore, sexual maturity and the onset of egg production are associated with increases in 1,25–dihydroxycholecalciferol

NATO ASI Series, Vol. H 48
Calcium Transport and
Intracellular Calcium Homeostasis
Edited by D. Pansu and F. Bronner

($1,25-(OH)_2-D_3$) production (Kenny, 1976) and in gonadal steroid. This paper describes the regulation of calbindin synthesis in the intestine and uterus of the laying hens and evaluates the role of transcriptional and post-transcriptional processes in the stimulation of calbindin D 28K synthesis associated with sexual maturity and shell formation. Moreover, we show that $1,25(OH)_2D_3$ production is controlled in hens by factors other than the sequences of events induced by the homeostatic regulation of blood calcium and that in the uterus, calbindin synthesis is controlled neither by vitamin D nor by gonadal steroids.

METHODS

ISA Brown hens were fed a diet containing 3–5% calcium and 2000 IU vitamin D/kg diet and were caged in an air-conditioned room, illuminated 14 h a day. Oviposition times for each hen were recorded daily using a computer recording system. Immature pullets were used at 12 weeks of age, were exposed to 6 h of light and were fed a 1% calcium diet with 2000 IU vitamin D/kg diet.

To dissociate the endocrine events associated with ovulation from the calcium requirement for shell formation, calcification of the egg-shell was suppressed for 7–8 days by premature explusion of the egg induced by injection of prostaglandins F2 α, 8 h after ovulation. Concentrations of calbindin m RNA were measured by dot blot hybridization using a ^{32}P cRNA probe (Mayel-Afshar et al, 1988) prepared from the recombinant plasmid consisting of the full length cDNA (Wilson et al, 1985). CaBP-28K levels were measured by immuno-electrophoresis using an antiserum specific to chick intestinal CaBP-28K. Blood $1,25-(OH)_2-D_3$ concentration were determined after high pressure liquid chromatography by radioimmunoassay (Bouillon et al, 1980).

RESULTS AND DISCUSSION

Regulation of intestinal calbindin synthesis

The injection of estradiol into immature pullets over 12 successive days led to an increase in blood $1,25-(OH)_2-D_3$ concentration (Table 1) but there was no change in intestinal calbindin levels. When testosterone was administered with estrogens, the concentration of $1,25-(OH)_2-D_3$ increased further and the concentration of calbindin and its mRNA rose in the intestine. In rachitic pullets, calbindin was not detectable in the duodenum. In mature laying hens, the intestinal concentrations of calbindin and its mRNA (Table 2) and the levels of 1,25-$(OH)_2$-D_3 in the blood were higher than in immature pullets. During the period of shell formation the levels of 1,25-$(OH)_2$-D_3 rose further (Fig. 1b) and the accumulation of calbindin mRNA increased slightly (Table 2) but calbindin concentration was not affected by the stage of shell formation as shown previously (Hurwitz, 1989).

These observations demonstrate that the synthesis of intestinal calbindin D 28K is stimulated a) at sexual maturity and then b) at the onset of egg production by 1,25-$(OH)_2$-D_3 and that the induction of calbindin synthesis is primarily at the transcriptional level. This conclusion is supported by the parallel decrease in blood concentration of 1,25-$(OH)_2$-D_3 and in intestinal accumulation of calbindin mRNA after suppression of the parathyroid glands in hens resumed laying hard shelled eggs (Table 3).

Suppression of shell formation depressed the concentration of 1,25-$(OH)_2$-D_3 6 h after the premature expulsion of the egg (Table 3) but did not change either the accumulation of the intestinal calbindin mRNA, or the CaBP-28K levels. However, the concentrations of intestinal CaBP-28K (Table 4) and the blood level of 1,25-$(OH)_2$-D_3, were lower during the period of shell formation (Fig. 1b) in hens laying shell-less egg for a week than those of hens laying hard shelled eggs. It is noteworthy that suppression of shell formation did not decrease in the duodenum the accumulation of calbindin mRNA relative to the level observed at the onset of shell formation in hens laying hard shelled egg (Table 4). It suggests, therefore, that the concentration of intestinal calbindin is also regulated by post-transcriptional processes.

Regulation of blood concentration of 1,25$(OH)_2$-D_3

The increase in intestinal calbindin concentration at sexual maturity and by the high demand for calcium during egg-shell deposition depends on the 1,25-$(OH)_2$-D_3 concentration. It is therefore of importance to know how fluctuation in blood concentration of 1,25-$(OH)_2$-D_3 are regulated in hens. Unlike hens laying shell-less eggs, the plasma concentration of 1,25-$(OH)_2$-D_3 increased during the period of shell formation (Fig. 1b), when the levels of ionized calcium were lower (Fig. 1a) and, consequently, stimulated the secretion of PTH as shown by Singh et al (1986). When the hens were fed on a low calcium diet (1%) there was a further decrease in blood ionized calcium associated with a large increase in blood levels of 1,25-$(OH)_2$-D_3. This stimulation in 1,25-$(OH)_2$-D_3

production results therefore from the calcium demands for shell formation and involves the sequence of responses observed in the homeostatic regulation of blood ionized calcium by the parathyroid glands.

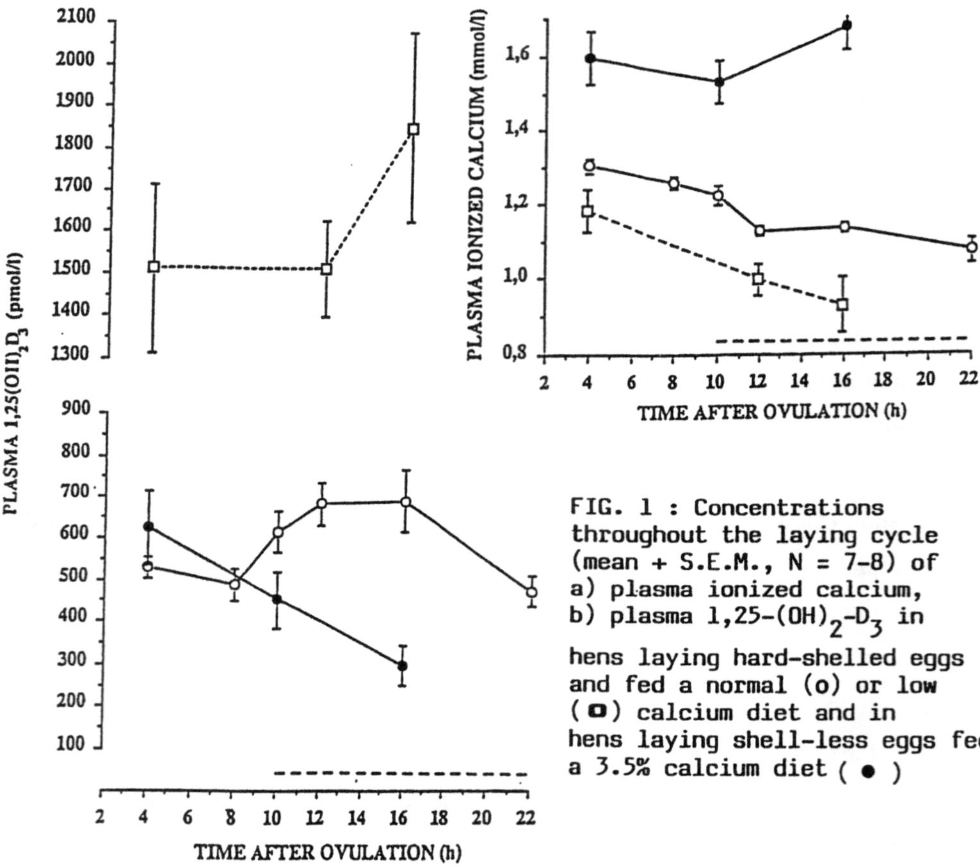

FIG. 1 : Concentrations throughout the laying cycle (mean + S.E.M., N = 7-8) of a) plasma ionized calcium, b) plasma 1,25-$(OH)_2$-D_3 in hens laying hard-shelled eggs and fed a normal (o) or low (◘) calcium diet and in hens laying shell-less eggs fed a 3.5% calcium diet (●)

Conversely, the second component involved in regulation of 1,25-$(OH)_2$-D_3 production was observed in hypercalcemic hens laying shell-less eggs and fed on a 3.5% calcium diet. The levels of 1,25-$(OH)_2$-D_3 just after ovulation increased 5 fold in hens laying shell less or hard-shelled eggs compared to the levels in immature pullets. Similar profiles of sex steroids were observed in both groups (Nys et al, 1986). It is therefore likely that this stimulation, observed even in hypercalcemic hens, does not involve the parathyroid glands and is associated with the endocrine events concomitant to ovulation.

Regulation of calbindin synthesis in the uterus

The growth of the oviduct and the uterine concentration of calbindin and its mRNA were stimulated by 12 daily injections of estradiol (Table 1). The observed increases were greater when testosterone was given with the estradiol. Estrogens also led to an increase in calbindin synthesis in the uterus of vitamin D deficient pullets, in which the duodenal CaBP–28K was undetectable. The stimulation of uterine calbindin at sexual maturity is, consequently, associated with the growth of the oviduct induced by sex steroids, is independent of vitamin D and is regulated in a tissue specific manner. In hens laying eggs which were undergoing calcification, the uterine concentration of calbindin mRNA was greatly increased relative to the levels observed just after ovulation (Table 2) or to those of hens laying shell–less eggs (Table 4). This increase in CaBP–28K synthesis in the uterus was maintained throughout the period of shell deposition (Table 4). Again, this effect of shell formation was not observed in the intestine.

The fluctuations in blood levels of gonadal steroids were similar during the ovulatory cycle in hens laying shell–less or hard–shelled eggs (Nys et al 1986). Consequently it is unlikely that the sex steroids are involved in the high stimulation in calbindin mRNA concentration observed at the time of shell calcification. This increase is not, however, associated with a concomitant or delayed increase in uterine calbindin concentration (Table 2, Bar and Hurwitz, 1973) suggesting variation in the turnover of mRNA or post–transcriptional regulation.

Suppression of the shell formation at the 12 h stage (Table 3) depressed the level of calbindin mRNA at 50% and 3.4% of its initial value 1 h and 6 H, respectively, after the premature expulsion of the egg. Calbindin D 28K concentration remained stable as expected from its long half–life (Norman et al, 1981) but was lower when the suppression of the shell was persued for 8 days (Table 4). Relatedly, at the time the hens resumed laying hard–shell eggs, the level of uterine CaBP–28K and its mRNA increased markedly during the formation of the first shell (Table 3). The genomic induction of calbindin was maintained in hens parathyroidectomized just before resumption of the first shell although the blood concentration of 1,25–$(OH)_2$–D_3 was depressed. Calbindin synthesis was maintained in the uterus but not in the intestine. PTH and 1,25–$(OH)_2$–D_3 are not, therefore directly involved in the stimulation in the uterine calbindin mRNA observed during shell formation. This daily formation is tissue specific and involved mRNA

Table 1 : Effects of gonadal steroids on duodenal and uterine calbindin of immature pullets fed on a vitamin D deficient or vitamin D repleted diet.

	Vitamin D repleted			Rachitic
	Control	Estrogen	Estrogen + Testosterone	Estrogen
Duodenal calbindin (µg/mg protein)	5a	8a	20b	nd
Duodenal calbindin mRNA (% control value)	100b	149b	489c	7.5a
Uterine calbindin (µg/mg protein)	nd	2.4	6.4	4.8
Uterinecalbindin mRNA/ uterus (% control value)	100a	850b	3750c	4172c
Plasma $1,25(OH)_2D_3$	106a	166b	225c	-

Values are means + SEM (n = 6). For any given variable, means without a common superscript are significantly different ($P < 0.05$, Newmans-Keuls test). nd = non detectable.

Table 2 : Increases in duodenal and uterine concentration of calbindin and its mRNA induced in hens by egg formation.

	Immature pullets	Mature hens 4 h after ovulation	Mature hens 12 h after ovulation
Duodenal calbindin (µg/mg protein)	5 ±2a	112±7b	82±4b
Duodenal calbindin mRNA (µg/mg duodenum)	49±12a	411±70b	556±89b
Uterine calbindin (µg/mg protein)	ND	54±6b	48±6b
Uterinecalbindin mRNA (ng/g uterus)	7±1a	22±4a	514±77b
Plasma $1,25(OH)_2D_3$	106±10a	610±89b	907±104c

See footnote table 1.

Table 3 : Effects of expulsion of the egg and resumption of shell formation and parathyroidectomy (PTX) on calbindin synthesis in the uterus and intestine of hens.

	Hens calcifying an egg	6 hours after egg expulsion	Resumption of the first shell after suppression of its formation		
Stage	12 h		8 h	24 h after ovulation	
			Control	sham PTX	PTX(1)
Duodenal calbindin (µg/mg protein)	138b	95ab	70a	95ab	76a
Duodenal calbindin mRNA (% initial value)	100a	83a	93a	126a	52b
Uterine calbindin (µg/mg protein)	60c	66c	9a	44b	31b
Uterine calbindin mRNA (% initial value)	100d	3.4a	17b	80cd	56c
Plasma 1,25-$(OH)_2$-D_3	764c	322b	460bc	948d	190a

See footnote table 1.

(1) The suppression of the parathyroid glands was carried out 16-18 h before oviposition

Table 4 : Increases in plasma 1,25$(OH)_2$-D_3 and in duodenal and uterine calbindin induced by shell formation.

Hens laying	shell-less eggs		Hard-shelled eggs	
Hours after ovulation	12 h	19 h	12 h	19 h
Duodenal calbindin (µg/mg protein)	74±11b	48±19a	138±12c	111±10c
Duodenal calbindin mRNA (% control, 12 h stage)	81a	83a	100ab	146b
Uterine calbindin (µg/mg protein)	25±2a	10±2a	60±11b	59±8b
Uterine calbindin mRNA (% control, 12 h stage)	13a	3.4a	100b	99.3b
Plasma 1,25$(OH)_2D_3$ (pmol/liter)	499±56b	305±15a	906±103c	615±78bc

See footnote table 1.

transcription within uterine tissue. Furthermore, the coincidence in calcium transport and stimulation in calbindin mRNA level within the uterus strongly suggests, at the cellular level, a calcium flux–dependent factor, which could act alone or in combination with vitamin D or sex steroids. Finally, it has been demonstrated that the timing of shell deposition is associated with the occurence of ovulation (Eastin and Spaziani, 1978) and that duration of shell formation is hormonaly controlled (Nys, 1987). Further studies are required to determine whether this postulated hormonal stimulant is involved in regulation of calbindin synthesis.

REFERENCES

Bar, A., and Hurwitz, S. (1973). Uterine calcium binding protein in the laying fowl. Comp. Biochem. Physiol. A45A, 579–586.

Bouillon, R., De Moor, P. Baggioni, E.G., and Uskokovic, M.R. (1980) A radioimmuno-assay for 1,25–dihychoxycholecalciferol. Clin. Chem., 26, 562–567.

Eastin, W.C., and Spaziani, E. (1978) On the control of calcium secretion in the avian shell gland (uterus). Biol. Reprod. &ç, 493–504.

Hurwitz, S. (1989) Calcium homeostasis in birds. Vitam. Horm. 45, 173–221.

Kenny, A.D. (1976) Vitamin D metabolism : physiological regulation in egg–laying quail. Am. J. Physiol. 230, 1609–1615.

Mongin, P., and Sauveur, B. (1979) The specific calcium appetite of the domestic fowl in food intake regulation in poultry. ed. Boorman K.N. and Freeman B.M. B.P.S. Edinburgh, 171–189.

Norman, A.W., Friedlander, E.J., and Henry, H.L. (1981) Determination of the rate of synthesis and degradation of vitamin D–dependent chick intestinal and renal calcium–binding proteins. Arch. Biochem. Biophys., 206, 305–317.

Nys, Y. (1987) Progesterone and testosterone elicit increases in the duration of shell formation in domestic hens. Brit. Poult. Sci. 28, 57–68.

Nys, Y., N'Guyen, T., Williams, J., and Etches, R.J. (1986) Blood levels of ionized calcium, inorganic phosphorus, 1,25–$(OH)_2D_3$ and gonadal hormones in hens laying hard–shelled or shell–less eggs. J. Endocrinol. 111, 151–157.

Mayel–Afshar, S., Lane, S.M., and Lawson, D.E.M. (1988) Relationship between the levels of calbindin synthesis and calbindin mRNA in chick intestine. Quantification of calbindin mRNA. J. Biol. Chem. 269, 4355–4361.

Singh, R., Joyner, C.J., Peddie, M.J., and Taylor, T.G. (1986). Changes in the concentration of parathyroid hormone and ionic calcium in the plasma of laying hens during the egg cycle in relation to dietary dificiencies of calcium and vitamin D. Gen. Comp. Endocrinol. 61, 50–28.

Wilson , P.W., Harding, M., and Lawson, D.E.M. (1985). Putative amino acid sequence of chick calcium–binding protein deduced from a complementary DNA sequence. Nucl. Acids Res. 13, 8867–8881.

CALBINDIN-D_{9k} LOCALIZATION AND GESTATIONAL CHANGES IN THE UTERO-PLACENTAL UNIT: A MODEL FOR MATERNAL-FETAL CALCIUM TRANSPORTS

M. Elizabeth Bruns and David E. Bruns
Department of Pathology
University of Virginia Health Sciences Center
Charlottesville, Virginia 22908

The calbindins are a class of E-F hand, high affinity calcium-binding proteins. In a variety of tissues, the synthesis of these proteins is known to be dependent on 1,25-dihydroxyvitamin D (1,25 $(OH)_2D$) (Christakos et al., Wasserman et al). Two major forms of calbindins have been isolated and characterized, namely a high molecular weight form of Mr 28,000 (calbindin-D_{28k}) first isolated from chick intestinal epithelium, and a low molecular weight form of Mr 9,000 (calbindin-D_{9k}) first identified in rat intestinal mucosa. These proteins show distinct differences in their distributions, both among species and among tissues. Although the molecular role of these proteins is as yet unclear, ample evidence indicates that calbindins are an excellent marker for epithelia that sustain high rates of transcellular calcium transport (Christakos et al., Wasserman et al.)

Calbindins have been shown to be present in key reproductive structures in both birds and mammals (Christakos et al., Wasserman et al). The avian protein is found in the egg shell gland and mammalian calbindin-D_{9k} is present in the feto-placental units of rats and mice. By biochemical and immunochemical means, high concentrations of calbindin-D_{9k} have been localized in pregnant mice and rats to epithelial cells of the uterus and yolk sac and to endodermal cells of the placenta termed intraplacental yolk sac (IPYS) (Mathieu et al). These cellular localizations of calbindin-D_{9k} in juxtaposed epithelial layers of uterus, yolk sac and intraplacental yolk sac strongly suggest that maternal-fetal calcium transport in rats and mice proceeds through a uterine luminal pathway similar to the bird pathway for egg shell formation. In lower mammals, this uterine luminal pathway through the yolk sac is important for

NATO ASI Series, Vol. H 48
Calcium Transport and
Intracellular Calcium Homeostasis
Edited by D. Pansu and F. Bronner

other nutrients such as immunoglobulins, amino acids, transferrin and vitamin B_{12}.

Delorme _et_ _al_. made the important observation that calbindin-D_{9k} was also present in the mature non-pregnant rat uterus. The uterine expression of calbindin-D_{9k} was not dependent upon 1,25(OH)$_2$D but rather on estrogen. Localization of uterine calbindin-D_{9k} was found in the myometrium and stromal cells of the uterus but not in the luminal cells. This muscle localization was a surprising observation, since in other organs, such as gut, kidney, placenta and yolk sac, calbindin-D_{9k} is located in calcium-transporting epithelia. So far, myometrium is the only muscle found to contain calbindin-D_{9k}.

Figure 1 summarizes the localization of calbindin-D_{9k} in the rodent uterine-fetal-placental unit and suggests a potential route of calcium transfer.

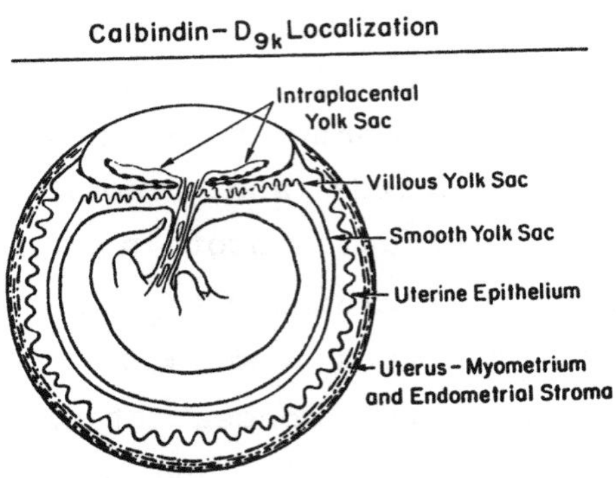

Figure 1. Immunochemical localization of calbindin-D in theutero-placental unit. Calbindin has been localized to the labeled structures: the uterus, the yolk sac, and the sinuses of Duvall of the placenta (intraplacental yolk sac) (Bruns _et al_.) The potential pathway for calcium transfer consists of transport across the uterine epithelial cells to uterine lumen, uptake by yolk sac and endodermal cells of placenta (intraplacental yolk sac), and transfer into the fetal circulation.

During gestation, the pregnant uterus undergoes two dynamic changes in calbindin content and localization, 1) induction of calbindin in the uterine epithelial layer and 2) biphasic fluctuation in muscle layers. Mathieu et al. in this laboratory traced the changes in calbindin-D_{9k} levels in rat uterus, yolk sac and placenta during gestation and found a coordinated induction of the protein in the uterine epithelial lining cells and the associated yolk sac visceral epithelium as well as in the intraplacental yolk sac epithelium. The time of this induction (day 17 to term) coincided with the time of maximal fetal accumulation of calcium, suggesting a role for calbindin-D_{9k} in maternal-fetal calcium transport via these juxtaposed epithelial layers. Dynamic changes also occurred in the calbindin-D_{9k} contents of the two layers of uterine smooth muscle (outer longitudinal and inner circular) during mid- and late gestation. During early pregnancy (days 0-4), calbindin-D_{9k} was present in the two smooth muscle layers. By mid-gestation (day 10), calbindin-D_{9k} had decreased by a factor of ten in these tissue layers. During late gestation calbindin-D_{9k} rebounded in the inner circular smooth muscle layer but not in the outer layer. These changes in uterine muscle calbindin content during early and mid-gestation appeared to be controlled by maternal rather than fetal factors as they were reproduced by the endocrine changes of pseudopregnancy. Progesterone appeared to be a good candidate for controlling the mid-gestational decrease of uterine muscle calbindin-D_{29k}, as it was shown to blunt estrogen's induction of the protein in the muscle layers and stroma of non-pregnant rats in a dose-dependent manner. As changes in myometrial calbindin-D_{9k} will alter intracellular Ca^{2+}, they may alter the potential for, and rates of, contraction.

In conclusion, during late pregnancy in rats, in association with maternal-fetal calcium transport and uterine muscle contraction, calbindin reappears in uterine muscle, as well as in the three epithelial transporting layers of the uterus, yolk sac, and placenta. Which endocrine systems control these late calbindin inductions is unknown. Previous work has indicated that $1,25(OH)_2D_3$ receptors appear in the rat uterus after estrogen treatment (Walters _et al_ and Levy _et al_). In addition, $1,25(OH)_2D_3$ receptors have been found in yolk sac and placenta. The study of calbindin-D_{9k} appearance and localization in the uterus-yolk sac-placenta provides an attractive model system to study the genetic interactions of the sex steroid hormones and $1,25(OH)_2D_3$. Additional studies are required to examine the function of calbindin in the myometrium and its role in normal and premature labor.

Bruns ME, Overpeck JG, Smith GC, Hirsch GN, Mills SE, Bruns DE (1988) Vitamin D-dependent calcium binding protein in rat uterus, differential effects of estrogen, tamoxifen, progesterone and pregnancy on accumulation and cellular localization. Endocrinology 122:2371-2378

Christakos S, Gabrielides C, Rhoten WB (1989) Vitamin D-dependent calcium binding proteins: chemistry, distribution, functional considerations, and molecular biology. Endocrine Reviews 10:3-26

Levy J, Zuili I, Yankowitz N, Shany S (1984) Induction of sytosolic receptors for 1,25-dihydroxyvitamin D_3 in the immature rat uterus by oestradiol. J Endocr 100:265-269

Mathieu CL, Burnett SH, Mills SE, Overpeck JG, Bruns DE, Bruns ME (1989) Gestational changes in calbindin-D_{9k} in rat uterus, yolk sac and placenta: Implications for maternal-fetal calcium transport and uterine muscle function. Proc Natl Acad Sci 86:3433-3437

Pike JW, Gooze LL, Haussler MR (1980) Biochemical evidence for 1,25-dihydroxyvitamin D receptor macromolecules in parathyroid, pancreatic, pituitary, and placental tissues. Life Sci 26:407-414

Stumpf WE, Sar M, Narbaitz R, Huang S, DeLuca HF (1983) Autoradiographic localization of 1,25-dihydroxyvitamin D_3 in rat placenta and yolk sac. Hormone Res 18:215-220

Walters MR, Cuneo DL, Jamison AP (1983) Possible significance of new target tissues for 1,25-dihydroxyvitamin D_3. Steroid Biochem 19:913-920

Wasserman RH, Fullmer CS (1982) Vitamin D-induced calcium-binding protein (Cheung Wy, ed.) Academic Press 2:175-215

POTENTIAL ROLE OF CALBINDINS IN ENAMEL CALCIFICATION

Alan N. Taylor
Department of Anatomy
Baylor College of Dentistry
Dallas, Texas, USA 75246

INTRODUCTION

Cellular elements residing between the source of calcium (blood) and the most highly mineralized substance in the body (enamel) regulate the flow of that divalent cation into the enamel matrix. Enamel formation is a complex two stage process involving secretion and maturation. In this presentation the cellular, subcellular and molecular parameters leading to enamel mineralization are reviewed and a model is presented which proposes a role for calbindin-D9K and calbindin-D28K in that process.

CELLULAR COMPONENTS

It has been established that the source of calcium for enamel mineralization is not from the capillaries in the pulp, i.e., the calcium does not traverse the odontoblast and dentin layers to reach the enamel (9). Therefore, calcium entering enamel must traverse the cells of the enamel organ situated between the capillaries and the future site of enamel. Prior to mineralization, inner enamel epithelial cells differentiate to become ameloblasts, the most intensly studied cells in attempts to understand enamel mineralization. Those studies in both unerupted molar teeth and continuously erupting rodent incisors have led to an understanding of the cell morphology involved. Pre-secretory ameloblasts (first formed) induce the adjacent mesenchymal tissue to form odontoblasts. After odontoblasts secrete the initial dentin, pre-secretory ameloblasts (now more

NATO ASI Series, Vol. H 48
Calcium Transport and
Intracellular Calcium Homeostasis
Edited by D. Pansu and F. Bronner

mature) secrete the first enamel (aprismatic enamel). Pre-secretory ameloblasts further differentiate to secretory ameloblasts which are characterized by their "picket fence" appearing distal ends or Tomes processes (Fig. 1).

The enamel matrix is secreted from the Tomes processes in a highly structured fashion but is only partially mineralized (30% inorganic). The framework for prismatic enamel, characteristic of mature enamel, is established. Secretory ameloblasts form the full thickness of enamel then differentiate to become maturation ameloblasts. These cells are responsible for removal of much of the organic component of the enamel matrix and introduction of the bulk of the inorganic component, i.e., composition changes from 30% to 96% inorganic. Maturation ameloblasts exist either as smooth-ended or ruffle-ended ameloblasts in well defined bands changing their morphology between the two types (8). The full significance of the modulation with regard to mineralization is yet to be clearly understood. All stages of ameloblast maturation can be observed (in sequential order) in histological sections of unerupted molars or continuously erupting incisors. A feature of these systems which make them useful for the study of mineralization is that under physiological conditions, there is no concurrent resorption to complicate the analysis as with bone mineralization (3).

SUBCELLULAR COMPONENTS (PRE-SECRETORY & SECRETORY STAGES)

In cells prior to enamel secretion there is no calbindin (13,14). Late pre-secretory ameloblasts, in the area where the aprismatic enamel is noted, exhibit the first presence of calbindin-D28K in some cells (13). As the ameloblasts mature and develop Tomes processes, i.e., become secretory ameloblasts, they all exhibit calbindin-D28K throughout their cytoplasm. Secretory ameloblasts are also known to have Ca-ATPase uniformly distributed over the entire plasma membrane (10). Tight junctions, which are impermeable to La and horseradish peroxidase, are present at the distal ends (12) and that

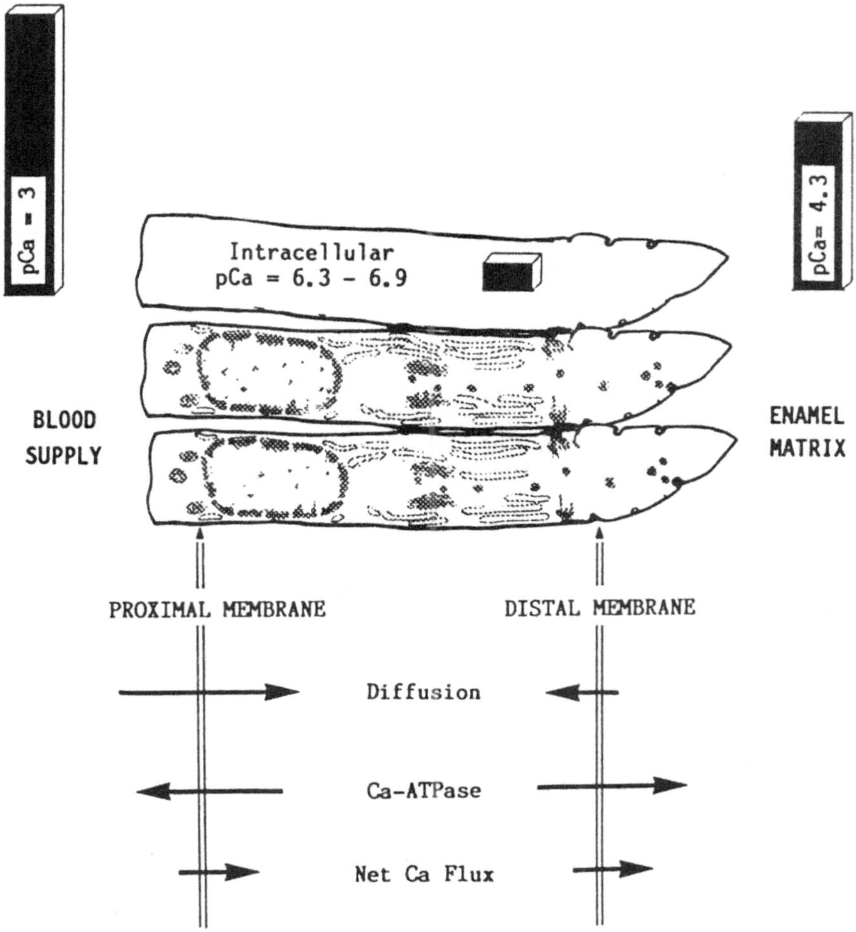

Figure 1. Ca transport parameters in secretory ameloblasts. Ca ion activities (top), morphology (center) and Ca fluxes (bottom). Also present, but not illustrated: membrane Ca-ATPase and cytoplasmic calbindin-D28K.

effectively separates the extracellular compartment from the developing enamel matrix compartment. Because of the relatively soft nature of the enamel associated with secretory ameloblasts (30 % inorganic), Aoba and Moreno (1) were able to determine the ionic composition of the matrix fluid from porcine teeth (Ca ion activity = 5×10^{-5} M). Total serum calcium was measured (2.9 mM), and if one assumes 50-60% bound, the Ca ion activity (10^{-3} M) in direct association with the extracelluar compartment is significantly higher than that in the developing enamel matrix (Fig. 1, upper half). That finding is in accord with the presence of tight junctions at the distal ends of secretory ameloblasts which would effectively prevent extracellular calcium from reaching the enamel matrix. Thus, calcium destined for enamel matrix must find an intracellular route where the calcium activity is assumed to be similar to that determined for other cell types, 1.2 - 4.9 x 10^{-7} M (2,4,16).

CALCIUM TRANSPORT BY SECRETORY AMELOBLASTS

Bawden (3) utilized the previously mentioned conditions to propose a model for calcium transport by secretory ameloblasts. In that model, the inward diffusional leak of calcium was greater at the proximal membrane, because of the steeper concentration gradient, than it was at the distal membrane. Those fluxes are indicated by the solid arrow lengths in the bottom of Figure 1. If one then subtracts the outwardly directed calcium flux due to Ca-ATPase distributed evenly in all parts of the membrane, a net inward flux results at the proximal membrane and a net outward flux is seen at the distal membrane (Fig. 1). In Bawden's model he assumes a transcellular flux of ionic Ca free of any association with calbindin-D28K, although the protein is known to be present. That assumption is apparently based on the slower rate of enamel mineralization in the secretory phase compared to the later maturation phase.

There are at least two observations, however, which support the involvement of calbindin-D28K in the transcellular transport of calcium in secretory ameloblasts. First, calbindin-D28K is initially noted in ameloblasts at the time of first enamel secretion (13). The protein then remains in all subsequent stages of enamel mineralization. Secondly, Feher et al. (6) have recently demonstrated (in vitro) that the diffusion coefficient for calcium increased linearly with increasing concentrations of calbindin-D9K. Calbindin-D28K possesses the same type of high affinity calcium-binding domains as does calbindin-D9K and earlier studies (5) with the larger protein suggest it would act in the same manner. Equations derived from the Feher system (6) illustrate how the presence of 0.15 mM calbindin-D9K (calculated enterocyte level from other studies) would increase the diffusive flux by a factor of nine. In that example intracellular free calcium is not significantly changed while the enhancement is accomplished by increasing the total intracellular calcium (now bound to calbindin).

Therefore, it is proposed that calcium transport by secretory ameloblasts is accomplished by: 1) the differential in inward directed calcium leaks down their respective concentration gradients at the proximal and distal poles, 2) the activity of membrane Ca-ATPase and 3) the increased diffusive flux of transcellular calcium provided by the presence of intracellular calbindin-D28K. The factors function to keep intracellular calcium within acceptable levels and at the same time allow a net flux from blood to enamel.

SUBCELLULAR COMPONENTS (MATURATION STAGE)

We have demonstrated both calbindin-D28K and calbindin-D9K in the cytoplasmic compartment of maturation ameloblasts (13,14). Although immunohistochemistry (at the LM level) is not a quantitative technique, the staining intensities noted for the two proteins in maturation ameloblasts would suggest that significant amounts of both are present. Although there are some conflicting reports in the literature, Takano and Akai (11) have been able to demonstrate the presence of a Quercetin sensitive Ca-ATPase activity in association with the ruffled border membrane (distal membrane), but not the proximal membrane of maturation ameloblasts.

Tight junctions exist between adjacent maturation ameloblasts with the location different in the two types of cells (close to distal pole in ruffle ended as diagramed in Fig 2, and close to proximal pole in smooth ended cells). There are conflicting reports, but, it has been shown that tight junctions effectively separate the extracellular compartment from the maturing enamel compartment (7). Since there are no reported direct measurements of calcium concentration in the maturing enamel matrix the same parameters which existed for secretory ameloblasts regarding electrolyte concentrations are assumed to exist for maturation ameloblasts (Fig. 2).

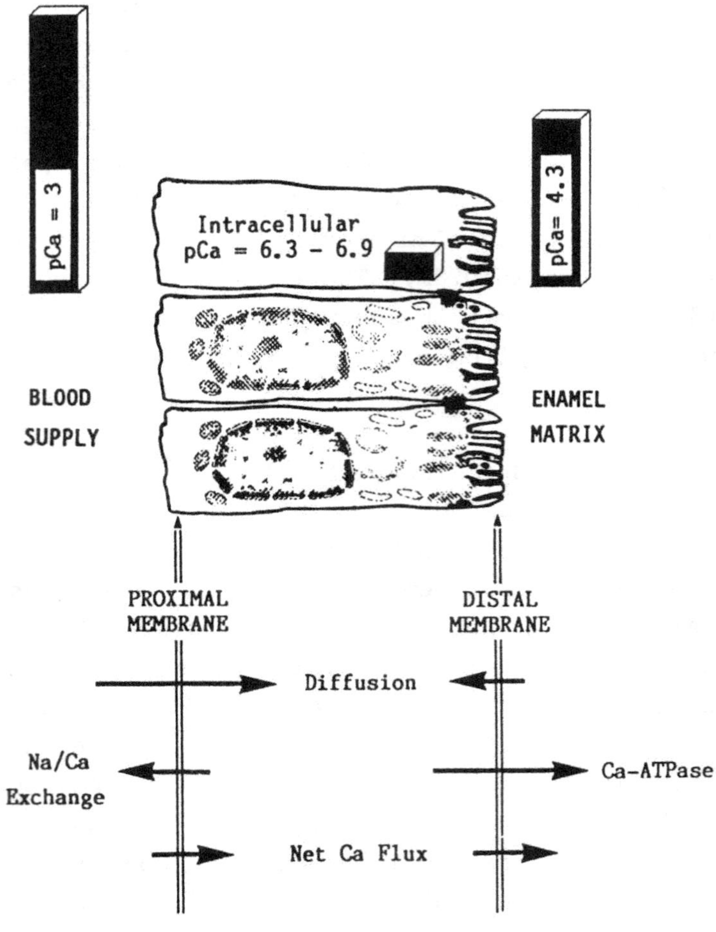

Figure 2. Ca transport parameters in maturation ameloblasts. Ca ion activities (top), morphology (center) and Ca fluxes (bottom). Also present, but not illustrated: both cytoplasmic calbindin-D9K and calbindin-D28K, and membrane Ca-ATPase in distal membrane only. The lengths of the solid arrows at the bottom of the figure are proportional to and directional for the individual calcium fluxes across the proximal and distal ameloblast membranes.

CALCIUM TRANSPORT BY MATURATION AMELOBLASTS

The same inward gradients of calcium diffusion exist for maturation ameloblasts as for secretory cells (Fig. 2), i.e., a larger gradient at the proximal membrane compared to the distal membrane. With Ca-ATPase only at the distal membrane, the efficiency of calcium removal from intracellular sites would be greater distally than proximally where the less efficient Na/Ca exchange (based on renal and intestinal studies) would presumably be the mechanism (15). With those ionic conditions, a larger transcellular calcium gradient could exist, thus requiring a more efficient intracellular movement of the ion in the direction of the enamel matrix. Thus, the proposed model for calcium transport by maturation ameloblasts is like that described for secretory ameloblasts except that with both calbindin-D9K and calbindin-D28K being present (13,14) there is an increase in the available efficiency for enhancing diffusability of intracellular calcium. That increased capability may be reflected in the fact that maturation ameloblasts are responsible for providing 66% of the inorganic component of enamel, whereas, the secretory ameoblasts were responsible for only 30% of the inorganic component of enamel.

SUMMARY

Model systems, based on Bawden's model (3), are presented for calcium transport from blood to enamel. This scheme builds on Bawden's proposal by including both calbindin-D9K and calbindin-D28K and suggests that the proteins function by increasing the diffusive calcium flux in ameloblasts.

ACKNOWLEDGMENTS

The technical and manuscript preparation help of J. Taylor, R. Spears and K. Hamilton is gratefully acknowledged. This study was supported by NIH grant DE07916.

REFERENCES

1. AOBA, T., and E.C. MORENO. The enamel fluid in the early secretory stage of porcine amelogenesis: chemical composition and saturation with respect to enamel mineral. Calcif. Tiss. Int. 41:86-94, 1987.
2. BARAN, D.T., and M.L. MILNE. 1,25 Dihydroxyvitamin D increases hepatocyte cytosolic calcium levels. J. Clin. Invest. 77:1622-1626, 1986.
3. BAWDEN, J.W. Calcium transport during mineralization. Anat. Rec. 224:226-233, 1989.
4. CHANDRA, S., D. GROSS, Y.C. LING, and G.H. MORRISON. Quantitative imaging of free and total intracellular Ca in cultured cells. Proc. Natl. Acad. Sci. USA 86:1870-1874, 1989.
5. FEHER, J.J. Facilitated Ca diffusion by intestinal calcium binding protein. Am. J. Physiol. 244:C303-C307, 1983.
6. FEHER, J.J., C.S. FULLMER, and G.K. FRITZSCH. Comparison of the enhanced steady-state diffusion of calcium by calbindin-D9K and calmodulin: possible importance in intestinal calcium absorption. Cell Ca. 10:189-203, 1989.
7. JOSEPHSEN, K. Lanthanum tracer study on permeability of ameloblast junctional complexes in maturation zone of rat incisor enamel organ. In: Tooth Enamel IV, ed. by R.W. Fearnhead, and S. Suga. Amsterdam: Elsevier Science Publ., 1984, p. 251-255.
8. JOSEPHSEN, K., and O. FEJERSKOV. Ameloblast modulation in the maturation zone of the rat incisor enamel organ. a light and electron microscopic study. J. Anat. 124:45-70, 1977.
9. REITH, E.J., and V.E. COTTY. Autoradiographic studies on calcification of enamel. Arch. Oral Biol. 7:365-372, 1962.
10. SASAKI, T., and P.R. GARANT. Ultracytochemical demonstration of ATP-dependent calcium pump in ameloblasts of rat incisor enamel organ. Calcif. Tiss. Int. 39:86-96, 1986.
11. TAKANO, Y., and M. AKAI. Demonstration of Ca-ATPase in the maturation ameloblast of rat incisor after vascular perfusion. J. Electron. Microsc. 36:196-203, 1987.
12. TAKANO, Y., H. OZAWA, and M.A. CRENSHAW. The mechanism of calcium and phosphate transport to the enamel. In: Mechanisms of Tooth Enamel Formation, ed. by S. Suga. Tokyo: Quintessence Publ., 1983, p. 49-64.
13. TAYLOR, A.N. Tooth formation and the 28,000-dalton vitamin D dependent calcium binding protein: an immunocytochemical study. J. Histochem. Cytochem. 32:159-164, 1984.
14. TAYLOR, A.N., W.A. GLEASON, and G.L. LANKFORD. Rat intestinal vitamin D dependent calcium binding protein: immunocytochemical localization in incisor ameloblasts. J. Dent. Res. 63:94-97, 1984.
15. VAN OS, C.H. Transcellular calcium transport in intestinal and renal epithelial cells. Biochim. Biophys. Acta 906:195-222, 1987.
16. WICKHAM, N.W.R., G.M. VERCELLOTTI, C.F. MOLDOW, M.R. VISSER, and H.S. JACOB. Measurement of intracellular calcium concentration in intact monolayers of human endothelial cells. J. Lab Clin. Med. 112:157-167, 1988.

SECTION VIII

DEFECTS OF CALCIUM TRANSPORT

INTESTINAL CALCIUM TRANSPORT AND BONE MINERALIZATION IN THE SPONTANEOUSLY DIABETIC BB RAT

R. Bouillon, J. Verhaeghe and M. Thomasset*

Laboratorium voor Experimentele Geneeskunde en Endocrinologie, Onderwijs & Navorsing, 3000 Leuven, Belgium; *Inserm U120, Le Vésinet, France

Diabetes mellitus is associated with a large number of abnormalities in calcium homeostasis: urinary calcium excretion in both human and experimental diabetes is increased; the duodenal calcium absorption is usually decreased in streptozotocin-induced diabetic rats and the duodenal calbindin concentration is similarly decreased (Table 1). The osteoblast function also seems to be impaired as suggested by decreased bone formation and possibly decreased bone mass in both experimental and human insulin-dependent diabetes mellitus (1). Moreover, the fetus or neonate of diabetic mothers have an increased risk for bone and calcium abnormalities, such as the caudal regression syndrome and neonatal hypocalcemia. The pathogenesis of these abnormalities is not well known. We, therefore, used the spontaneously diabetic BB rat, a model of human autoimmune diabetes, to study the effect of insulin deficiency on vitamin D and calcium homeostasis.

TABLE 1 : SYNOPTIC VIEW ON DUODENAL CALCIUM ABSORPTION AND CALBINDIN CONCENTRATION IN DIABETIC RATS

	Type of diabetes	Diabetes duration	Calcium absorption	Duodenal CaBP-9k
Schneider (9)	alloxan	6 d	decreased	-
Schneider (10)	streptozotocin	5 d	-	decreased
Schneider (11)	streptozotocin	4 d	decreased	-
Charles (3)	streptozotocin	5 d	decreased	normal
Hough (4)	streptozotocin	48 d	increased	-
Wood (18)	streptozotocin	9 d	decreased	-
Verhaeghe (13)	BB rat	24 d	decreased	-
Nyomba (6)	BB rat	24 d	decreased	decreased
Verhaeghe (17)	BB rat	12 w	decreased	decreased

NATO ASI Series, Vol. H 48
Calcium Transport and
Intracellular Calcium Homeostasis
Edited by D. Pansu and F. Bronner

The BB rat is a strain of Wistar rats, initially detected at the Bio-Breeding laboratories (Ottawa, Canada) which develops an autoimmune type of diabetes. As in human type 1 diabetes, an "insulitis" is followed by a complete destruction of the endocrine β-cells at or after puberty (70-110 d), and is characterized by severe insulinopenia, hyperglycemia, polyuria, weight loss and ultimately keto-acidosis and death unless insulin treatment is started. The genetic mechanism involves the rat equivalent of the major histocompatibility complex but environmental factors (e.g. diet) also influence the frequency of diabetes (5).

Materials and Methods

The BB rat colony in Leuven is derived from the original Ottawa colony and is further inbred since 1983 (BB/pfd or BB/Leuven) and is kept in standard laboratory conditions as described previously (13).

Serum 1,25-$(OH)_2D_3$, osteocalcin and the vitamin D-binding protein (DBP) were measured by RIA or RID (2, 16). The vitamin D-dependent calbindins were also measured by RIA (12) in the duodenal mucosa and kidney. Moreover, duodenal calcium absorption (7) and a 3-day calcium balance study was performed in some animals.

Results

Serum 1,25-$(OH)_2D_3$ concentrations were markedly decreased in all diabetic animals but so were the concentrations of its transport protein, DBP. Therefore, the physiologically active or free concentration of 1,25-$(OH)_2D_3$ was actually increased in the short-term diabetic animals (diabetes duration of 4 weeks) and slightly in the long-term diabetic animals (diabetes duration of 12 weeks) (Fig. 1).

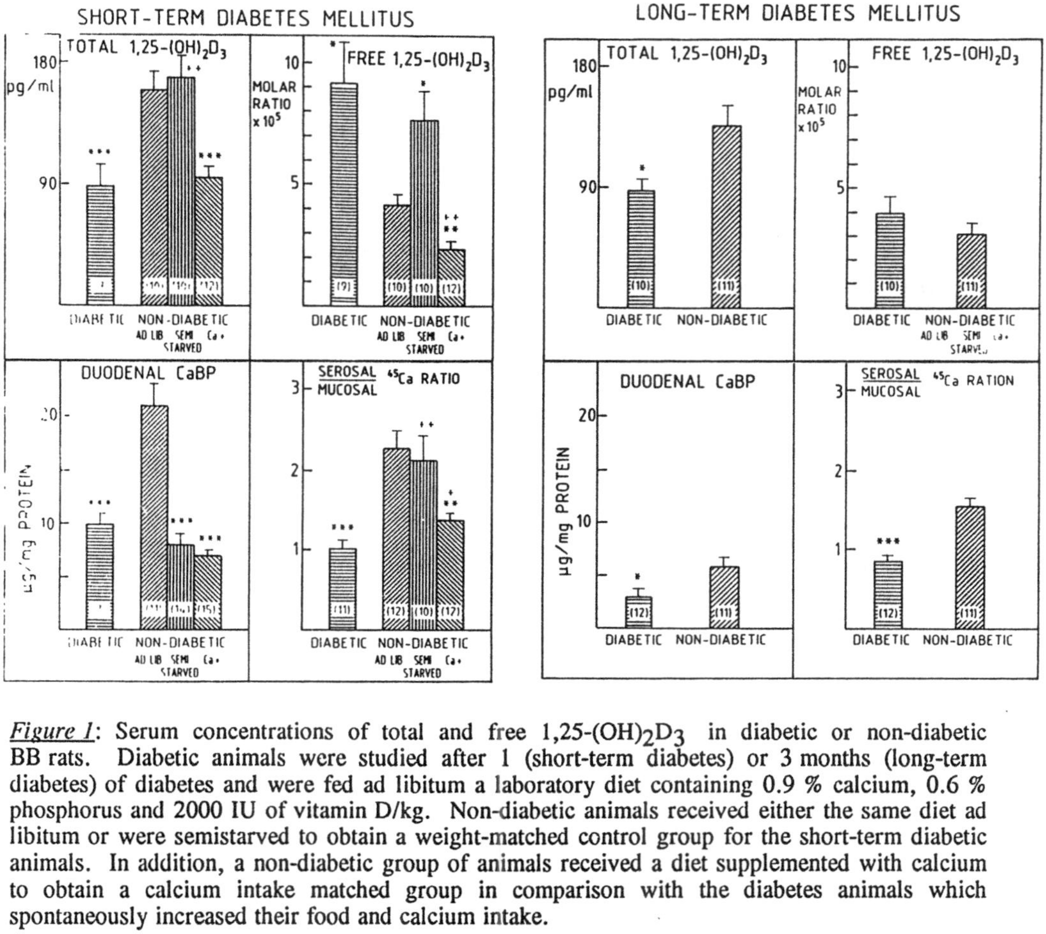

Figure 1: Serum concentrations of total and free 1,25-$(OH)_2D_3$ in diabetic or non-diabetic BB rats. Diabetic animals were studied after 1 (short-term diabetes) or 3 months (long-term diabetes) of diabetes and were fed ad libitum a laboratory diet containing 0.9 % calcium, 0.6 % phosphorus and 2000 IU of vitamin D/kg. Non-diabetic animals received either the same diet ad libitum or were semistarved to obtain a weight-matched control group for the short-term diabetic animals. In addition, a non-diabetic group of animals received a diet supplemented with calcium to obtain a calcium intake matched group in comparison with the diabetes animals which spontaneously increased their food and calcium intake.

The duodenal concentrations of CaBP-9k significantly decreased in all diabetic animals and so was the active duodenal calcium absorption. The serosal/mucosal ^{45}Ca ratio was, indeed, not different from one, indicating a total loss of active duodenal calcium absorption in diabetic animals. Moreover, the duodenal mucosa contained but half of the normal concentration of the vitamin D receptor, as measured by binding studies and Scatchard plot analysis. In non-diabetic rats, semistarved to obtain a final body weight similar to that of diabetic animals, the duodenal calbindin concentration was also reduced but the duodenal calcium absorption remained normal. In non-diabetic rats, given a calcium supplement so that their calcium intake was similar to that of (polyphagic) diabetic rats, a decrease in both intestinal calbindin concentration and duodenal calcium absorption was observed (Fig. 1) but the active calcium absorption, although decreased, remained positive. During a 3-day calcium balance study in short-term diabetic animals a normal net calcium balance was observed despite an increased urinary calcium excretion and decreased active duodenal calcium absorption. This can only be explained by their increased calcium intake and, consequently, an increased passive intestinal calcium diffusion (6, 15).

The urinary calcium excretion was always higher in diabetic animals than in controls irrespective of diabetes duration, calcium intake or pregnancy (Table 2). When given a low calcium diet, however, a substantial reduction of calciuria was observed but the calciuria was still at least fourfold higher than in non-diabetic animals.

TABLE 2 : URINARY CALCIUM EXCRETION (mg/d) IN DIABETIC AND NON-DIABETIC BB RATS

	Diabetic	Non-diabetic
Short-term diabetes (1 m)		
Non-pregnant (≥ 7 w) - female:		
0.9 % Ca diet ad libitum	6.1 ± 0.7***	0.46 ± 0.05
0.2 % Ca diet ad libitum	1.3 ± 0.2**	0.3 ± 0.1
Pregnant:		
0.9 % Ca diet ad libitum	9.4 ± 1.3***	0.7 ± 0.1
0.2 % Ca diet ad libitum	2.8 ± 0.7**	0.3 ± 0.04
Long-term diabetes (12 w) - male		
0.9 % Ca diet ad libitum	5.6 ± 0.5***	0.5 ± 0.04

** $p < 0.01$; *** $p < 0.001$

The calcium transfer across the osteoblast for new bone formation also dramatically decreased. Indeed, the osteoblast surface, osteoid surface and bone mineral apposition rate (as calculated from double calcein labelling in the tibial metaphysis and the lumbar vertebrae) were severely impaired in both short-term and long-term diabetic BB rats (Table 3). This abrupt decrease in bone formation was severe (≤ 20 % of normal) and would explain the gradual occurrence of osteoporosis. After 3-4 weeks of diabetes, the relative (trabecular and cortical) bone volume in tibial metaphysis/lumbar vertebrae remained normal, as well as several parameters of bone strength; after 12 weeks of diabetes, however, both the relative trabecular bone volume and bone strength were significantly reduced, indicating a true osteoporosis (17).

TABLE 3 : BONE MINERAL APPOSITION RATE (μm/day) AND SERUM OSTEOCALCIN (ng/ml) IN DIABETIC AND NON-DIABETIC BB RATS

	Bone mineral apposition rate (μm/day)		Serum osteocalcin (ng/ml)	
	Diabetic	Non-diabetic	Diabetic	Non-diabetic
Short-term diabetes (male)	1.0 ± 0.4*	5.6 ± 0.6	24 ± 2*	108 ± 10
Non-pregnant (female):				
0.9 % calcium diet			40 ± 3*	68 ± 6
0.2 % calcium diet	-	-	54 ± 13*	133 ± 7
Pregnant (day 21):				
0.9 % calcium diet	-	-	40 ± 4*	129 ± 11
0.2 % calcium diet	-	-	54 ± 7*	285 ± 27
Long-term diabetes (male):				
0.9 % calcium diet	0.3 ± 0.1*	2.2 ± 0.3	23 ± 3*	62 ± 4

* $p < 0.001$; mean ± SE

The net transport of <u>calcium across the placenta</u> of diabetic rats was also impaired as can be indirectly derived from the fetal calcium content and the ossification of fetal bone. Indeed, the number of ossification centra was decreased in fetuses of diabetic rats and their total calcium content was decreased both in absolute amount and after correction of relative fetal size (14). Interestingly, placental calbindin concentrations were lower in diabetic rats than in non-diabetic rats (530 ± 30 vs 670 ± 20 ng/mg protein). Moreover, osteocalcin concentrations were also decreased (Fig. 2). This suggests a primary osteoblast dysfunction with secondary decreased fetal calcium requirements.

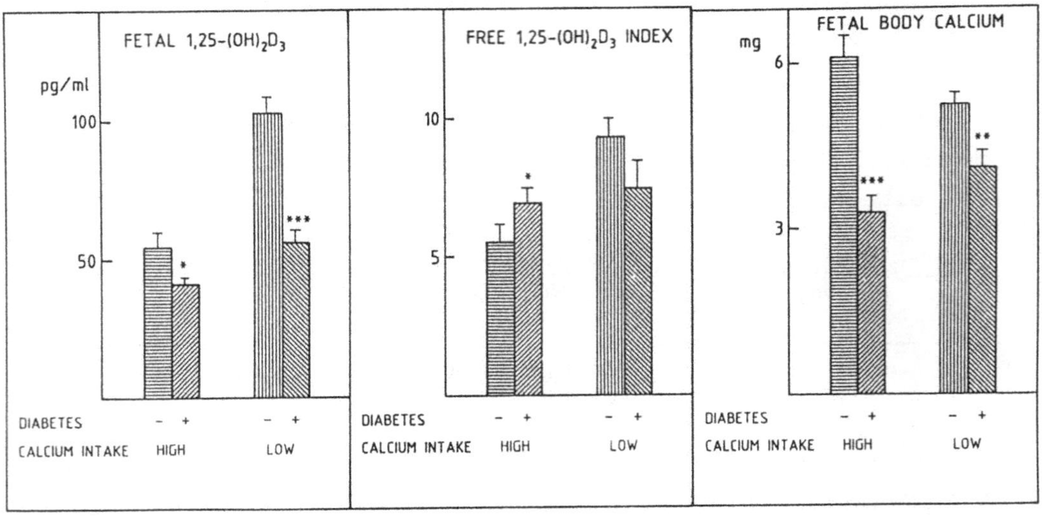

Figure 2: Serum concentrations of 1,25-$(OH)_2D_3$ and osteocalcin and total body calcium content of fetuses from diabetic and non-diabetic BB rats.

Discussion

In diabetic BB rats, there are marked alterations in the calcium transfer across the classical vitamin D target tissues. First, the duodenal active calcium absorption is impaired or even abolished but this is completely compensated for (and probably caused) by the increased calcium intake and passive calcium diffusion. Second, the urinary calcium excretion is increased severalfold. This cannot simply be explained by increased glomerular filtration or a slight increase of the serum concentration of diffusible calcium; thus, the renal reabsorption of calcium must also be impaired.

Third, the mineralization of bone is decreased by decreased osteoblast number or function. Fourth, the calcium transport across the placenta of diabetic rats is impaired as the total body calcium of the fetuses is significantly decreased. Finally, the calcium excretion in the milk of lactating diabetic rats is also decreased but this is largely due to their decreased milk production (13). Although the total amount of calcium crossing all critical membranes (duodenum, renal tubuli, osteoblasts, placenta and mammary gland) is clearly decreased it is doubtful whether this is due to a defect in cellular calcium transport. A different experimental approach is necessary to address this question. The described phenomena could also be explained as adaptations to decreased calcium demands or other abnormalities: increased calcium intake and low duodenal CaBP levels due to abnormalities in vitamin D metabolism and action, decreased bone formation of adult rats and decreased bone maturation of the fetus, and decreased milk production in lactating diabetic rats.

In conclusion: untreated or poorly insulin-treated animals have a markedly decreased concentration of vitamin D-dependent proteins (CaBP-9k and osteocalcin) despite increased free concentration of 1,25-$(OH)_2D_3$. This suggests a state of vitamin D resistance and can be partially explained by a decreased vitamin D receptor concentration. The decreased concentration of the vitamin D-dependent proteins corresponds to a decreased physiological function as both active duodenal calcium absorption and bone formation are similarly impaired. All abnormalities can be reversed by a continuous insulin infusion but injections or an infusion of 1,25-$(OH)_2D_3$ to normalize its total concentration resulted in normalization of duodenal calbindin concentration and an increase in active duodenal calcium absorption but was unable to increase the serum osteocalcin concentration. The different target tissues of diabetic rats are not equally resistant to the vitamin D action.

References

1. Auwerx J, Dequeker J, Bouillon R, Geusens P, Nijs J (1988) Mineral metabolism and bone mass at peripheral and axial skeleton in diabetes mellitus. Diabetes 37:8-12
2. Bouillon R, De Moor P, Baggiolini EG, Uskokovic MR (1980) A radioimmunoassay for 1,25-dihydroxycholecalciferol. Clin Chem 26:562-567
3. Charles MA, Tirunaguru P, Zolock DT, Morissey RL (1981) Duodenal calcium transport and calcium binding protein levels in experimental diabetes mellitus. Miner Electrolyte Metab 5:15-22
4. Hough S, Avioli AV (1984) Alterations of bone and mineral metabolism in diabetes. In: Recent advances in diabetes, pp. 223-229. Nattrass M, Santiago JV (eds). Edinburgh: Churchill Livingstone
5. Like AA (1985) Spontaneous diabetes in animals. In: Volk B, Arquilla ER. The diabetic pancreas. 2nd ed.: 385-413
6. Nyomba BL, Verhaeghe J, Thomasset M, Lissens W, Bouillon R (1989) Bone mineral homeostasis in spontaneously diabetic BB rats. I. Abnormal vitamin D metabolism and impaired active intestinal calcium absorption. Endocrinology 124:565-572
7. Pansu D, Bellaton C, Roche C, Bronner F (1983) Duodenal and ileal calcium absorption in the rat and effects of vitamin D. Am J Physiol 244:G659-G700
8. Schedl HP, Heath H, Wenger J (1978) Serum calcitonin and parathyroid hormone in experimental diabetes: effects of insulin treatment. Endocrinology 103:1368-1373
9. Schneider LE, Schedl HP (1972) Diabetes and intestinal calcium absorption in the rat. Am J Physiol 223:1319-1323
10. Schneider LE, Wilson HD, Schedl HP (1974) Effects of alloxan diabetes on duodenal calcium-binding protein in the rat. Am J Physiol 227:832-838
11. Schneider LE, Omdahl J, Schedl HP (1976) Effects of vitamin D and its metabolites on calcium transport in the diabetic rat. Endocrinology 99:793-799
12. Thomasset M, Parkes CO, Cuisinier-Gleizes P (1982) Rat calcium-binding proteins: distribution, development, and vitamin D dependence. Am J Physiol 243:E483-E488
13. Verhaeghe J (1988) Calcium metabolism during reproduction in the diabetic BB rat. Thesis K.U. Leuven, pp. 1-154
14. Verhaeghe J, Thomasset M, Bréhier A, Van Assche FA, Bouillon R (1988) $1{,}25(OH)_2D_3$ and Ca-binding protein in fetal rats: relationship to the maternal vitamin D status. Am J Physiol 254:E505-E512
15. Verhaeghe J, Suiker AMH, Nyomba BL, Visser WJ, Einhorn TA, Dequeker J, Bouillon R (1989) Bone mineral homeostasis in spontaneously diabetic BB rats. II. Impaired bone turnover and decreased osteocalcin synthesis. Endocrinology 124:573-582
16. Verhaeghe J, Van Herck E, Van Bree R, Van Assche FA, Bouillon R (1989) Osteocalcin during the reproductive period in normal and diabetic rats. J Endocr 120:143-151
17. Verhaeghe J, Van Herck E, Visser WJ, Suiker AMH, Thomasset M, Einhorn TA, Faierman E, Bouillon R (1990) Bone and mineral metabolism in BB rats with long-term diabetes. Diabetes, in press
18. Wood RJ, Allen LH, Bronner F (1984) Regulation of calcium metabolism in streptozotocin-induced diabetes. Am J Physiol 247:R120-R123

CALCIUM TRANSPORT IN THE STREPTOZOTOCIN DIABETIC RAT: STUDIES WITH BRUSH BORDER MEMBRANE VESICLES

H. P. Schedl, K. Christensen, W. Ronnenberg
Medical Service, Veterans Administration
Medical Center and Department of Medicine
University of Iowa
Iowa City, Iowa 52242
U.S.A.

Background

Streptozotocin (Sz) diabetes in the rat is an experimental animal model for examination of pathophysiology of diabetes mellitus. When Sz-diabetes is induced in the young, rapidly growing rat, abnormalities of calcium homeostasis are to be anticipated. Sz-diabetes leads to growth arrest and hyperphagia, i.e. the major pathway of calcium utilization, bone growth, is blocked, in the setting of excessive intake of dietary calcium. Although growth arrest characterizes the organism as a whole, the alimentary tract, particularly the small intestine, grows (Jervis and Levin, 1966; Schedl and Wilson, 1971a, 1971b; Schedl et al, 1982). During the first 2-3 days after Sz injection, food intake is decreased. Subsequently, with restoration of food intake and development of hyperphagia, the alimentary tract grows. Crypt cell proliferation rate of jejunum and ileum in the Sz-diabetic rat is double that of the control (Miller, et al, 1977). This increased rate of cell division is present in Sz-diabetics pair-fed with controls, and is minimally increased by the hyperphagia of the ad lib fed diabetic, although mucosal mass of the entire small intestine is significantly greater in the ad lib as compared with the pair-fed Sz-diabetic (Schedl, et al, 1982).

Transport Studies

Interpretation of results of intestinal transport studies is dependent upon mucosal growth status of the Sz-diabetic as compared with the control. Early in the course of Sz-diabetes, i.e. approximately 4-8 days after Sz-injection, size of small intestine of the diabetic is comparable to that of

NATO ASI Series, Vol. H 48
Calcium Transport and
Intracellular Calcium Homeostasis
Edited by D. Pansu and F. Bronner

control. At 12 days post Sz-injection, increased mucosal growth is well established in Sz-diabetes, and the effect of the increased growth on transport must be considered.

For example, the role of the mucosal mass of the entire small intestine as the determinant of *in vivo* transport rate has been examined for amino acid absorption (Lal and Schedl, 1974). Early (5 days) after induction of diabetes, transport of L-leucine and L-lysine is increased per unit weight of mucosa, as well as per unit length of intestine. At 12 days, transport per intestine remains increased because of the greater mucosal growth, but transport per unit of mucosa is identical in control and diabetic. Thus, both absorption per unit of mucosa (transport specific activity) and absorption per intestine (transport per unit length) must be considered in interpreting results of transport studies. In fact, since in long-term Sz-diabetes mucosal weight becomes twice as great and the small intestine becomes over 1/3 longer (Schedl and Wilson, 1971a and 1971b), mucosal mass must ultimately be a major regulatory factor in transport. The impact on total body homeostasis of a direct measurement of transport specific activity must be interpreted in terms of mucosal mass, i.e. for this paper, uptake per unit brush border protein must be related to total mucosa.

Calcium transport in the diabetic rat has been studied *in vivo* by *in situ* luminal perfusion of intestinal segments and *in vitro*, with the everted duodenal sac. Both of these techniques examine the net result of the multiple steps in transport across the enterocyte, i.e. the entry process at the brush border, transit across the cell, and extrusion at the basolateral membrane. In the case of *in vivo* perfusion, transported calcium enters the circulation directly from the interstitial fluid. Hence, *in vivo* transport mimics the mechanisms expected to be encountered under physiologic conditions. In contrast, with the everted sac, calcium must traverse the submucosa, muscularis mucosae, muscularis propria and serosa for the transport process to be measured. The everted duodenal sac performs uphill transport of calcium across all of these barriers. Mucosal growth in diabetes is associated with increased full-wall thickness of the duodenum including mucosal hyperplasia and hypertrophy of the muscle layers. Hence, *in vitro* transport must be studied early in diabetes, where resistance of the tissue to transport is similar in the diabetic and control.

The first evidence for an abnormality of intestinal calcium transport in chemically induced diabetes was the finding of decreased duodenal calcium transport in the alloxan diabetic rat at 5, 7, or 10 days post-injection (Schneider and Schedl, 1972). *In vivo*, *in situ* perfusion of the most proximal 10 cm segment of small intestine showed net calcium absorption per unit dry weight of mucosa in the diabetic to be half that of the control because of decreased lumen-to-plasma flux. In the most distal 15 cm segment of small intestine, transport was the same in diabetic and control.

Subsequent studies in the Sz-diabetic rat confirmed these results (Walker and Schedl, 1979). Using the same segment sites, absorption was measured at luminal calcium concentrations below (0.8 mM) and above (3.4 mM) plasma ionized calcium. Absorption was measured at 4 and 11 days post injection to define the effect of time course of diabetes and intestinal growth on calcium transport. Calcium transport specific activity of duodenum was lower in diabetics at both concentrations, and the depression in specific activity was greater at the longer duration (11 days) of diabetes. Calcium transport specific activity of ileum was depressed only at 11 days and only for 3.4 mM calcium. Transport per unit segment length was decreased in duodenum despite mucosal growth, but was not decreased in ileum in diabetes. The decreased net calcium absorption was secondary to decreased lumen-to-plasma flux.

In vivo in situ segment perfusion was also used to compare effects of Sz-diabetes on strontium and magnesium absorption with the response of calcium absorption. In accord with the concept that strontium is the divalent cation that shares most closely the physical, chemical and biological properties of calcium, transport response of strontium to diabetes mimicked that of calcium (Miller and Schedl, 1976a). Duodenal transport specific activity for strontium in the Sz-diabetic rat was half that of the control at 5 and 12 days post injection. Despite 60% more mucosa in the duodenum of the diabetic at 12 days, transport per cm was also lower in the diabetic. Effects of diabetes on ileal strontium transport were minimal, localizing the defect in strontium as well as calcium transport in Sz-diabetes to the proximal segment.

Unlike strontium, transport specific activity and transport per unit length for magnesium were the same in Sz-diabetics and controls at 5 and 12 days (Miller and Schedl, 1976b). This is consistent with the differing biology of magnesium as compared with calcium.

Mechanism of Transport Depression

The first evidence for a possible mechanism for the depression of calcium transport was the finding of decreased circulating 1,25-dihydroxycholecalciferol [$1,25-(OH)_2D_3$] in the Sz-diabetic rat (Schneider, et al, 1977a). The decreased circulating $1,25-(OH)_2D_3$ was restored to normal by insulin treatment. *In vivo* duodenal calcium transport also increased in response to treatment with insulin (Schneider, 1977b).

The possibility that decreased $1,25-(OH)_2D_3$ was the determinant of the depressed duodenal calcium transport in the diabetic was further substantiated by the response of duodenal calcium transport to treatment with vitamin D_3 metabolites. These studies were performed *in vitro* with the everted duodenal sac in the 5 d diabetic rat. The duodenum is the only site in the small intestine that transports calcium against a concentration gradient *in vitro*. Active calcium transport is decreased in the everted sac from the diabetic as compared with the control rat.

Treatment of the Sz-diabetic rat with extract of *Solanum malacoxylon*, a plant containing a glycoside of $1,25-(OH)_2D_3$, increased calcium transport in diabetes (Schneider et al, 1975). Treatment of the diabetic rat with vitamin D_3 or 25-hydroxycholecalciferol ($25-OH-D_3$) had no effect on duodenal calcium transport (Schneider, et al, 1976). In contrast, treatment with $1,25-(OH)_2D_3$ or 1α-hydroxycholecalciferol ($1\alpha-OH-D_3$) restored calcium transport in the Sz-diabetic rat to control levels. The response to 1α-OH-D_3 and lack of response to 25-OH-D_3 is consistent with depression of 1α-hydroxylation of 25-OH-D_3. These findings with respect to vitamin D_3 metabolism have been confirmed by _in vivo_ studies of vitamin D_3 metabolism (Spencer, et al, 1980).

Depressed _in vivo_ calcium transport in diabetes has been demonstrated in duodenum (Schneider and Schedl, 1972), cecum and colon (Petith and Schedl, 1979). The decreased calcium transport in the proximal small intestine is associated with decreased vitamin-D-dependent calcium binding protein (CaBP, Calbindin-D 9K) (Schneider et al, 1974).

Brush Border

The site or sites in the enterocyte determining the decreased transcellular transport in Sz-diabetes have not been identified. Brush border function is altered in Sz-diabetes: disaccharidase activities are increased in mucosa (Younoszai and Schedl, 1972) and brush border (Olsen and Rogers, 1971) and mucosal hexose uptake is enhanced (Flores and Schedl, 1968). Down-regulation of brush border calcium uptake, such as that associated with vitamin D deficiency, would explain decreased transcellular calcium transport in Sz-diabetes. However, a difference in chemical and physical properties of the brush border determined by its differing chemical composition could also decrease calcium uptake.

Calcium uptake by brush border membrane vesicles (BBMV) comprises saturable and non-saturable components of nearly equal magnitude (Schedl et al, 1985). The saturable process shows typical characteristics of carrier mediation: a) competitive inhibition by strontium; b) non-competitive inhibition by magnesium; and c) countertransport of medium calcium by intravesicular strontium (but not by intravesicular magnesium) (Wilson, et al, 1989). Characteristics of non-saturable calcium uptake are consistent with at least two components of nearly equal magnitude: a) a component inhibited by both strontium and magnesium in the uptake medium, probably representing electrostatic binding of calcium to the BBMV membrane; b) an uninhibited component representing diffusion of calcium into the BBMV down its concentration gradient. Saturable uptake is probably the biologically significant component of calcium transport.

The key requirement in calcium uptake studies is to define V_{max} for the saturable process, i.e. separate this component of transmembrane movement from other uptake processes. We prepared BBMV from proximal small intestine of the control and 5 day Sz-diabetic rat (Table) (Schedl, et al, 1989a). BBMV were characterized a) morphologically by electron microscopy; b) biochemically by measurement of enzyme activity; and c) in terms of capability for active transport by sodium-driven glucose uptake. Electron micrographs showed closed vesicles. Sucrase activity was greater in mucosal homogenate and BBMV of the diabetic than the control (Table), but degree of enrichment (i.e. 26 fold) was the same in both. At initial rate (3 sec), Na^+ driven D(+)-glucose uptake by BBMV from the diabetic was greater than from control (4.94 vs 1.60; after subtracting uptake in absence of Na^+ of 0.60; all data nmole/mg protein per 3 sec at 2 mM D-glucose conc).

TABLE

Sucrase, units‡		Ca uptake, nmole per 3 sec x 10^3‡		K_T,
Homogenate	BBMV	V_{max}	K_D**	mM
C 0.077	1.95	211	324	0.051
D 0.102*	2.74*	129*	199*	0.038

‡per mg protein. *Differ, $p<0.05$. **at 1 mM (Ca^{++}).

Both saturable and non-saturable calcium uptake were depressed in BBMV from the diabetic as compared to the control (Table). The 40% decrease in both components of calcium uptake could explain the depressed transcellular transport found by _in vivo_ and _in vitro_ studies, since mucosal mass is the same in control and diabetic at 5 days. Further studies using BBMV from 12 day diabetics will be necessay to compare the effects of mucosal growth on BBMV uptake of calcium with _in vivo_ transport response (Walker and Schedl, 1979).

It is recognized that V_{max} values (Table) compare carrier specific activity in diabetic and control, not carrier number. In BBMV from the normal rat, V_{max} for calcium can be altered by changing membrane lipid without changing carrier number (Schedl, et al, 1989b). Therefore, the relationship of findings with BBMV to decreased 1,25-$(OH)_2D_3$ in diabetes is not established by these studies. Diabetes alters not only membrane lipids but also content of brush border integral membrane proteins such as sucrase-isomaltase (Table) and the hexose carrier. Even if the decreased V_{max} in the diabetic rat is secondary to decreased 1,25-$(OH)_2D$, the decreased K_D could be the consequence of altered membrane composition. Further experiments, including comparing treatment with 1,25-$(OH)_2D_3$ and insulin are necessary.

LITERATURE REFERENCES

Flores P, Schedl HP (1968) Intestinal transport of 3-0-methyl-D-glucose in the normal and alloxan-diabetic rat. Am J Physiol 214:725-729

Jervis EL, Levin RJ (1966) Anatomic adaptation of the alimentary tract of the rat to the hyperphagia of chronic alloxan diabetes. Nature 210:391-393

Lal D, Schedl HP (1974) Intestinal adaptation in diabetes: amino acid absorption. Am J Physiol 227:827-831

Miller DL, Hanson W, Schedl HP, Osborne JW (1977) Proliferation rate and transit time of mucosal cells in small intestine of the diabetic rat. Gastroenterol 73:1326-1332

Miller DL, Schedl HP (1976a) Effects of experimental diabetes on intestinal strontium absorption in the rat. Proc Soc Exp Biol Med 152:589-592

Miller DL, Schedl HP (1976b) Effects of diabetes on intestinal magnesium absorption in the rat. Am J Physiol 231:1039-1042

Olsen WA, Rogers L (1971) Jejunal sucrase activity in diabetic rats. J Lab Clin Med 77:838-842

Petith MM, Schedl HP (1979) Effects of diabetes on cecal and colonic calcium transport in the rat. Am J Physiol 235:E699-E702

Schedl H, Christensen K, Ronnenberg W (1989a) Intestinal calcium (Ca) transport in the streptozotocin (Sz) diabetic rat: studies with brush border membrane vesicles (BBMV). FASEB J 3:A1155

Schedl HP, Wilson HD (1971a) Effects of diabetes on intestinal growth in the rat. J Exp Zool 176:487-495

Schedl HP, Wilson HD (1971b) Effects of diabetes on intestinal growth and hexose transport in the rat. Am J Physiol 220:1739-1745

Schedl HP, Wilson HD, Mathur S, Murthy S, Field J (1989b) Effects of phospholipid or cholesterol enrichment of rat intestinal brush border membrane on membrane order and transport of calcium. Metabolism 38:1164-1169

Schedl HP, Wilson HD, Miller T, Ogesen B (1985) Calcium uptake by intestinal brush border membrane vesicles, comparison with in vivo calcium transport. J Clin Invest 76:1871-1878

Schedl HP, Wilson HD, Ramaswamy K, Lichtenberger L (1982) Gastrin and growth of alimentary tract in the streptozotocin diabetic rat. Am J Physiol 242:G460-G463

Schneider LE, Nowosielski LM, Schedl HP (1977b) Insulin-treatment of diabetic rats: effects on duodenal calcium absorption. Endocrinol 100:67-73

Schneider LE, Omdahl JH, Schedl HP (1976) Effects of vitamin D and its metabolites on calcium transport in the diabetic rat. Endocrinol 99:793-799

Schneider LE, Schedl HP (1972) Diabetes and intestinal calcium absorption in the rat. Am J Physiol 223:1319-1323

Schneider LE, Schedl HP, McCain T, Haussler MR (1977a) Experimental diabetes reduces circulating 1,25-Dihydroxy vitamin D in the rat. Science 196:1452-1454

Schneider LE, Wasserman RH, Schedl HP (1975) Depressed duodenal calcium absorption in the diabetic rat: restoration by Solanum Malacoxylon. Endocrinol 97:649-653

Schneider LE, Wilson HD, Schedl HP (1974) Effects of alloxan diabetes on duodenal calcium binding protein in the rat. Am J Physiol 227:832-838

Spencer EM, Khalil M, Tobiassen O (1980) Experimental diabetes in the rat causes an insulin-reversible decrease in renal 25-hydroxyvitamin D_3-1α-hydroxylase activity. Endocrinol 107:300-305

Walker B, Schedl HP (1979) Small intestinal calcium absorption in the rat with experimental diabetes. Proc Soc Exp Biol Med 161:149-152

Wilson HD, Schedl HP, Christensen K (1989) Calcium uptake by brush-border membrane vesicles from the rat: effects of strontium and magnesium. Am J Physiol 257:F446-F453

Younoszai MK, Schedl HP (1972) Effects of diabetes on intestinal disaccharidase activity. J Lab Clin Med 79:579-586

EPITHELIAL CALCIUM TRANSPORT IN THE SPONTANEOUSLY HYPERTENSIVE RAT (SHR)

Uta Hennessen, Bernard Lacour and Tilman B. DRÜEKE
Unité 90 de l'INSERM and Département de Néphrologie
Hôpital Necker
161 rue de Sèvres, 75743 Paris Cedex 15, France

During recent years, numerous clinical and experimental studies have provided evidence in favor of a relationship between disturbed calcium metabolism and arterial hypertension. The possibility has not been definitely excluded that the calcium abnormalities were only an epiphenomenon. However, the increasing number of reports in animals and in man showing defects of calcium handling at the level of the entire organism, of organs, cells and subcellular structures point to a fundamental role of these abnormalities in the pathogenesis of hypertension (McCarron 1985, Young et al 1988).

Increased blood pressure values in rats have been found to be associated with elevated intracellular levels of free cytosolic Ca^{2+} in platelets (Resnick 1989, Brushi et al 1985) and lymphocytes (Brushi et al 1985). However, other investigators found no relation between blood pressure and free cytosolic Ca^{2+} in platelets (Zimlichman et al 1986). In kidney tubular epithelium of spontaneously hypertensive rats, free cytosolic Ca^{2+} has been found normal (Llibre 1988) or even decreased (Jacobs et al 1990). Moreover, perturbations of calcium handling by the vascular smooth muscle cell in the hypertensive state are probably more complex in that free cytosolic Ca^{2+} may be normal in the resting state (Bukoski in press, Stern et al 1984), but cellular Ca^{2+} transients upon stimulation abnormally high (Bukoski in press). Since the Ca^{2+} ion plays a major role in vascular smooth muscle contraction and relaxation, it is tempting to speculate that the numerous abnormalities of Ca^{2+} concentration in body fluids and cells and of the handling of the Ca^{2+} ion by cellular and subcellular membranes in experimental animals with arterial hypertension are reflective of some fundamental perturbations associated with high blood pressure.

Alterations of calcium metabolism have been particularly well described in the spontaneously hypertensive rat of the Okamoto-Aoki strain (SHR) and in its normotensive control, the Wistar-Kyoto rat (WKY). They include disturbances of free Ca^{2+} concentration in the extracellular and intracellular space, abnormalities of the hormonal regulation of cellular Ca^{2+} handling and hence profound perturbations of Ca^{2+} metabolism at the level of the intestine, the skeleton and the kidney (McCarron 1985, Young et al 1988).

Disturbed intestinal transport of calcium. Studies examining intestinal Ca^{2+} transport in the SHR have led to conflicting results. This could in part be due to the different methods used for exploration, but also in part to factors such as age, gender, strain, diet and duration of fast.

NATO ASI Series, Vol. H 48
Calcium Transport and
Intracellular Calcium Homeostasis
Edited by D. Pansu and F. Bronner

Everted gut sac and early in vivo studies. In studies using the everted gut sac technique, duodenal Ca^{2+} transport has been found unchanged in the 5- and 12-week-old male SHR (Stern et al 1984), increased in the 12-week-old male SHR (Toraason et al 1981) and diminished in 5- and 12-week-old male SHR (Llibre et al 1988, Jacobs et al 1990). Early *in vivo* perfusion studies of Ca^{2+} absorption showed either a decrease (Schedl et al 1984 and 1986) or an increase (Toraason et al 1981) in the 12-week-old male SHR and an increase in the 50-week-old female SHR (Lau et al 1984). It was difficult to reconcile such discrepant results. Therefore, experimental models were required that allowed a more precise assessment of Ca^{2+} transport at the tissue or cellular level.

Ussing chamber. We and others have used the modified Ussing chamber technique which provides a direct measurement of active Ca^{2+} fluxes across isolated intestinal segments under electrically controlled, short-circuited conditions (Lucas et al 1986, McCarron et al 1985, Gafter et al 1986). We observed in the male SHR at 12-14 weeks of age a decrease in absorptive Ca^{2+} flux in the duodenum (directed from the mucosa to the serosa), with no change of the secretory Ca^{2+} flux (in the opposite direction). The sum of both fluxes resulted in a significant decrease in net Ca^{2+} absorption. Both the SHR and the WKY were able to increase net Ca^{2+} flux in response to a low (0.1%) calcium diet. However, the extent to which the SHR increased its absorptive Ca^{2+} flux was less marked than that achieved by the control rat. In the 20- to 24-week-old male SHR and the WKY of similar age, basal net active Ca^{2+} transport was near zero. In response to a low (0.1%) calcium diet, however, only the normotensive rat strain was capable of increasing absorptive Ca^{2+} flux. In the 12-14-week-old animals, it was interesting to note that the per cent enhancement of duodenal active Ca^{2+} transport was paralleled by a similar per cent increase in circulating calcitriol levels. Calcium transport has also been found less responsive to exogenous vitamin D in the SHR than in the WKY by Wilson et al. (Wilson et al 1988). Similar results of active Ca^{2+} transport have been obtained by others in the duodenum and colon in 24-week-old SHR and WKY using the Ussing chamber method (Gafter et al 1986).

In situ loop perfusion. Schedl et al. (1988) have recently examined SHR and WKY rats from different suppliers, using the in situ loop perfusion technique. They found that the SHR had always lower Ca^{2+} transport rates than the WKY, independent of the source from which the strains had been purchased. In contrast, circulating calcitriol levels were unpredictive and unrelated to Ca^{2+} fluxes.

Isolated enterocyte. We have also used the isolated duodenal enterocyte model in order to gain further insight into the mechanism of the intestinal Ca^{2+} transport defect at the cellular level. With this technique and using a simplified model of analysis which postulated a linear Ca^{2+} uptake during the first 15 min, the 12-week-old male SHR had a decrease in Ca^{2+} influx rate into duodenal epithelial cells (Fig. 1), compared with that of the WKY (Lucas et al 1988). Moreover, Ca^{2+} efflux rate constant (i.e., the flux directed from the cell interior to

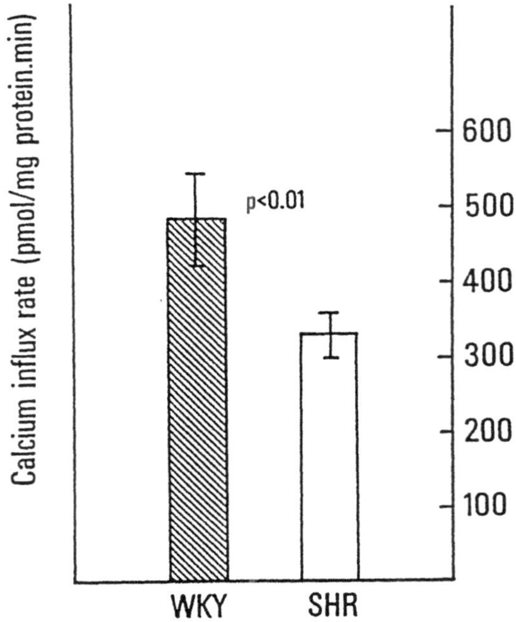

Fig. 1. Duodenal enterocyte Ca^{2+} uptake (from 1 to 15 min) in WKY and SHR expressed as pmol Ca^{2+} per mg cell protein and per min, at a incubation medium Ca^{2+} concentration of 1.0 mM. The rate of Ca^{2+} uptake was significantly reduced in the SHR (n=14) compared with the WKY (n=14).

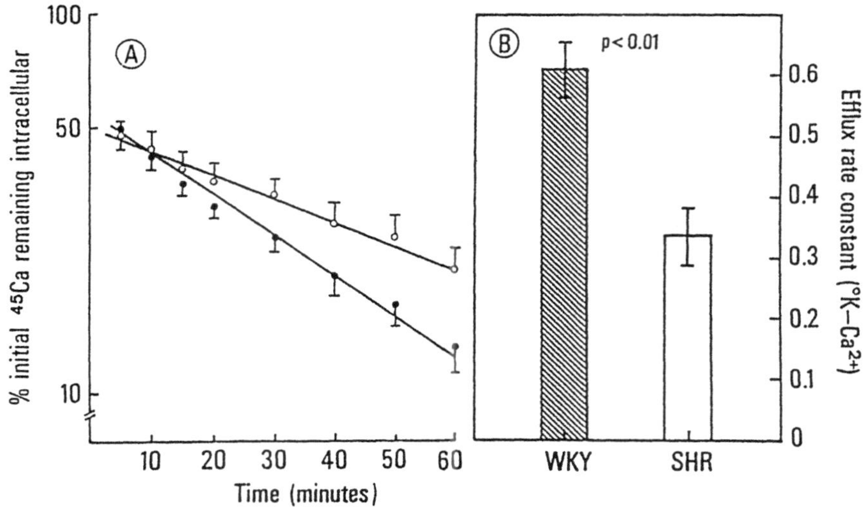

Fig. 2. (A) Efflux rate constant of Ca^{2+} from isolated duodenal enterocytes : Percentage tracer remaining in intracellular fraction as a function of time. Solid circles (●) represent WKY, open circles (o) SHR. Note that the scale of the ordiante is logarithmic. (B) Efflux rate constant for calcium was reduced in SHR compared with WKY enterocytes (n=11 pairs).

outside) was also significantly decreased in the male SHR of same age (Fig. 2). In a subsequent study with isolated enterocytes from SHR and WKY rats (Roullet C et al in press), we applied the three-compartment, double exponential model of analysis of Ca^{2+} fluxes described by Borle (Borle 1987). These experiments showed that intracellular Ca^{2+} flux was reduced in the SHR compared with the WKY for both young (12-week-old) and mature (24-week-old) animals on a normal (1%) Ca diet. On a high (2%) Ca diet, this strain difference disappeared in the young rats. In mature SHR, the high Ca diet paradoxically stimulated cellular Ca^{2+} flux whereas it lowered it in mature WKY resulting in a similar flux in both strains. Young SHR had a lower intracellular Ca pool than WKY, and this defect was corrected by the high Ca diet. Interestingly, diet-induced changes of plasma calcitriol levels in the SHR did not parallel the observed changes of intestinal Ca^{2+} fluxes. Thus in this study duodenal enterocytes from SHR appeared to have an intrinsic alteration of Ca^{2+} transport that could be corrected in part by a higher Ca diet. Taken together, these experiments indicate that transcellular Ca^{2+} fluxes of the adolescent SHR are profoundly disturbed. It is however still unknown whether the primary defect is located at the site of the brush border membrane or the baso-lateral membrane. Experiments are presently in progress in our laboratory which are designed to explore Ca^{2+} uptake velocity by isolated brush borders from SHR and WKY enterocytes.

Balance studies. Finally, balance studies have been performed to determine whether net intestinal Ca^{2+} absorption was disturbed at the level of the intact animal. These studies have also led to conflicting results. Increased absorption of Ca^{2+} was found in the very young (3.5-week-old) male SHR (Lau et al 1986), compared with the WKY. No difference was observed in male 6- to 10-week-old male rats (Stern et al 1984, Bindels et al 1987) whereas it was decreased in 10-week-old male SHR in another balance study (Lau et al 1984). The female SHR appears to hyperabsorb Ca^{2+} at all ages studied, namely 3.5, 25, and 50 weeks (Lau et al 1984 and 1986) but apparently the female SHR has as yet only been studied by one single group of workers.

Net Ca^{2+} retention or loss in the whole animal depends mainly on the amount of Ca^{2+} absorbed or excreted at the level of the intestine and much less on that excreted by the kidney. However, measurements of fecal Ca excretion are relatively imprecise, in contrast to determinations of urinary Ca concentrations. Therefore, balance studies are difficult to interpret and relatively long fecal collection periods are needed in order to avoid too large fluctuations of Ca concentration measurements in feces. This may explain several discrepant results published in the literature, the Ca balance being increased, similar or decreased in the SHR compared with the WKY. We have performed a balance study with 10-day fecal collection periods in 5-week-old rats and repeated the balance in the same animals at the age of 12 weeks (Bourgouin et al, in press). Whereas Ca balance was slightly (though not significantly) greater in the very young SHR, it was smaller at the age of 12 weeks, compared with the WKY of same age and raised under identical conditions (Table 1).

Altered renal tubular transport of calcium. At least two recent studies were devoted to the measurement of intracellular free Ca^{2+} concentration in the epithelium of isolated proximal renal tubules of the SHR and WKY (Llibre et al 1988, Jacobs et al 1990). One group of authors found no difference between both strains (Llibre et al 1988) whereas the other found even a decrease in the SHR (Jacobs et al 1990) but in the latter study, only young SHR and WKY rats (age 8 weeks and less) were studied. To the best of our knowledge, no studies of renal Ca^{2+} transport have as yet been performed in the SHR at the level of the isolated kidney tubule, the epithelial cell or subcellular structures. Studies of renal Ca^{2+} handling have all been performed in the intact animal. Urinary excretion of Ca^{2+} has been variously reported as decreased (Lau et al 1986, Bindels et al 1987), not different (Lau et al 1984, Schedl et al 1988, McCarron et al 1981), or increased in the very young SHR (Lau et al 1984, Hsu et al 1984 and 1986). It appears to be constantly increased in the mature SHR. This finding is intriguing, in face of a lower filtered Ca^{2+} load and a state of secondary hyperparathyroidism (Merke et al 1989). Hypercalciuria persists on a low Ca^{2+} diet in mature male SHR (Grady et al 1983) providing evidence for a renal Ca^{2+} leak. However, others found a decrease of urinary Ca^{2+} excretion after fasting in 8- to 14-week-old male SHR and 23-week-old female SHR whereas it was increased in the postabsorptive state (Lau et al 1984). They therefore hypothesized that the hypercalciuria of the SHR was of the absorptive, not of the renal type. It is probable that a renal type of hypercalciuria would also have been found by this group if older rats had been studied.

In order to localize the precise site or sites of the underlying transport abnormality, studies of tubular epithelial Ca^{2+} transport will have to be carried out at the cellular and subcellular level in the SHR. We have recently examined the epithelium of the proximal renal tubule of male SHR and WKY rats by electron microscopy. In the epithelium of the 12-week old SHR, a shortening and even a patchy loss of microvilli was observed which did not exist in the WKY (Drüeke et al, in press). Moreover, irregular spaces were found between the basal aspects of cells and the underlying basement membrane. Similar ultrastructural abnormalities were also apparent in the absorptive epithelium of the duodenum of SHR, but not of WKY rats.

Disturbances of Ca regulating hormones. Disturbances of Ca^{2+} transport may be either intrinsic or due to changes in the secretion and/or the cellular action of Ca^{2+} regulating hormones. Several endocrine abnormalities have been reported in the mature male SHR for the three major hormones, namely calcitriol, parathyroid hormone and calcitonin (McCarron 1985, Young et al 1988, Merke et al 1989, Bindels et al 1987). Thus decreased plasma calcitriol levels due to insufficient renal calcitriol production prevail in the male hypertensive strain at maturity. Such low circulating calcitriol levels are clearly inappropriate

Table 1 : Calcium balance (means±SEM) at age 4-5 and 13-14 weeks

	Mineral intake coefficient++ (mg/day)	Fecal excretion (mg/day)	Mineral absorption* (mg/day)	Percent absorption** (%)	Urinary excretion (mg/day)	Retained amount+ (mg/day)	Retention (%
Age 4-5 weeks							
WKY (n=6)	130±2.4	63.3±2.9	66.5±2.5	51.3±1.9	8.82±0.35	57.7±2.4	44.4±1.3
SHR (n=6)	129±2.1	61.3±3.4	67.7±3	52.5±2.4	5.31±0.14	62.3±2.6	48.4±2.1
P	NS	NS	NS	NS	< 0.02	NS	NS
Age 13-14 weeks							
WKY (n=5)	201±13	132±12	69.9±2.8	35.1±2.3	3±0.57	66.9±2.7	33.6±2.4
SHR (n=6)	207±2.3	149±4.8	57.5±4.8	27.8±2.2	4.3±0.64	53.2±4.6	25.7±2.2
P	NS	NS	NS	< 0.05	NS	< 0.05	< 0.05

* absorbed = absolute amount ingested minus absolute fecal amount excreted
** absorbed amount as per cent of amount ingested
\+ retained = absorbed minus excreted in urine
\++ retention of nutrient as per cent of amount ingested

since the adult male SHR has low serum Ca^{2+} and phosphorus concentrations and elevated serum immunoreactive PTH levels, together with a defect of skeletal mineralization rate and secondary hyperparathyroidism. Serum calcitriol increments are subnormal in response to PTH and phosphorus depletion, and the response to Ca-deficient diets has been either normal or low. Finally, serum calcitonin levels have been found to be inapproppriately increased.

Conclusion. The metabolism of Ca^{2+} is profoundly disturbed in the spontaneously hypertensive rat (SHR), compared with its normotensive control, the WKY. Numerous perturbations of Ca^{2+} handling by the hypertensive animal have been observed in balance studies as well as at the level of epithelial tissues, cells and subcellular structures. It is not yet known at present whether the initial anomaly in the SHR is an epithelial (and a non-epithelial) defect of Ca^{2+} handling by the cell or whether the perturbations of Ca^{2+} transport are secondary to some circulating non-hormonal or hormonal factor including calcitriol and parathyroid hormone. Altered cellular Ca^{2+} permeability and a compromised ability of the cell to remove or sequester Ca^{2+} may be relevant to the pathogenesis and maintenance of hypertension.

References

Bindels RJM, van den Brock LAM, Jongen MJM, Hackeng WHL, Löwick CW, van Os CH (1987) Increased plasma calcitonin levels in young spontaneously hypertensive rats : Role in disturbed phosphate homeostasis. Pflügers Arch 408 : 395-400

Borle A (1987) Kinetic analyses of calcium movements in cell cultures. III. Effects of calcium and parathyroid hormone in kidney cells. J Gen Physiol 55 : 163-186

Bourgouin P, Lucas P, Roullet C et al (in press) Developmental changes of Ca^{2+}, PO_4 and Calcitriol metabolism in spontaneously hypertensive rats. Am J Physiol

Brushi G, Brushi ME, Caroppo M, Orlandini G, Pavarani C, Cavatorta A (1984) Intracellular free [Ca^{2+}] in circulating lymphocytes of spontaneously hypertensive rats. Life Sci 35 : 535-542

Brushi G, Brushi ME, Caroppo M, Orlandini G, Spaggiari M, Cavatorta A (1985) Cytoplasmic free [Ca^{2+}] is increased in platelets of spontaneously hypertensive rats and essentail hypertensive patients. Clin Sci 68 : 179-184

Bukoski RD (to be published) Intracellular Ca^{2+} metabolism of isolated resistance arteries and cultured vascular myocytes of spontaneously hypertensive and Wistar Kyoto normotensive rats. J Hypertens

Drüeke T, Lucas PA, Nabarra B, Ben Nasr L, Hennessen U, Dang P, Thomasset M, Lacour B, Coudrier E, McCarron DA (to be published) Ultrastructural and functional abnormalities of intestinal and renal epithelium in the spontaneously hypertensive rat. Kidney Int

Gafter U, Kathpalia S, Zikos D, Lau K (1986) Ca fluxes across duodenum and colon of spontaneously hypertensive rats : Effect of $1,25(OH)_2D_3$. Am J Physiol 251 : F278-F282

Grady JR, Dorow J, McCarron DA (1983) Urinary calcium excretion and cAMP response of the spontaneously hypertensive rat to Ca^{2+} deprivation (abstract). Clin Res 31 : 330A

Hsu CH, Chen PS, Caldwell RM (1984) Renal phosphate excretion in spontaneously hypertensive and Wistar Kyoto rats. Kidney Int 25 : 789-795

Hsu CH, Chen PS, Smith DE, Yang CS (1986) Pathogenesis of hyper-calciuria in spontaneously hypertensive rats. Miner Electrolyte Metab 12 : 130-141

Jacobs WR, Ferrari C, Brazy PC, Mandel LJ (1990) Cytosolic free calcium regulation in renal tubules from spontaneously hypertensive rat. Am J Physiol 258 : F175-F182

Lau K, Chen S, Eby B (1984) Evidence for an intestinal mechanism in hypercalciuria of the spontaneously hypertensive rat. Am J Physiol 247 : E625-E633

Lau K, Langman CB, Gafter U, Dudeja PK, Brasitus TA (1986) Increased calcium absorption in hypertensive spontaneously hypertensive rat : role of serum 1,25-dihydroxyvitamin D3 levels and intestinal brush border membrane fluidity. J Clin Invest 78 : 1803-1090

Llibre J, LaPointe M, Battle DC (1988) Free cytosolic calcium in renal proximal tubules from the spontaneously hypertensive rat. Hypertension [Dallas] 12 : 399-404

Lucas PA, Brown RC, Drüeke T, Lacour B, Metz G, McCarron DA (1986) Abnormal vitamin D metabolism, intestinal calcium transport, and bone calcium status in the spontaneously hypertensive rat compared with its genetic control. J Clin Invest 78 : 221-227

Lucas PA, Roullet C, Ben Nasr L, Lacour B, Nabarra B, Dang P, McCarron DA, Drüeke T (1989) Decreased duodenal enterocyte calcium fluxes in the spontaneously hypertensive rat. Am J Hypertens 2 : 86-92

McCarron DA, Yung NN, Ugoretz BA, Krutzik S (1981) Disturbances of calcium metabolism in the spontaneously hypertensive rat : attenuation of hypertension by calcium supplementation. Hypertension [Dallas] 3 (suppl I) : I-162 - I-167

McCarron DA (1985) Is calcium more important than sodium in the pathogenesis of essential hypertension ? (review) Hypertension [Dallas] 7 : 607-627

McCarron DA, Lucas PA, Shneidman J, Lacour B, Drüeke T (1985) Blood pressure development of the spontaneously hypertensive rat following concurrent manipulations of dietary Ca^{2+} and Na^{+} : relation to intestinal Ca^{2+} fluxes. J Clin Invest 76 : 1147-1154

Merke J, Lucas PA, Szabo A, Cournot-Witmer G, Mall G, Bouillon R, Drüeke T, Mann J, Ritz E (1989) Hyperparathyroidism and abnormal calcitriol metabolism in the spontaneously hypertensive rat. Hypertension [Dallas] 13 : 233-242

Resnick LM (1989) Calcium metabolism in the pathophysiology and treatment of clinical hypertension. Am J Hypertens 2 : 179S-185S

Roullet C, Young E, Roullet JB, Lacour B, Drüeke T, McCarron DA (to be published) Calcium uptake by duodenal enterocytes isolated from young and mature SHR and WKY rats : influence of dietary calcium. Am J Physiol

Schedl HP, Miller DL, Paper JM, Horst RL, Wilson HD (1984) Calcium and sodium transport and vitamin D metabolism in the spontaneously hypertensive rat. J Clin Invest 73 : 980-986

Schedl HP, Miller DL, Horst RL, Wilson HD, Natarajan K, Conway T (1986) Intestinal calcium transport in the spontaneously hypertensive rat : response to calcium depletion. Am J Physiol 250 : G412-G419

Schedl HP, Wilson HD, Horst RL (1988) Calcium transport and vitamin D in three breeds of spontaneously hypertensive rats. Hypertension [Dallas] 12 : 310-316

Stern N, Lee DBN, Silis V, Beck FW, Deftos L, Manolagas SC, Sowers JR (1984) Effects of high calcium intake on blood pressure and calcium metabolism in young SHR. Hypertension [Dallas] 6 : 639-646

Toraason MA, Wright GL (1981) Transport of calcium by duodenum of spontaneously hypertensive rat. Am J Physiol 241 : G344-G347

Wilson HD, Schedl HP, Horst RL (1988) Calcium transport in the spontaneously hypertensive rat is responsive to vitamin D. Proc Soc Exp Biol Med 198 : 141-146

Young EW, Bukoski RD, McCarron DA (1988) Calcium metabolism in experimental hypertension. Proc Soc Exp Biol Med 187 : 123-141

Zimlichman R, Goldstein DS, Zimlichman S, Keiser HR (1986) Cytosolic free calcium in platelets of spontaneously hypertensive rats. Hypertension [Dallas] 4 : 283-287

INTERACTION BETWEEN CALCITRIOL AND THYROID HORMONE: EFFECTS ON INTESTINAL CALCIUM TRANSPORT AND BONE RESORPTION

H. S. Cross, E. Lehner*, N. Fratzl-Zelman*, K. Klaushofer * and M. Peterlik
Department of General and Experimental Pathology
University of Vienna Medical School,
Währingerstraße 13
A-1090 Vienna
Austria

Since disorders of calcium and inorganic phosphate (Pi) metabolism are frequently observed in diseases of the thyroid gland, it is evident that thyroid hormones must play an important role in mineral metabolism (for review, see Perry 1989). Triiodothyronine (T_3) can be considered a "calciotropic" hormone insofar as its direct actions on bone, kidney and intestine resemble those of 1,25-dihydroxycholecalciferol (calcitriol). Similar to the steroid hormone, T_3 has been shown to stimulate osteoclastic bone resorption in cultured rodent bone (Mundy and Raisz1979; Klaushofer et al. 1989), to enhance renal tubular absorption of Pi (Espinosa 1984) and to increase Na^+/Pi cotransport in cultured embryonic chick small intestine (Cross et al. 1986). In addition, a particular role of T_3 as a calcium-regulating hormone is emerging from observations that T_3, though it has no effect by itself on intestinal calcium absorption, nevertheless influences calcium uptake from the intestinal lumen considerably through potentiation of the genomic effect of calcitriol on the calcium absorptive mechanism (Cross and Peterlik 1988). In this study we extend this observation by showing that T_3 is able to modulate the stage-specific induction of calcium transport by calcitriol in embryonic small intestine. Thereby, the thyroid hormone increases the sensitivity of enterocytes to the action of the sterol. A similar pattern of interaction of the two hormones was seen when their combined effects on osteoclastic bone resorption were evaluated in cultured neonatal mouse calvaria. We therefore propose that thyroid hormones are involved in the homeostatic control of calcium metabolism by potentiating the effects of calcitriol in its classic target organs, intestine and bone.

---

*Ludwig Boltzmann Research Unit for Clinical and Experimental Osteology, 4th Medical Department, Hanusch Hospital, Heinrich Collin-Str. 30, A-1140 Vienna, Austria

NATO ASI Series, Vol. H 48
Calcium Transport and
Intracellular Calcium Homeostasis
Edited by D. Pansu and F. Bronner

MATERIALS AND METHODS

Organ Culture of Embryonic Chick Jejunum. Fertilized eggs were obtained from a local poultry farm and kept in an incubator with forced ventilation at 70 % relative humidity and 38° C for 15 - 20 days. The proximal segments of the jejunum were excised, slit open and cultured mucosa-side up on stainless steel grids in serum-free McCoy's 5A modified medium (Corradino 1973a). Culture time was 48 h. Synthetic calcitriol or T_3 were added to cultures in ethanolic solution so that the final ethanol concentration did not exceed 0.15 %. Ethanol only was added to control cultures.

Uptake measurements. Transport rates were evaluated from measurements of tissue uptake of radiocalcium as described earlier. Briefly, eight cultured intestines were incubated for 30 min in 5.0 ml of a "low sodium" mannitol buffer containing 0.25 mM Ca^{2+} (cf. Corradino 1973a). Radioisotope concentration was 0.5 mCi/l. Uptake was terminated by transferring the guts onto filter paper in a rapid filtration apparatus, where they were rinsed thoroughly under continuous suction with 3 x 10 ml ice-cold 0.9% NaCl . The guts were then blotted lightly on tissue paper, weighed. and processed for liquid scintillation counting.

Organ culture of bone. Calvariae were dissected from 4-6 day-old mice (strain HIM:OF1 Swiss, SPF, Institute for Experimental Animal Research of the University of Vienna, Himberg, Austria) and cultured as described elsewhere (Stern and Krieger 1983): Briefly, bones were immersed in 2.0 ml culture medium and incubated in rotating stoppered glass tubes, which had been gassed with a $O_2/N_2/CO_2$ (50/45/5 %) mixture. Culture medium was changed after 24 h, and fresh treatments were added. Total culture period was 72 h. The culture medium was prepared from Dulbecco's modified Eagle's medium (DMEM, without L-glutamine, M.A, Bioproducts, Walkersville, Md.) by addition of 1.4% L-glutamine, 15% heat-inactivated (56° C, 45 min) horse serum (Gibco-Europe), 10 U/ml heparin and 100 U/ml Na-penicillin G, and was sterilized by filtration through 0.22 μm membrane filters.

Measurement of bone resorption. Bone resorption was evaluated by determination of calcium release from cultured bones. Medium calcium concentrations were determined in 0.2 ml aliquots at 0, 24, 48, and 72 h by fluorescence titration with a Corning 940 calcium analyzer.

Materials. $^{45}CaCl_2$ was obtained from the Radiochemical Centre Amersham, England. calcitriol was a generous gift from Hoffmann-LaRoche, Basle, Switzerland. Triiodothyronine was donated by Sanabo, Vienna, Austria.

RESULTS

At day 20 of embryonic development, gut segments cultured for 48 h in serum-free medium in the presence of 10^{-10} - 10^{-7} M calcitriol exhibited increased uptake of calcium (Fig. 1). In accordance with previous results, T_3 (10^{-8} M) had no influence whatsoever on calcium uptake by cultured jejuna (Cross et al. 1986, Cross and Peterlik 1988). However, when the steroid and the thyroid hormone were both present in the culture medium, T_3 was able to further raise levels of calcitriol-induced calcium uptake (Fig. 1). Similarly, a potentiating effect of T_3 had also been observed on induction of Na+/Pi and Na+/D-glucose transport by calcitriol in cultured embryonic chick jejunum cultured on day 20 of embryonic development (Cross and Peterlik 1988; Cross et al. 1988). Studies utilizing brush-border membrane vesicles isolated from cultured embryonic chick jejunum confirmed the notion that the effect of the two hormones on tissue uptake reflects their interaction on activation of Na^+ gradient driven Pi and D-glucose transport across the brush-border membrane of enterocytes (Cross et al. 1990).

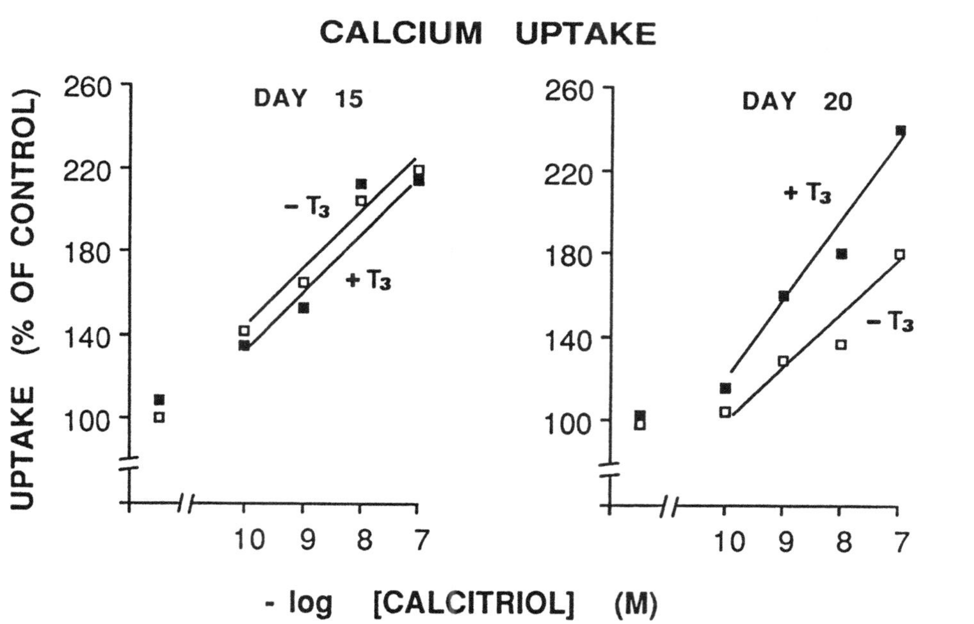

Figure 1: Influence of T_3 on the induction by calcitriol of calcium transport in embryonic chick small intestine cultured on day 15 (left panel) and day 20 (right panel) of embryonic development. Culture time 48 h. T_3 (■) concentration in culture medium: $1x10^{-7}$ M. Data are means from two separate experiments (n ≥ 12 guts).

We had shown previously that the expression of calcitriol actions in cultured embryonic small intestine depends very much on the degree of epithelial cell differentiation (Cross and Peterlik 1982; cf. also Cross et al 1986). Thus, the peak value of calcium uptake induced by calcitriol was observed on day 15, when the epithelium is comprised only of undifferentiated absorptive cells (Black and Moog 1977). The sterol-related increment then gradually declined until day 18-19, but tended to rise again until hatching on day 20. In contrast, in embryonic small intestine cultured between day 15-17, calcitriol was totally ineffective in inducing Na^+/Pi uptake. A significant stimulation of Na^+/Pi transport by calcitriol could be observed only on day 20. Similarly, D-glucose uptake in response to calcitriol increased from day 15 to 20 in parallel with advancing differentiation of the small intestinal epithelium (unpublished data).

Fig. 1 shows that at embryonic day 15, T_3 did not further increase the already high rate of calcitriol-mediated calcium transport. However, T_3 allowed the precocious induction of Na^+/Pi transport by calcitriol in day 17 embryonic jejunum (Cross and Peterlik 1988) and also enhanced low rates of calcitriol-related Na+/D-glucose transport in immature day 15 embryonic jejunum (unpublished data).

Since osteoclastic bone resorption can be stimulated by both calcitriol and T_3, we examined the possibility that an interaction of the steroid and the thyroid hormone similar to that observed in the intestine could be revealed in bone. Fig. 2 shows that calcitriol ($5x10^{-11}$ M) significantly stimulated release of calcium from neonatal mouse calvariae during a 72 h culture period. We had shown previously that the bone resorbing activity of T_3 becomes apparent in this system above 10^{-9} M (Klaushofer et al. 1989). Thus, in the experiment shown in Fig. 2, T_3 at $5x10^{-9}$ M had only a marginal effect by itself, but apparently potentiated the bone resorbing action of the sterol.

When we examined the influence of T_3 on the dose-response relation of the calcitriol effect on bone resorption, it became apparent, that the thyroid hormone enhanced the bone resorbing activity in particular of low sterol concentrations ($1x10^{-12}$ - $5x10^{-11}$ M !) (Fig. 3).

DISCUSSION

Since the effects of calcitriol on uptake of calcium as well as on Na^+-dependent transport of Pi and D-glucose result from receptor-mediated changes in the activity of genes coding for components of the respective transport systems (Corradino 1973b; Franceschi and

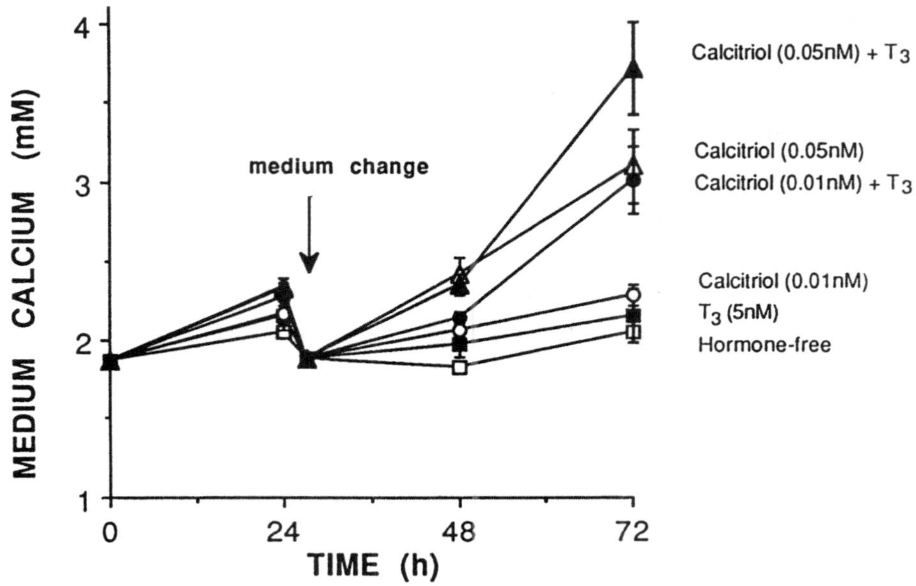

Figure 2: Effect of calcitriol and T_3 on resorption of cultured neonatal mouse calvariae. Data from a typical experiment are shown as means ± S.E.M. (n = 5 bones per time point and treatment). Arrow indicates change of culture medium (see Methods).

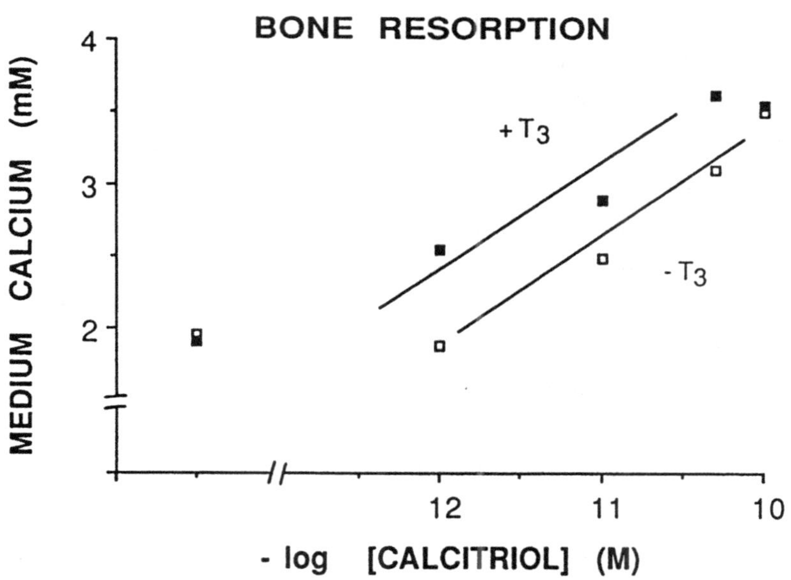

Figure 3: Influence of T_3 on calcitriol-induced resorption of cultured calvarial bone. Culture time 72 h. T_3 (■) concentration in medium was $5x10^{-9}$ M. Data are means from three separate experiments (n = 15 bones per data point).

DeLuca 1981; Peterlik 1978; Peterlik et al. 1981), we conclude that the potentiation of calcitriol actions by thyroid hormone probably occurs during the multi-step process of activation and control of gene expression. Although it cannot be excluded that T_3 increases the binding affinity of the vitamin D receptor or augments its synthesis, an interaction of the two hormones at the DNA level has to be considered a likely possibility. A logical basis for this assumption is provided by the high degree of sequence homology between the respective nuclear hormone receptors (Evans 1988). Steroid hormone-sensitive genes apparently contain clusters of overlapping "hormone response elements" in their 5' flanking region, and can be activated therefore in a cooperative fashion by receptor-mediated binding of one hormone molecule in combination with another member of the steroid/thyroid hormone family (Schüle et al. 1988). In this respect it must be noted that calbindin-9k, the calcium-binding protein, an integral component of the transcellular calcium transfer process (Bronner et al. 1986), is coded for by a calcitriol-sensitive gene which contains also a thyroid hormone responsive element in its 5' flanking region (Perret et al. 1989). Cooperativity between the two hormones may further enhance expression of the gene when its activation by calcitriol alone is restricted at a certain stage of epithelial differentiation (cf. Fig. 1). This assumption seems to be particularly valid for activation of genes coding for elements of the Na^+ gradient-driven transport mechanisms of Pi or D-glucose (Cross and Peterlik 1988; unpublished data).

It is interesting that the potentiating effect of thyroid hormone on calcitriol actions is not only limited to the small intestine but can also be observed in another target organ of the two hormones, namely in bone (Figs. 2, 3). It is however completely unknown at which step in the induction of osteoclastic bone resorption the two hormones could interact. Since both hormones play an important role in cell differentiation, recruitment of osteoclasts from monocyte-like precursors is likely to be influenced by the combination of steroid and thyroid hormones. In addition, T_3 as well as calcitriol are able to stimulate, at least to some extent, the endogenous synthesis of prostaglandins in bone (Hoffman et al. 1987; Klaushofer et al. 1989) and may thus indirectly stimulate recruitment and/or activation of osteoclasts. This could provide also a basis for the observed synergistic action of the two hormones on resorption of cultured calvarial bone.

In any case, through its potentiating effect on calcitriol actions, T_3 increases the sensitivity of bone and also of the small intestine to low sterol concentrations. It is therefore tempting to speculate that also *in vivo* the presence of thyroid hormone could be required for optimal calcitriol action in its classic target tissues. It is for example conceivable that induction of osteoclastic bone resorption by calcitriol, which is postulated to occur mainly at elevated sterol concentrations, may be facilitated by thyroid hormones even at normal circulating calcitriol levels. Furthermore, synergistic effects of T_3 and calcitriol could contribute to increased bone resorption associated with hyperthyroidism. This phenomenon might also explain the fact that in a number of hyperthyroid patients malabsorption of calcium from the small intestine is absent despite low circulating levels of calcitriol (for discussion, see Cross et al. 1986). Consequently, the permissive effect of T_3 on the genomic action of calcitriol implies an important role of thyroid hormones in the regulation of calcium and phosphate metabolism.

ACKNOWLEDGMENTS

These investigations were supported by Grant No. 73/86 and 61/85 of the Anton Dreher-Gedächtnisschenkung. The authors thank Mrs. Helga Gazda and Mrs. Renate Koreny for skillful technical assistance.

REFERENCES

Black BL, Moog F (1977) Goblet cells in embryonic intestine: accelerated differentiation in culture. Science 197:368-370

Bronner F, Pansu D, Stein WD (1986) An analysis of intestinal calcium transport across the intestine. Am J Physiol 250:G561-G569

Corradino RA (1973a) Embryonic chick intestine in organ culture: A unique system for the study of intestinal calcium absorptive mechanism. J Cell Biol 58: 64-78

Corradino RA (1973b) 1,25-Dihydroxycholecalciferol: Inhibition of action in organ-cultured intestine by actinomycin D and α-amanitin. Nature 243:41-43

Cross HS, Peterlik M (1982) Differential response of enterocytes to vitamin D during embryonic development: induction of intestinal inorganic phosphate, D-glucose and calcium uptake. Horm Metab Res 14: 649- 652

Cross HS, Peterlik M (1988) Calcium and inorganic phosphate transport in embryonic chick intestine: Triiodothyronine enhances the genomic action of 1,25-dihyroxy-cholecalciferol. J Nutrition 118: 1529-1534

Cross HS, Pölzleitner D, Peterlik M (1986) Intestinal phosphate and calcium absorption: joint regulation by thyroid hormones and 1,25-dihydroxyvitamin D_3. Acta Endocrinol (Copenh) 113: 96-103.

Cross HS, Debiec H Peterlik M (1988) Thyroid Hormone Regulation of Calcitriol-mediated Intestinal D-Glucose transport: Whole Tissue and Brush-border Membrane Vesicle Studies. Biol Chem Hoppe-Seyler 369:803

Cross HS, Debiec H, Peterlik M (1990) Thyroid hormone enhances the genomic action of calcitriol in the small intestine. In: Peterlik M, Bronner F, ed. Molecular and Cellular Regulation of Calcium and Phosphate Metabolism. New York: Alan R. Liss Inc.,

1990:163-180

Espinosa RE, Keller MJ, Yusufi ANK, Dousa TP (1984) Effect of thyroxine administration on phosphate transport across renal cortical brush border membrane. Am J Physiol 246:F133-F139

Evans RM (1988) The steroid and thyroid hormone receptor superfamily. Science 240:889-895.

Franceschi RT, DeLuca, HF (1981) The effect of inhibitors of protein and RNA synthesis on 1,25-dihydroxyvitamin D_3-dependent calcium uptake in cultured embryonic chick duodenum. J Biol Chem 256:3848-3852.

Hoffmann O, Klaushofer K, Czerwenka E, Koller K, Leis H-J, Gleispach H, Peterlik M (1987) Calcitriol-stimulated bone resorption in neonatal mouse calvaria: effects of indomethacin and gamma interferon. J Bone Min Res 2 (Suppl 1): 455

Klaushofer K, Hoffmann O, Gleispach H, Leis H-J, Czerwenka E, Koller K, Peterlik M (1989) Bone-Resorbing Activity of Thyroid Hormone Is related to Prostaglandin Production in Cultured Neonatal Mouse Calvaria. J Bone Min Res 4:305-312

Mundy GR, Raisz LG (1979) Thyrotoxicosis and calcium. Metabolism 2:285 -292

Perret C, Bréhier A, l'Horset F, Gouhier N, Thomasset M (1989) Analysis of rat calbindin-9k promotor, regulation by 1,25(OH)2D3. J Bone Min Res 4 (Suppl 1): 700

Perry III HM. Thyroid hormones and mineral metabolism. In: Peck WA, ed. Bone and Mineral Research/6, Amsterdam New York Oxford: Elsevier 1989, 113 - 137

Peterlik M (1978) Phosphate transport by embryonic chick duodenum: stimulation by vitamin D_3. Biochim Biophys Acta 514:164-171.

Peterlik M, Fuchs R, Cross, HS (1981) Stimulation of D-glucose transport: a novel effect of vitamin D on intestinal membrane transport. Biochim Biophys Acta 649: 138-142.

Schüle R, Müller M, Kaltschmidt C, Renkawitz R. (1988) Many Transcription Factors interact Synergistically with Steroid Receptors. Science 242:1418-1420

Stern PH, Krieger NS (1983) Comparison of fetal rat limb bones and neonatal mouse calvaria: effects of parathyroid hormone and 1,25-dihydroxyvitamin D_3. Calcif Tissue Int 35:172-176

LEAD POISONING AND CALCIUM TRANSPORT

T. J. B. Simons
Biomedical Sciences Division
King's College London,
Strand, London WC2R 2LS,
U. K.

The blood of normal human subjects contains 10 - 20 µg Pb/100 ml (equivalent to 0.5 - 1 µmol/l.). The blood levels are higher in Pb poisoning. Haematological and neurological symptoms appear when blood Pb is in the range 5 - 10 µmol/l. (Hernberg, 1980). Nearly all the Pb in the blood is in the red cell fraction: the percentage of Pb in serum rises from about 1% when the blood Pb content is 1 µmol/l. to 2% at 7 µmol/l. (Manton & Cook, 1984). The aim of the present research is to understand the factors that determine the distribution of Pb across the erythrocyte membrane, under physiological and pathological conditions, when the erythrocyte Pb content is in the range 1 - 20 µmol/l.cells, and the serum Pb concentration is in the range 5 - 200 nmol/l..

Earlier studies have shown that there are two main pathways for Pb^{2+} movement across the erythrocyte membrane (Fig. 1). When erythrocytes are suspended in Pb-buffered media containing 10^{-8} to 10^{-6} M Pb^{2+}, Pb^{2+} uptake is HCO_3^-- and Cl^--dependent, and DIDS-sensitive (Simons, 1986a). The anion exchanger is implicated in Pb uptake, and appears to catalyse an exchange of $PbCO_3Cl^-$ for Cl^- (Simons, 1986b). Binding by the erythrocyte membrane is negligible, compared with uptake into the cytoplasm (Simons, 1986a). Pb^{2+} efflux can occur passively, catalysed by the anion exchanger, but is also brought about by an ATP-dependent and vanadate-sensitive pathway. This pathway has been characterised in experiments with resealed ghosts (Simons, 1988). It was shown that Pb efflux occurs against a concentration gradient for Pb^{2+}, and has a V_{max} of 14 mmol/(l.cells x hr) and K_m of 5 x 10^{-8} M intracellular Pb^{2+} (Fig. 2). The IC_{50}s for vanadate inhibition of Ca^{2+} efflux and Pb^{2+} efflux are the same. Intracellular Ca^{2+} inhibits Pb^{2+} efflux, but is

NATO ASI Series, Vol. H 48
Calcium Transport and
Intracellular Calcium Homeostasis
Edited by D. Pansu and F. Bronner

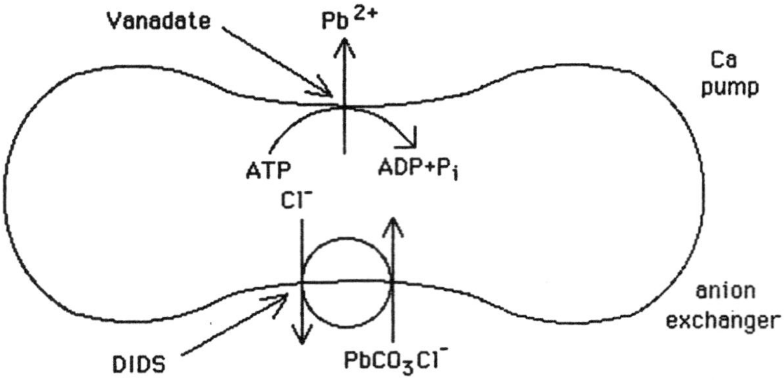

Figure 1 Pb^{2+} can cross the erythrocyte membrane by two pathways. The anion exchanger allows bidirectional movement by exchange of $PbCO_3Cl^-$ for Cl^- (inwards Pb^{2+} movement shown for convenience). The Ca^{2+} pump catalyzes ATP-dependent extrusion of Pb^{2+}, which can occur against an electrochemical gradient.

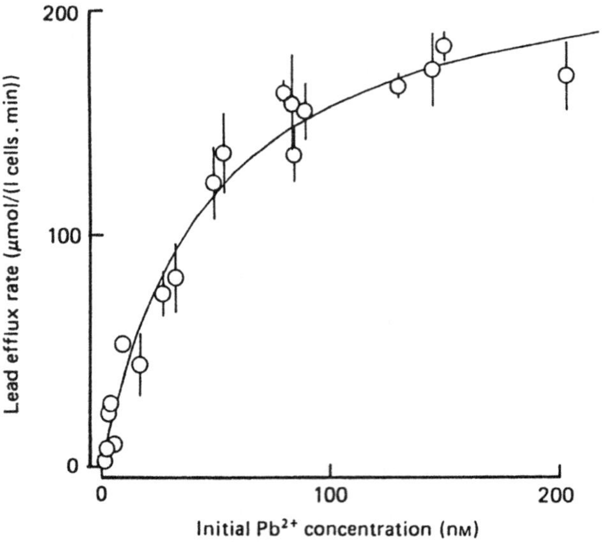

Figure 2 Variation of lead efflux rate with Pb^{2+} concentration. Each point gives the inital rate of efflux, ± S.D., plotted against the initial Pb^{2+} concentration. The line is a rectangular hyperbola, and corresponds to a V_{max} of 230 ± 14 μmol/(l.cells.min) and a K_m of 47 ± 8 nM Pb^{2+}. Taken, with permission, from Simons (1988).

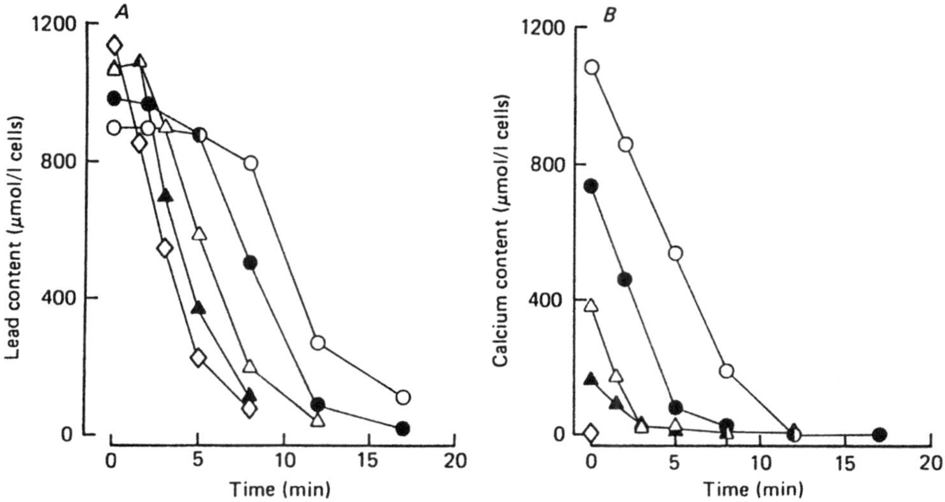

Figure 3 Effect of Ca^{2+} on Pb^{2+} efflux. Each ghost preparation was made to contain 3 mM Tiron (which chelates Pb^{2+} but not Ca^{2+}) and varying quantities of Pb^{2+} and Ca^{2+}. The initial free concentrations of Pb^{2+} and Ca^{2+}, measured with ion-sensitive electrodes, were: Pb^{2+} (nM): 150 (◇), 170 (▲), 140 (△), 110 (●) and 100 (○); Ca^{2+} (μM): 3.5 (◇), 54 (▲), 140 (△), 290 (●) and 460 (○). Taken, with permission, from Simons (1988).

also pumped out of the ghosts, and Pb^{2+} efflux is delayed until after the Ca^{2+} has been pumped out (Fig. 3). The conclusion is that Pb^{2+} is transported by the erythrocyte Ca^{2+} pump, with a similar V_{max} to Ca^{2+}, but a higher affinity. It is not possible to perform quantitative competition experiments between Pb^{2+} and Ca^{2+} with this preparation, so details of the effects of Pb^{2+} and Ca^{2+} on each other's effluxes are lacking.

The presence of an active pump which extrudes Pb^{2+} from erythrocytes makes it hard to explain the distribution of Pb^{2+} between erythrocytes and plasma. One possible explanation for the 100-fold accumulation of Pb^{2+} in erythrocytes under physiological circumstances might be the presence of an intracellular ligand with a very high affinity for Pb^{2+}. Only small quantities of a molecule with a K_d for Pb^{2+} of 10^{-11} or 10^{-12} M would be needed. It has not yet been possible to test this hypothesis. An alternative explanation might be that the pathways for Pb^{2+} transport shown in Fig. 1 are not operative under physiological conditions. The intracellular Pb^{2+} concentration may be too low to allow significant Pb^{2+} efflux via the Ca^{2+} pump, and Pb^{2+} uptake may occur via other pathways that become significant at extracellular Pb^{2+} concentrations below 10^{-9} M. These possibilities were tested in a new series of experiments.

Tracer ^{203}Pb uptake was measured into human erythrocytes suspended in artificial media containing a Pb-nitrilotriacetate (NTA) buffer, which maintains the extracellular Pb^{2+} concentration close to a fixed value in the range 10^{-12} to 10^{-9} M. Fig. 4 shows a sample experiment, in which extracellular [Pb^{2+}] was fixed at 2×10^{-11} M. There was no significant ^{203}Pb uptake during 4 hours incubation at 37°C of a bicarbonate-free erythrocyte suspension. Inclusion of 25 mM bicarbonate (replacing chloride) stimulates ^{203}Pb uptake. The cellular ^{203}Pb level plateaus at about 2 µmol/l.cells after 3 hours incubation. Addition of 20 µM DIDS, which inhibits the anion exchanger, abolishes ^{203}Pb uptake, in the presence of bicarbonate. On the other hand,

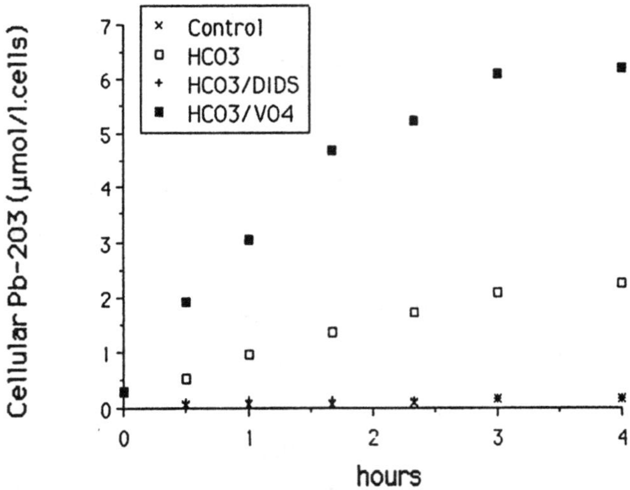

Figure 4 Uptake of ^{203}Pb by human erythrocytes at 37°C. In the control conditions, the cells were suspended at 5% haematocrit in 145 mM KCl buffered with 15 mM Hepes (pH 7.4), and containing 1 mM NTA and 20 µM $^{203}Pb(NO_3)_2$ (free Pb^{2+} 20 pM) and 5 mM glucose. In the other conditions, 25 mM $KHCO_3$ replaced an equivalent quantity of KCl, and DIDS (20µM) and Na vanadate (4 mM) were added as required. The uptake of radioactivity was measured after centrifuging the cells through silicone fluid.

addition of 4 mM vanadate, which inhibits the Ca^{2+} pump, further boosts ^{203}Pb uptake to a level of 6 µmol/l. cells, in the presence of bicarbonate. These results are consistent with the model of Pb^{2+} transport shown in Fig. 1. Pb^{2+} uptake occurs via the anion exchanger, and is stimulated by bicarbonate and inhibited by DIDS. The cellular accumulation of ^{203}Pb depends upon the balance between influx and efflux. Vanadate stimulates ^{203}Pb accumulation by inhibiting the efflux pathway.

Fig. 4 also shows that the quantity of ^{203}Pb taken up by the erythrocytes when suspended in a 25 mM-bicarbonate medium is similar to the Pb content of human erythrocytes from normal subjects. This prompted an investigation of the variation in steady-state ^{203}Pb accumulation with extracellular Pb^{2+} concentration. The combined results of 4 experiments are shown in Fig. 5. Physiological levels of erythrocyte Pb are achieved by incubating cells in artifical media containing 1 - 2 x 10^{-11} M buffered Pb^{2+}, while pathological levels, of 10 µmol/l. cells or more, require extracellular Pb^{2+} to be 10^{-10} M or higher.

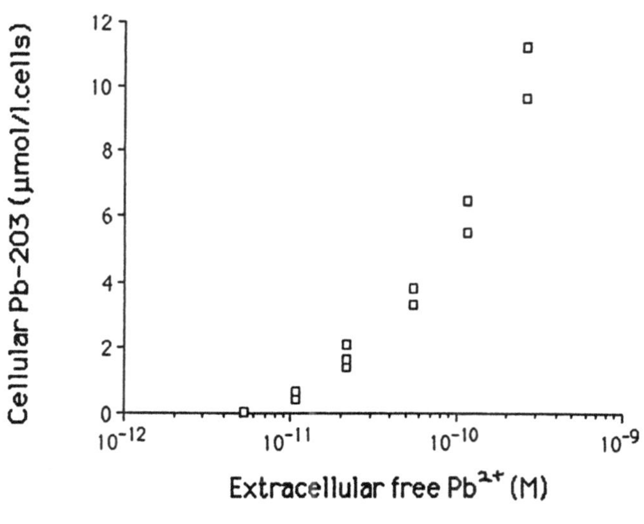

Figure 5 Dependence of ^{203}Pb uptake on extracellular Pb^{2+} concentration. Each point gives the ^{203}Pb content of cells after 3 hr incubation at 37°C in the presence of 25 mM HCO_3^- (as in Fig. 4, without inhibitors) plotted against the extracellular Pb^{2+} concentration.

The conclusion to be drawn from this work is that the anion exchange and Ca^{2+} pump pathways for Pb^{2+} transport across the erythrocyte membrane operate down to physiological erythrocyte Pb levels, and down to an extracellular free Pb^{2+} concentration of 10^{-11} M. It does not follow, however, that the free Pb^{2+} concentration in human serum is about 10^{-11} M. Although artifical media containing 10^{-11} M Pb^{2+} appear to be in a steady-state with erythrocytes containing a physiological quantity of Pb, it is possible that plasma contains additional, unidentified substances that stimulate Pb^{2+} influx into erythrocytes. Additional influx pathways would then allow erythrocytes to be in equilibrium with a concentration of plasma Pb^{2+} lower than 10^{-11} M in vivo. The most important low molecular weight ligand for Pb^{2+} in plasma is cysteine (May, Linder & Williams, 1977; al-Modhefer, Bradbury & Simons, unpublished). One pathway for Pb^{2+} uptake could be as Pb-cysteine, transported by an amino-acid carrier. This possibility was tested in the experiment illustrated in Fig. 4. Inclusion of 1 mM cysteine, in the absence of bicarbonate, failed to stimulate ^{203}Pb uptake above control levels (not shown on graph). This makes it unlikely that cysteine is involved in the membrane transport of Pb^{2+}, but other possibilities remain, such as transport with other amino-acids or as a Pb-protein complex.

This work was supported by the Wellcome Trust (UK)

REFERENCES

Hernberg S (1980) Biochemical and clinical effects and responses as indicated by blood concentration. In: Singhal RL, Thomas JA (eds) Lead Toxicity. Urban & Schwarzenbach, Baltimore Munich, p. 367

Manton WI, Cook JD (1984) High accuracy (stable isotope dilution) measurements of lead in serum and cerebrospinal fluid. Brit J Ind Med 41:313-319

May PM, Linder PW, Williams DR (1977) Computer simulation of metal-ion equilibria in biofluids: models for the low-molecular-weight complex distribution of calcium (II), magnesium (II), manganese (II), iron (III), copper (II), zinc (II) and lead (II) ions in human blood plasma. J Chem Soc Dalton Trans 588-595

Simons TJB (1986a) Passive transport and binding of lead by human red blood cells. J Physiol Lond 378:267-286

Simons TJB (1986b) The role of anion transport in the passive movement of lead across the human red cell membrane. J Physiol Lond 378:287-312

Simons TJB (1988) Active transport of lead by the calcium pump in human red cell ghosts. J Physiol Lond 405:105-113

TUMORAL PARATHYROID HORMONE-RELATED PROTEIN AND CALCIUM TRANSPORT

J.-Ph. Bonjour and R. Rizzoli
Division of Clinical Pathophysiology
Department of Medicine
University Hospital of Geneva
1211 Geneva 4
Switzerland

The main mechanisms involved in the pathogenesis of hypercalcemia observed in cancer patients comprise increased net bone resorption and enhanced tubular reabsorption of calcium (Ca). Recently, a protein derived from various human and animal tumors associated with hypercalcemia and interacting with parathyroid hormone receptors at the level of bone and kidney has been identified (3,11,12). In our laboratory we have shown by using renal cell cultured epithelia that this parathyroid hormone-related protein (PTHrP) exerts many of the in vitro effects of PTH, such as stimulation of adenylate cyclase (6,7), inhibition of inorganic phosphate (Pi) transport (6,7) and decrease of sodium-proton exchange (4). We also observed striking homologies with the action of PTH in vivo (10). When perfused over a 6 day period to thyroparathyroidectomized (TPTX) rats by intraperitoneal osmotic minipumps, the synthetic aminoterminal fragment of the protein [PTHrP(1-34)] enhances the mobilization of Ca from bone (10). In collaboration with M. Arlot and P. Meunier (Unité INSERM 234, Lyon) we observed that this effect was associated with an increased number of osteoclasts in the tibial long bone (9). Renal reabsorption of Ca and magnesium were increased by

NATO ASI Series, Vol. H 48
Calcium Transport and
Intracellular Calcium Homeostasis
Edited by D. Pansu and F. Bronner

PTHrP(1-34), while that of Pi was depressed (10). However, for similar effects on plasma Ca, PTHrP(1-34) stimulated bone formation less than PTH (9). This difference in bone formation-resorption coupling could be due to the attenuated increase in the plasma level of 1,25-dihydroxyvitamin D3 monitored in PTHrP(1-34) as compared to PTH(1-34) infused animals (10). Thus, these experiments indicate that both an increase in bone Ca resorption, that maybe associated with a reduced coupling with bone Ca deposition, and a stimulation in tubular Ca reabsorption contribute to the rise in calcemia observed under PTHrP(1-34) administration. Use of inhibitors of either bone Ca resorption or tubular Ca reabsorption demonstrate that about two thirds of the PTHrP(1-34) induced increase in calcemia was due to the increment in renal Ca flux (10). These findings, which are similar to those we previously reported in rats bearing the hypercalcemic Leydig cell tumor (5,8) emphasizes the importance of the tubular Ca reabsorption in the pathogenesis of tumoral hypercalcemia (2). They are also directly relevant to clinical observations (1,2) made in some patients with hypercalcemia of malignancy in whom the elevated plasma Ca level cannot be corrected despite effective inhibition of bone resorption.

References

1. Bonjour JP, Philippe J, Guelpa G, Bisetti A, Rizzoli R, Jung A, Rosini S, Kanis JA (1988) Bone and renal components in hypercalcemia of malignancy and responses to a single infusion of clodronate. Bone 9:123-130

2.Bonjour JP, Rizzoli R (1989) Pathophysiological aspects and therapeutic approaches of tumoral osteolysis and hypercalcemia. In: Bisphosphonates and Tumor Osteolysis. Ed.K.W. Brunner, H. Fleisch, H.J. Senn. Recent Results in Cancer Research, vol.116. Springer-Verlag, Berlin, Heidelberg, pp.29-39

3.Burtis WJ, Wu T, Bunch C, Wysolmerski JJ, Insogna KL, Weir EC, Broadus AE, Stewart AF (1987) Identification of a novel 17,000-Dalton parathyroid hormone-like adenylate cyclase-stimulating protein from a tumor associated with humoral hypercalcemia of malignancy. J Biol Chem 262:7151-7156.

4.Caverzasio J, Rizzoli R, Martin TJ, Bonjour JP (1988) Tumoral synthetic parathyroid hormone related peptide inhibits amiloride-sensitive sodium transport in cultured renal epithelia. Europ. J. Physiol. Pflügers Arch 413:96-98

5.Hirschel-Scholz S, Caverzasio J, Rizzoli R, Bonjour JP (1986) Normalization of hypercalcemia associated with a decrease in renal calcium reabsorption in Leydig cell tumor-bearing rats treated with WR-2721. J Clin Invest 78:319-322

6.Pizurki L, Rizzoli R, Caverzasio J, Mundy G, Bonjour JP,(1988) Factor derived from human lung carcinoma associated with hypercalcemia mimics the effects of parathyroid hormone on phosphate transport in cultured renal epithelia. J Bone Min Res 2:233-239

7.Pizurki L, Rizzoli R, Moseley J, Martin TJ (1988) Effect of synthetic tumoral PTH-related peptide on cAMP production and Na-dependent Pi transport. Am J Physiol 255:F957-F961

8.Rizzoli R, Caverzasio J, Fleisch H, Bonjour JP (1986) Parathyroid hormone-like changes in renal calcium and phosphate reabsorption induced by Leydig cell tumor in thyroparathyroidectomized rats. Endocrinology 119:1004-1009

9.Rizzoli R, Arlot M, Caverzasio J, Meunier P, Bonjour JP (1989) Blunted stimulation of bone formation (BF) in parathyroid hormone-related protein (PTHrP)-infused thyroparathyroidectomized (TPTX) rats. J Bone Min Res 4 suppl 1:S265

10.Rizzoli R, Caverzasio J, Chapuy MC, Martin TJ, Bonjour JP, (1989) Role of bone and kidney in parathyroid hormone-related peptide-induced hypercalcemia in rats. J Bone Miner Res 4:759-765

11.Strewler GJ, Stern PH, Jacobs JW, Eveloff J, Klein RF, Leung SC, Rosenblatt M, Nissenson RA (1987) Parathyroid hormonelike protein from human renal carcinoma cells. J Clin Invest 80:1803-1807

12.Suva LJ, Winslow GA, Wettenhall REH,Hammonds RG, Moseley JM, Diefenbach-Jagger H, Rodda CP, Kemp BE, Rodriguez H, Chen EY, Hudson PJ, Martin TJ, Wood WI (1987) A parathyroid hormone-related protein implicated in malignant hypercalcemia: Cloning and expression. Science 237:893-896

SECTION IX

IN RECOGNITION

FELIX BRONNER

FORTY YEARS DEVOTED TO CALCIUM HOMEOSTASIS AND TRANSPORT

FELIX BRONNER - FORTY YEARS DEVOTED TO CALCIUM HOMEOSTASIS AND TRANSPORT

D. Pansu and C. Bellaton
Laboratoire de Physiologie des Echanges Mineraux
Ecole Pratique des Hautes Etudes
Hopital E. Herriot
69437 Lyon Cedex 03
France

The Lyon Workshop took place at a time that coincided with Felix Bronner becoming professor emeritus after 40 years of professional activity, largely in the field of calcium metabolism. It seemed appropriate to us, therefore, to mark this milestone in the career of our colleague and teacher by looking at his manifold accomplishments and contributions.

Felix Bronner's scientific career began at a time when radioactive tracers had become available for biological research. He was among the first to study calcium metabolism in humans, first in boys and later in women. Early studies dealt with the effect of immobilization on calcium metabolism, a problem that space travel has rendered acute, as well with postmenopausal osteoporosis. He then turned his attention to detailed quantitative studies, with the rat as the model. This enabled him, together with his friend J.-P. Aubert of the Pasteur Institute, to construct a mathematical model of mammalian calcium homeostasis (Aubert and Bronner, 1965).

The discovery in 1967 of the intestinal calcium-binding protein (CaBP, calbindin) by Wasserman (Wasserman and Taylor, 1966) made it possible for investigators of calcium metabolism to focus on molecular aspects of regulation. Felix Bronner and his associates began studying the nutritional regulation of this protein.

He and his group found that high levels of CaBP were associated with low calcium intakes and vice versa (Freund and Bronner, 1975). It was not until many years later that in our joint studies we were able to show a clear linear relationship between active transcellular calcium transport and the CaBP content of the transporting cell (Bronner et al, 1986).

Another interesting discovery was that the administration of vitamin D in the form of 1,25-$(OH)_2$-D_3 to vitamin D-replete rats on a high calcium diet that therefore had low levels of CaBP led to a very rapid increase in CaBP, an increase that did not seem to be accompanied by a corresponding increase in mRNA-CaBP (Buckley and Bronner, 1980). This was interpreted as suggesting a post-transcriptional regulatory step in CaBP biosynthesis (Singh and Bronner, 1980).

NATO ASI Series, Vol. H 48
Calcium Transport and
Intracellular Calcium Homeostasis
Edited by D. Pansu and F. Bronner

Our own collaboration with Felix Bronner began when we measured calcium absorption by duodenal loops in rats that had been fed a diet enriched in lactose. We found that the addition of lactose enhanced total calcium absorption, but led to depressed CaBP levels and explained this as due to the down-regulation of CaBP, equivalent to that caused by high calcium intakes (Pansu et al., 1979). In subsequent studies, we measured the V_m of active transport directly and correlated it with CaBP levels under a variety of physiological, nutritional and developmental conditions (Pansu et al., 1981, 1983a, 1983b). Quantitative analysis of the three steps of transcellular calcium transport - entry across the brush border, intracellular diffusion and extrusion across the basolateral membrane - led to the conclusion that intracellular movement of calcium is rate-limiting and that it is the near millimolar concentration of CaBP that permits intracellular calcium flux to rise to the rate required by the experimentally arrived values of transcellular calcium transport (Bronner et al. 1986). Subsequently Felix, in collaboration with W.D. Stein, extended this analysis to the distal convoluted tubule of the kidney (Bronner and Stein, 1987). Inhibition of active calcium transport by theophylline has now been shown to be due to the inhibition of calcium-binding by CaBP (Pansu et al., 1989).

Felix's papers illustrate how he has worked, utilizing carefully gathered experimental results - his own or the literature's - to develop a unifying conceptual and theoretical framework. But Felix's strength in analysis and synthesis extends to his art of communication. As editor of the important series on Mineral Metabolism (Comar and Bronner, 1960-2), on Disorders of Mineral Metabolism (Bronner and Coburn, 1981-82) and of the 27 volumes in the Current Topics in Membranes and Transport series (Bronner and Kleinzeller, 1970-1987; Bronner, 1984-1988; Kleinzeller and Bronner, 1982-6), he taught us, brought us up to date and helped diffuse the knowledge gathered by so many authors and contributors from throughout the world. In addition, he organized numerous meetings, from the first Gordon Research Conference on Bones and Teeth in 1954 to the three predecessor meetings of this one, the Vienna workshops on Calcium and Phosphate Transport Across Biomembranes.

This is by no means, however, an end to Felix's career. He is continuing his scholarly work of editing, analyzing and searching for new insights. In recent years, he has also begun a career as a painter who has already received some recognition. We salute you, Felix, and wish you many more creative years.

REFERENCES:

Aubert JP, Bronner F (1965) A symbolic model for the regulation by bone metabolism of the blood calcium level in rats. Biophys. J., 5: 349-358.

Bronner F ed. (1984-88) Current Topics in Membranes and Transport. Vols. 21, 24, 25, 29, 32.

Bronner F, Coburn JW eds. (1981-2) Disorders of Mineral Metabolism. Academic Press, New York. Vols. I-III.

Bronner F, Kleinzeller A eds. (1970-1987) Current Topics in Membranes and Transport. Vols. 1-15, 17, 19, 23, 28.

Bronner F, Pansu D, Stein WD (1986) An analysis of intestinal calcium transport across the rat intestine. Amer. J. Physiol.,250: G561-G569.

Bronner F, Stein WD (1988) $CaBP_r$ facilitates intracellular diffusion for Ca pumping in distal convoluted tubule. Am J. Physiol, 255: F558-F562.

Buckley M, Bronner F (1980) Calcium binding protein biosynthesis in the rat: regulation by calcium and 1,25-dihydroxyvitamin D_3. Arch. Biochem. Biophys.,202: 235-241.

Comar CL, Bronner F eds. (1961-1964, 1969) Mineral Metabolism-An Advanced Treatise, Academic Press, New York, Vols. IA, IB, IIA, IIB, III.

Freund T, Bronner F (1975) Regulation of intestinal calcium-binding protein by calcium intake in the rat. Am. J. Physiol., 228: 861-869.

Kleinzeller A, Bronner F eds. (1982-6) Current Topics in Membranes and Transport, Vols. 16, 20, 26.

Pansu D, Bellaton C, Bronner F (1979) Effect of lactose on duodenal calcium-binding protein and calcium absorption. J. Nutr. 109: 509-512.

Pansu D, Bellaton C, Bronner F (1981) The effect of calcium intake on the saturable and non-saturable components of duodenal calcium transport. Am. J. Physiol., 240: G32-G37.

Pansu D, Bellaton C, Bronner F (1983a) Developmental changes in the mechanisms of duodenal calcium transport in the rat. Am. J. Physiol., 244: G20-G26.

Pansu D, Bellaton C, Roche C, Bronner F (1983b) Duodenal and ileal calcium absorption in the rat and effects of vitamin D. Am. J. Physiol., 244: G695-G700.

Pansu D, Bellaton C, Roche C, Bronner F (1989) Theophylline inhibits transcellular Ca transport in intestine and Ca-binding by CaBP. Am. J. Physiol., 257: G935-G943.

Singh RP, Bronner F (1980) Vitamin D acts posttranscriptionally. In vitro studies with the vitamin D-dependent calcium-binding protein of rat duodenum. In: Calcium Binding Proteins: Structure and Function. (Siegel FL, Carafoli RH, Kretsinger RH, MacLennan, DH, Wasserman RH, eds), Elsevier-North Holland, New York, pp 379-383.

INDEX

NATO ASI Series H

Vol. 1: **Biology and Molecular Biology of Plant-Pathogen Interactions.** Edited by J.A. Bailey. 415 pages. 1986.

Vol. 2: **Glial-Neuronal Communication in Development and Regeneration.** Edited by H.H. Althaus and W. Seifert. 865 pages. 1987.

Vol. 3: **Nicotinic Acetylcholine Receptor: Structure and Function.** Edited by A. Maelicke. 489 pages. 1986.

Vol. 4: **Recognition in Microbe-Plant Symbiotic and Pathogenic Interactions.** Edited by B. Lugtenberg. 449 pages. 1986.

Vol. 5: **Mesenchymal-Epithelial Interactions in Neural Development.** Edited by J.R. Wolff, J. Sievers, and M. Berry. 428 pages. 1987.

Vol. 6: **Molecular Mechanisms of Desensitization to Signal Molecules.** Edited by T.M. Konijn, P.J.M. Van Haastert, H. Van der Starre, H. Van der Wel, and M.D. Houslay. 336 pages. 1987.

Vol. 7: **Gangliosides and Modulation of Neuronal Functions.** Edited by H. Rahmann. 647 pages. 1987.

Vol. 8: **Molecular and Cellular Aspects of Erythropoietin and Erythropoiesis.** Edited by I.N. Rich. 460 pages. 1987.

Vol. 9: **Modification of Cell to Cell Signals During Normal and Pathological Aging.** Edited by S. Govoni and F. Battaini. 297 pages. 1987.

Vol. 10: **Plant Hormone Receptors.** Edited by D. Klämbt. 319 pages. 1987.

Vol. 11: **Host-Parasite Cellular and Molecular Interactions in Protozoal Infections.** Edited by K.-P. Chang and D. Snary. 425 pages. 1987.

Vol. 12: **The Cell Surface in Signal Transduction.** Edited by E. Wagner, H. Greppin, and B. Millet. 243 pages. 1987.

Vol. 13: **Toxicology of Pesticides: Experimental, Clinical and Regulatory Perspectives.** Edited by L.G. Costa, C.L. Galli, and S.D. Murphy. 320 pages. 1987.

Vol. 14: **Genetics of Translation. New Approaches.** Edited by M.F. Tuite, M. Picard, and M. Bolotin-Fukuhara. 524 pages. 1988.

Vol. 15: **Photosensitisation. Molecular, Cellular and Medical Aspects.** Edited by G. Moreno, R.H. Pottier, and T.G. Truscott. 521 pages. 1988.

Vol. 16: **Membrane Biogenesis.** Edited by J.A.F. Op den Kamp. 477 pages. 1988.

Vol. 17: **Cell to Cell Signals in Plant, Animal and Microbial Symbiosis.** Edited by S. Scannerini, D. Smith, P. Bonfante-Fasolo, and V. Gianinazzi-Pearson. 414 pages. 1988.

Vol. 18: **Plant Cell Biotechnology.** Edited by M.S.S. Pais, F. Mavituna, and J.M. Novais. 500 pages. 1988.

Vol. 19: **Modulation of Synaptic Transmission and Plasticity in Nervous Systems.** Edited by G. Hertting and H.-C. Spatz. 457 pages. 1988.

Vol. 20: **Amino Acid Availability and Brain Function in Health and Disease.** Edited by G. Huether. 487 pages. 1988.

NATO ASI Series H

Vol. 21: **Cellular and Molecular Basis of Synaptic Transmission.**
Edited by H. Zimmermann. 547 pages. 1988.

Vol. 22: **Neural Development and Regeneration. Cellular and Molecular Aspects.**
Edited by A. Gorio, J.R. Perez-Polo, J. de Vellis, and B. Haber. 711 pages. 1988.

Vol. 23: **The Semiotics of Cellular Communication in the Immune System.**
Edited by E.E. Sercarz, F. Celada, N.A. Mitchison, and T. Tada. 326 pages. 1988.

Vol. 24: **Bacteria, Complement and the Phagocytic Cell.**
Edited by F.C. Cabello und C. Pruzzo. 372 pages. 1988.

Vol. 25: **Nicotinic Acetylcholine Receptors in the Nervous System.**
Edited by F. Clementi, C. Gotti, and E. Sher. 424 pages. 1988.

Vol. 26: **Cell to Cell Signals in Mammalian Development.**
Edited by S.W. de Laat, J.G. Bluemink, and C.L. Mummery. 322 pages. 1989.

Vol. 27: **Phytotoxins and Plant Pathogenesis.**
Edited by A. Graniti, R.D. Durbin, and A. Ballio. 508 pages. 1989.

Vol. 28: **Vascular Wilt Diseases of Plants. Basic Studies and Control.**
Edited by E.C. Tjamos and C.H. Beckman. 590 pages. 1989.

Vol. 29: **Receptors, Membrane Transport and Signal Transduction.**
Edited by A.E. Evangelopoulos, J.P. Changeux, L. Packer, T.G. Sotiroudis, and K.W.A. Wirtz. 387 pages. 1989.

Vol. 30: **Effects of Mineral Dusts on Cells.**
Edited by B.T. Mossman and R.O. Bégin. 470 pages. 1989.

Vol. 31: **Neurobiology of the Inner Retina.**
Edited by R. Weiler and N.N. Osborne. 529 pages. 1989.

Vol. 32: **Molecular Biology of Neuroreceptors and Ion Channels.**
Edited by A. Maelicke. 675 pages. 1989.

Vol. 33: **Regulatory Mechanisms of Neuron to Vessel Communication in Brain.**
Edited by F. Battaini, S. Govoni, M.S. Magnoni, and M. Trabucchi. 416 pages. 1989.

Vol. 34: **Vectors as Tools for the Study of Normal and Abnormal Growth and Differentiation.**
Edited by H. Lother, R. Dernick, and W. Ostertag. 477 pages. 1989.

Vol. 35: **Cell Separation in Plants: Physiology, Biochemistry and Molecular Biology.**
Edited by D.J. Osborne and M.B. Jackson. 449 pages. 1989.

Vol. 36: **Signal Molecules in Plants and Plant-Microbe Interactions.**
Edited by B.J.J. Lugtenberg. 425 pages. 1989.

Vol. 37: **Tin-Based Antitumour Drugs.** Edited by M. Gielen. 226 pages. 1990.

Vol. 38: **The Molecular Biology of Autoimmune Disease.**
Edited by A.G. Demaine, J-P. Banga, and A.M. McGregor. 404 pages. 1990.

Vol. 39: **Chemosensory Information Processing.** Edited by D. Schild. 403 pages. 1990.

Vol. 40: **Dynamics and Biogenesis of Membranes.**
Edited by J.A.F. Op den Kamp. 367 pages. 1990.

Vol. 41: **Recognition and Response in Plant-Virus Interactions.**
Edited by R.S.S. Fraser. 467 pages. 1990.

NATO ASI Series H

Vol. 42: Biomechanics of Active Movement and Deformation of Cells.
Edited by N. Akkaş. 524 pages. 1990.

Vol. 43: Cellular and Molecular Biology of Myelination.
Edited by G. Jeserich, H. H. Althaus, and T. V. Waehneldt. 565 pages. 1990.

Vol. 44: Activation and Desensitization of Transducing Pathways.
Edited by T. M. Konijn, M. D. Houslay, and P. J. M. Van Haastert. 336 pages. 1990.

Vol. 45: Mechanism of Fertilization: Plants to Humans.
Edited by B. Dale. 710 pages. 1990.

Vol. 46: Parallels in Cell to Cell Junctions in Plants and Animals.
Edited by A. W. Robards, W. J. Lucas, J. D. Pitts, H. J. Jongsma, and D. C. Spray. 296 pages. 1990.

Vol. 47: Signal Perception and Transduction in Higher Plants.
Edited by R. Ranjeva and A. M. Boudet. 357 pages. 1990.

Vol. 48: Calcium Transport and Intracellular Calcium Homeostasis.
Edited by D. Pansu and F. Bronner. 456 pages. 1990.

GPSR Compliance
The European Union's (EU) General Product Safety Regulation (GPSR) is a set of rules that requires consumer products to be safe and our obligations to ensure this.

If you have any concerns about our products, you can contact us on

ProductSafety@springernature.com

In case Publisher is established outside the EU, the EU authorized representative is:

Springer Nature Customer Service Center GmbH
Europaplatz 3
69115 Heidelberg, Germany

www.ingramcontent.com/pod-product-compliance
Lightning Source LLC
Chambersburg PA
CBHW061153240926
51846CB00006B/999
* 9 7 8 3 6 4 2 8 3 9 7 9 5 *